电子科学与技术导论

闫小兵　编著

科学出版社

北　京

内 容 简 介

本书是一本旨在介绍电子科学与技术前沿发展的综合性教材。本书内容涵盖了电子科学与技术的概述、阻变存储器的研究及其性能提升、忆阻器的多种应用、基于新型二维材料及莫特绝缘体的忆阻器、柔性铁电器件的研究现状和应用、石墨烯场效应晶体管研究进展、氮化物激光器技术、太阳能电池技术的最新进展以及集成电路技术与仿真等内容。

本书提供了从基础理论到最新应用的全方位知识，可作为高等院校电子科学与技术专业本科生及相关专业研究生的教材或参考书，也可作为相关领域工程技术人员的参考资料。

图书在版编目（CIP）数据

电子科学与技术导论 / 闫小兵编著. --北京：科学出版社，2025.1
ISBN 978-7-03-079781-0
Ⅰ. TN
中国国家版本馆 CIP 数据核字第 2024WB4170 号

责任编辑：赵敬伟　赵　颖 / 责任校对：彭珍珍
责任印制：张　伟 / 封面设计：无极书装

科 学 出 版 社 出版
北京东黄城根北街 16 号
邮政编码：100717
http://www.sciencep.com

北京华宇信诺印刷有限公司印刷
科学出版社发行　各地新华书店经销
*
2025 年 1 月第 一 版　开本：720×1000　1/16
2025 年 1 月第一次印刷　印张：21 1/4
字数：420 000
定价：149.00 元
(如有印装质量问题，我社负责调换)

前　言

在信息化和智能化快速发展的当代，电子科学与技术作为基础学科和关键技术，在全球科技进步中扮演着核心角色。尤其是半导体技术的飞速发展，不仅极大地推动了信息技术的革命，也在能源、材料科学、生物医学等多个交叉科学领域中显示出广泛的应用潜力。本书旨在为读者提供一种全面系统性的认识和深入理解电子科学与技术的最新发展，尤其是半导体技术领域的前沿动态。

随着微电子技术的进步，从微型化、低功耗到高性能，半导体器件和材料的创新直接推动了整个信息技术领域的革新。当前，随着云计算、物联网和人工智能等技术的发展，对半导体领域提出了新的挑战与机遇。本书正是在这样的背景下，为读者提供了一个关于电子科学与技术的深入探讨和学习的平台。

本书共 10 章，全面覆盖电子科学与技术的多个关键组成部分，每一章都力求深入浅出，提供最新的研究成果和技术进展。第 1 章概述电子科学与技术的基本概念和发展历程，为后续章节奠定基础，该章由王博平撰写。第 2 章聚焦于阻变存储器的研究及性能提升，介绍阻变存储器的材料、阻变机制和性能提升的方法，该章由王博平撰写。第 3 章深入探讨忆阻器在神经仿生、神经网络、模式识别和电路等领域中的应用，该章由孙凯旋撰写。第 4 章介绍基于新型二维材料的忆阻器，包括基于不同二维材料的忆阻器以及基于二维材料的忆阻器的开关机制，该章由赵乾龙和侯其锐撰写。第 5 章介绍基于莫特绝缘体的忆阻器，包括莫特绝缘体材料、莫特转变的驱动方法和基于莫特绝缘体的忆阻器应用，该章由冉云峰撰写。第 6 章关注于柔性铁电器件的研究现状和应用，包括柔性铁电器件的制备方法和应用，该章由贾晓彤撰写。第 7 章详述石墨烯场效应晶体管的研究进展，包括石墨烯的概述、表征方法、制备方法及功能化方法，石墨烯场效应晶体管的基本结构及原理、制备工艺、特性分析和在生化物质传感中的应用，该章由王中荣撰写。第 8 章探讨了氮化物激光器技术新进展，包括氮化物激光器的概述、基本特性、生长及测试分析方法和应用，该章由杨杰撰写。第 9 章介绍太阳能电池的研究进展，包括太阳能电池的种类和光伏原理，该章由王中荣和杨杰撰写。第 10 章介绍集成电路的设计与仿真，包括集成电路的概述、数字集成电路、模拟集成电路、Bi-CMOS 型集成电路和能量回收型集成电路，该章由师建英撰写。参与本书编写工作的还有赵桢、郭振强、张伟峰、王永睿、贾刚、王雅红、高加洋、白建婷、赵鹏丽、高冰洋、何畅、薛喆、杨文琦、王嘉成、贾永庆等。

书中彩图可扫描封底二维码查看。

目　录

前言
第1章　绪论 ·· 1
　1.1　电子科学与技术的发展史 ··· 1
　1.2　电子科学与技术的研究方向 ··· 3
　1.3　电子科学与技术遭遇的瓶颈及解决方案 ······················· 4
　　参考文献 ··· 7
第2章　阻变存储器研究及性能提升 ··· 8
　2.1　引言 ·· 8
　　2.1.1　存储器的发展 ··· 8
　　2.1.2　阻变存储器的分类 ··· 10
　2.2　阻变存储器的材料 ··· 12
　　2.2.1　氧化物阻变存储器 ··· 12
　　2.2.2　非氧化物无机材料阻变存储器 ························ 17
　　2.2.3　有机材料阻变存储器 ······································· 19
　　2.2.4　二维材料阻变存储器 ······································· 20
　2.3　阻变存储器的阻变机制 ··· 24
　　2.3.1　电化学金属化导电细丝 ··································· 24
　　2.3.2　价态变化存储器 ·· 24
　　2.3.3　相变阻变存储器 ·· 26
　　2.3.4　铁电阻变存储器 ·· 26
　2.4　阻变存储器的性能提升 ··· 27
　　2.4.1　利用二维材料提升器件性能 ··························· 27
　　2.4.2　利用量子点提升器件性能 ······························· 28
　　2.4.3　利用外围电路提升器件性能 ··························· 29
　　2.4.4　通过杂质掺杂提升器件性能 ··························· 29
　　2.4.5　特殊器件形貌提升器件性能 ··························· 30
　2.5　小结 ·· 32
　　参考文献 ··· 32

第 3 章　忆阻器的应用 ·· 42
3.1　引言 ·· 42
3.2　忆阻器在神经仿生中的应用 ··· 44
3.2.1　生物突触和人工突触 ·· 44
3.2.2　导电丝忆阻器人工突触 ·· 46
3.2.3　相变忆阻器人工突触 ·· 46
3.2.4　光电忆阻器人工突触 ·· 46
3.3　忆阻器神经网络 ·· 49
3.3.1　人工神经网络 ·· 49
3.3.2　深度神经网络 ·· 52
3.3.3　脉冲神经网络 ·· 55
3.4　忆阻器的应用 ·· 57
3.4.1　忆阻器在生物学上的应用 ·· 57
3.4.2　忆阻器在模式识别中的应用 ··· 59
3.4.3　忆阻器在电路中的应用 ·· 62
3.5　小结 ·· 64
参考文献 ·· 65

第 4 章　基于新型二维材料的忆阻器 ·· 75
4.1　引言 ·· 75
4.2　基于二维材料的忆阻器 ·· 77
4.2.1　石墨烯和衍生物 ·· 77
4.2.2　过渡金属二硫化物 ·· 79
4.2.3　其他二维材料 ·· 84
4.3　基于二维材料忆阻器的开关机制 ··· 89
4.3.1　细丝 ··· 89
4.3.2　原子空位 ··· 92
4.3.3　电子的捕获和释放 ·· 96
4.4　现有的挑战 ·· 97
4.5　小结 ·· 98
参考文献 ·· 99

第 5 章　基于莫特绝缘体的忆阻器 ·· 119
5.1　引言 ·· 119
5.2　莫特绝缘体转变和莫特绝缘体材料 ··· 120
5.3　莫特转变的驱动方法 ·· 121
5.3.1　压力驱动的莫特转变 ·· 121

5.3.2 温度驱动的莫特转变 ········· 122
5.3.3 电压驱动的莫特转变 ········· 125
5.4 基于莫特绝缘体的忆阻器应用 ········· 126
5.4.1 非易失性特性 ········· 126
5.4.2 选择器和神经元的挥发性特性 ········· 130
5.5 小结 ········· 136
参考文献 ········· 138

第 6 章 柔性铁电器件的研究现状和应用 ········· 151
6.1 引言 ········· 151
6.2 制备柔性铁电器件的途径 ········· 152
6.2.1 通过范德瓦耳斯异质外延制备柔性钙钛矿铁电薄膜 ········· 152
6.2.2 通过蚀刻法制备柔性钙钛矿铁电薄膜 ········· 156
6.2.3 二维柔性铁电材料 ········· 163
6.3 柔性铁电器件的应用 ········· 169
6.3.1 柔性铁电存储器 ········· 169
6.3.2 柔性铁电传感器 ········· 175
6.3.3 柔性铁电光伏器件 ········· 177
6.3.4 柔性铁电能量采集器 ········· 178
6.4 柔性铁电器件应用的潜在研究领域 ········· 180
6.5 小结 ········· 182
参考文献 ········· 183

第 7 章 石墨烯场效应晶体管研究进展 ········· 198
7.1 石墨烯的发现、基本结构及性能 ········· 198
7.2 石墨烯的表征方法 ········· 200
7.2.1 显微镜法 ········· 200
7.2.2 光谱法 ········· 202
7.3 石墨烯的制备方法 ········· 206
7.4 石墨烯场效应晶体管基本结构及原理 ········· 215
7.4.1 背栅石墨烯场效应晶体管 ········· 215
7.4.2 顶栅石墨烯场效应晶体管 ········· 216
7.4.3 液栅石墨烯场效应晶体管 ········· 216
7.5 石墨烯场效应晶体管的制备工艺 ········· 217
7.6 石墨烯场效应晶体管中石墨烯的功能化方法 ········· 219
7.6.1 共价功能化方法 ········· 219
7.6.2 非共价功能化方法 ········· 220

7.7 石墨烯场效应晶体管在生化物质传感中的应用 ································ 222
    7.7.1 石墨烯场效应晶体管生化传感基本理论 ································ 222
    7.7.2 G-SgFET 在生化物质传感领域的应用 ································ 223
参考文献 ································ 231

第 8 章 氮化物激光器技术新进展 ································ 238
8.1 引言 ································ 238
    8.1.1 氮化物激光器的应用 ································ 238
    8.1.2 氮化物激光器发展历程 ································ 239
8.2 氮化物激光器基本特性 ································ 241
    8.2.1 氮化物激光器工作原理 ································ 241
    8.2.2 氮化物激光器工作特性 ································ 243
    8.2.3 氮化物激光器调制特性 ································ 248
8.3 氮化物白光激光技术 ································ 249
    8.3.1 荧光转化激光白光技术 ································ 250
    8.3.2 RGB 三基色复合激光白光技术 ································ 253
    8.3.3 白光 LD 光通信技术 ································ 256
参考文献 ································ 260

第 9 章 太阳能电池研究进展 ································ 267
9.1 引言 ································ 267
9.2 太阳能电池的种类和光伏原理 ································ 267
    9.2.1 pn 结型太阳能电池 ································ 267
    9.2.2 肖特基势垒太阳能电池 ································ 268
    9.2.3 染料敏化太阳能电池 ································ 269
    9.2.4 聚合物太阳能电池 ································ 270
    9.2.5 新型太阳能电池 ································ 273
参考文献 ································ 275

第 10 章 集成电路设计与仿真 ································ 277
10.1 引言 ································ 277
    10.1.1 集成电路的过去、现状和未来 ································ 277
    10.1.2 集成电路的分类及主流工艺 ································ 280
10.2 数字集成电路 ································ 281
    10.2.1 数字集成电路的设计方法 ································ 281
    10.2.2 数字集成电路的基本单元电路 ································ 283
    10.2.3 数字集成电路中的时序电路 ································ 295
10.3 模拟集成电路 ································ 301

  10.3.1 模拟集成电路的设计方法 ··· 301
  10.3.2 模拟集成电路的主要电路单元 ····································· 302
10.4 Bi-CMOS 型集成电路 ··· 318
  10.4.1 Bi-CMOS 型集成电路的设计方法 ································· 318
  10.4.2 Bi-CMOS 型集成电路的典型电路的设计及应用 ···················· 319
10.5 能量回收型集成电路的设计 ·· 322
  10.5.1 能量回收型集成电路的原理 ······································ 322
  10.5.2 能量回收型电路的结构分类 ······································ 324
参考文献 ··· 328

第1章 绪　　论

1.1　电子科学与技术的发展史

电子科学与技术专业是国家级一流本科专业建设点、国家工程教育认证专业、国家"十二五"高等学校专业综合改革试点专业、国家特色专业建设点、教育部"卓越工程师教育培养计划"试点专业，下设了电路与系统、电磁场与微波技术、微电子学与固体电子学、物理电子学四个二级学科。其中，电路与系统主要研究电路与系统的理论、分析、测试、设计和物理实现；电磁场与微波技术主要研究天线与微波技术的理论与应用；微电子学与固体电子学主要研究各种微电子器件、集成电路、集成电子系统的分析、设计、制造、测试和应用；物理电子学主要研究电子工程和信息科学技术领域内的理论与应用。

电子科学与技术专业中微电子技术和光电子技术的前身是半导体专业和激光专业。

早在中国古代，我国就通过烽火台进行远程信号传输。1820 年，奥斯特发表了著名的奥斯特实验，第一次揭示了电流能够产生磁的现象；1831 年，迈克尔·法拉第发现了电磁感应定律；1873 年，麦克斯韦提出电磁场理论，预测了电磁波的存在；1888 年，德国科学家赫兹成功地在导线周围测得电磁场，证实了电磁波的存在，成为无线电通信的先驱，为无线电通信奠定了坚实的理论基础。自赫兹证实了电磁波的存在之后，人们就产生了用电磁波传递信息的想法。1895 年，亚历山大·斯捷潘诺维奇·波波夫发明了第一台"雷电指示器"的无线电接收机；第二年，成功表演了 250m 的无线电通信，开创了无线电技术的新时代，意义深远。

随着第二次世界大战的爆发，在军事上对电子器件的要求逐步提高。1880 年，爱迪生发现将加入到灯泡里的电极连接到钨丝的电源后，被加热的钨丝会向电极产生放电现象。1904 年，英国科学家 J. A. Fleming 基于这一现象发明了用于无线电通信中检波器的真空二极管，将其取名为 Bulb，或称 Valve。1906 年，美国科学家德福雷斯特发明了真空三极管。1946 年，第一台电子计算机 ENIAC 诞生，其共使用了 18800 个真空管，占地 $1500\text{ft}^2 (1\text{ft}^2 = 0.0929\text{m}^2)$，重达 30t。ENIAC 标志着人类社会从此进入计算机时代。1948 年，美国贝尔实验室威廉·肖克利、约翰·巴丁和沃尔特·布拉顿等发明了晶体三极管，开创了固体电子时代，真正地带来了"固态革命"，推动了全球范围内半导体事业的发展，标志着集成电路时代的到来。1958 年，杰克·基尔比 (Jack Kilby) 在面积不超过 4mm^2 的晶片上集成了 20 余个元件，并申请了专利，将这种由半导体元件构成的微型固体组合件称为"集成电路"。随后，罗伯特·诺伊斯 (Robert Noyce) 提出了一种"半导体器件与铅结构"模型，为集成电路的大规模生产提供了关键技术。1960 年，仙童半导体公司采用罗伯特·诺伊斯的方案制造出第一块可以实际使用的集成电路，杰克·基尔比和罗伯特·诺伊斯被人们公认为集成电路的共同发明者。1967 年，大规模集成电路 (LSI) 出现，集成度迅速提高；1977 年，超大规模集成电路面世，一个硅晶片中已经可以集成 15 万个以上的晶体管；1988 年，16M 动态随机存储器 (DRAM) 问世，1cm^2 大小的硅片上集成有 3500 万个晶体管，标志着进入超大规模集成 (VLSI) 电路阶段；1997 年，300MHz 奔腾 II 问世，采用 $0.25\ \mu\text{m}$ 工艺，奔腾系列芯片的推出让计算机的发展如虎添翼，发展速度让人惊叹，至此，超大规模集成电路的发展又到了一个新的高度。2009 年，Intel 酷睿 i 系列全新推出，创纪录地采用了领先的 32nm 工艺，并且下一代 22nm 工艺正在研发。现如今，集成电路的发展已经经历了小规模集成 (SSI)、中规模集成 (MSI)、大规模集成 (LSI) 以及超大规模集成等阶段。随着集成度的不断提高，器件尺寸也在持续减小，先进的半导体制造技术已经可以实现 7nm 甚至更小的制程。除此之外，在人工智能方面也有新的发展，2011 年，IBM 公司通过模拟大脑结构，首次研制出两个具有感知认知能力的硅芯片原型，可以像大脑一样具有学习和处理信息的能力。2014 年 8 月，IBM 公司推出名为"真北"的第二代类脑芯片，它采用 28nm 硅工艺制作，包括 54 亿个晶体管和 4096 个处理核，相当于 100 万个可编程神经元，以及 2.56 亿个可编程突触。

电子科学与技术对于国家经济发展、科技进步和国防建设具有重要的战略意义。今天，面对电子科学与技术的迅猛发展，世界上许多发达国家，像美国、德国、日本、英国、法国等，都竞相将微电子技术和光电子技术引入国家发展计划。我国根据国外电子器件的进程，于 1956 年提出了"向科学进军"计划，将半导体技术列为重点发展的领域之一。至 20 世纪 90 年代，由于微型计算机、通信、家电等信息产业的发展和普及，对集成电路芯片的需求量越来越大，此外几场局部战争让全世界接受了电子战、信息战的高科技战争的理念。我国也越发重视集成电路的发展，先后在多项国家级战略性科技计划，如 863 计划、973 计划、国家攻关计划中，微电子技术 (集成电路技术) 和光电子技术 (激光技术) 均有立项；1995 年，原电子工业部提出了"九五"集成电路发展战略，并实施了"909 工程"；国家自然科学基金委员会在 1996 年底立项开展"光子学与光子技术发展战略研究"；在"九五"和"十五"期间，国家自然科学基金委员会在重大项目、重点项目和国家杰出青年科学基金中对电子科学与技术方面的立项给予了足够的重视和支持；近年，更是大力支持电子科学与集成电路，国务院发布《新时期促进集成电路产业和软件产业高质量发展的若干政策》来大力支持该领域的发展。国家的大力支持推动了我国在电子科学领域的快速发展。

1.2 电子科学与技术的研究方向

电子科学与技术专业涵盖的学科范围极其广泛，它以电子和光电子器件为核心，由物理、材料、工艺、器件、系统构成一个完整的学科体系。器件的物理、材料和制作工艺构成了电子和光电子器件的技术支撑，形成了多个紧密关联的学科群，器件的封装工艺和测试技术的不断发展，确保了电子和光电子器件在器件应用、器件集成和系统集成等方面的应用不断提升，形成了以系统带动器件、以器件带动材料的良性循环发展。其主要分为以下几个研究方向。

微电子和光电子技术方向。微电子技术主要指的是以集成电路技术为核心的学科，它涉及微小型电子元器件和电路的设计、制造及封装。这些技术和工艺共同实现了电子系统的高级功能。1947 年，美国贝尔实验室发明了晶体管，开创了固体电子技术时代。在随后的 10 年中，晶体管技术不断进步，德州仪器公司和贝尔实验室在 1954 年分别推出晶体管收音机和全晶体管计算机，并且 1958 年，美国第一颗轨道卫星"探险者 1 号"也首次使用了晶体管技术。随着近几年通信产业的迅速发展，微电子技术成为热门方向，芯片技术也成为各国争相抢占的热门技术。现如今，5G、6G 通信技术领域是各国研究的热点，芯片战争已经打响。因此，未来几年微电子方向的人才需求会进一步扩大。

在微电子技术蓬勃发展的同时，人们发现可以利用光电各自的优势来为我们服务，即通过光子激发电子或者电子跃迁产生光子，实现光能与电能的转换，比如激光器、光电探测器、太阳能电池等。这就是早期的光电子学。随着光电子学的发展，人们开始研究完全利用光来处理信息，于是诞生了光子学。所以可以说，先有了光电子学，又有了光子学。而最终的发展会是光电的再次统一，即更高一个层次上的光电子学。目前光电子技术涉及以下内容：激光技术、波导技术、光子检测技术、光计算和信息处理技术、光存储技术、光子显示技术、光子加工与光子生物技术等。国际公认：21 世纪是光电子与微电子紧密结合发挥作用的时代，以光通信为龙头的信息光电子产业将成为 21 世纪的明星产业和支柱产业。信息产业已成为我国第一支柱产业，光电子产品在信息产业中占有很大比重，未来光电子产业还将继续蓬勃发展。

半导体材料和电子器件方向。第一代半导体是元素半导体，以硅 (Si)、锗 (Ge) 为代表，主要应用于低压、低频、中功率晶体管以及光电探测器中。第二代半导体材料是化合物半导体，以砷化镓 (GaAs)、磷化铟 (InP) 为代表，主要用于制作高速、高频、大功率以及发光电子器件。但是 GaAs、InP 材料资源稀缺，价格昂贵，并且还有毒性，能污染环境，这些缺点使得第二代半导体材料的应用具有很大的局限性。第三代半导体材料是宽禁带半导体(禁带宽度大于或等于 2.3 eV)，以 SiC、GaN、金刚石等为代表，与第一代、第二代半导体材料相比，第三代半导体材料具有高热导率、高击穿场强、高饱和电子漂移速率和高键合能等优点，可以适应高温、高压、高功率、高频以及抗辐射等特殊环境，最大限度地提高了电子元器件的内在性能。随着第三代半导体材料的发展，宽禁带技术的进步，材料工艺与器件工艺的逐步成熟，宽禁带半导体材料作为一种新型的半导体材料非常具有研究前景，其重要性将逐渐显现，在高端领域已经逐步取代第一代、第二代半导体材料，成为电子信息产业的主宰。

电磁场与电磁波方向。电磁场与微波专业方向是由 20 世纪 70 年代的无线电通信与电子系统学科演变而来的，该方向以电磁场理论、光导波理论、光器件物理及微波电路理论为基础，并与通信系统、微电子系统、计算机系统等相结合。随着当代物理、数学、技术科学的不断进步，电磁场和微波技术得到日新月异的发展，在通信、雷达、激光和光纤、遥感、卫星、微电子、高能技术、生物和医疗等高新技术领域中起着重要作用。

物理电子技术方向。物理电子技术专业起源于 20 世纪 50 年代清华大学、华中科技大学、西安电子科技大学等院校设立的电真空技术专业。1985 年本科专业调整，物理电子技术整合了电子物理技术、真空电子技术、气体电子学与激光等 11 个专业。物理电子技术是一个宽口径的专业方向，与近代物理学、电子学、光学、光电子学、量子电子学及相关技术交叉与融合，形成了真空电子学与技术、微波电子学与技术、光电子学与技术、纳米电子学与技术、超导电子学与技术等专业，并形成了若干新的科学技术增长点，如光波与光子技术、信息显示技术与器件、高速光纤通信与光纤网等。

1.3 电子科学与技术遭遇的瓶颈及解决方案

微电子学技术及超大规模集成电路的发展使得人类在计算机、电子通信、航空航天等领域取得了重大突破，逐渐成为当代各行各业智能工作的基石。但是随着集成度的提高，微电子学技术的进一步发展却受到了极大的限制。① 光刻技术：光刻是集成电路加工设备的核心，其直接决定了单个晶体管器件的物理尺寸。②材料质量：集成度越高，芯片中晶体管的尺寸越小，对制作集成电路的半导体单晶硅材料的纯度要求就越高。③ 制造工艺：随着光刻精度的提高，基板和光刻掩模版的表面平整度也需相应提高(对于数十纳米的最小线宽制程，表面平整度几乎是原子尺度)。④ 能耗和散热：随着集成电路芯片中晶体管数量的增多，芯片工作时产生的热量也大幅度增加。光电子技术是继微电子技术之后近 30 年来

1.3 电子科学与技术遭遇的瓶颈及解决方案

迅猛发展的综合性高新技术。光电子器件性能的提升依赖于模拟设计、材料生长、器件制备、封装测试四个过程。能否突破这些限制,是微电子与光电子技术发展所面临的极大挑战。

从 2013 年起,中国进口集成电路的价值就超过 2000 亿美元,2018 年创下历史新高 (进口价值为 3120.8 亿美元),而 2017 年原油进口 2402.6 亿美元,远低于进口芯片价值,由此产生的贸易逆差也创下历史新高[1],主要原因是高端芯片领域国产集成电路的占有率很低,国内自主芯片产品结构处于中低端,如表 1.1 所示。因此要想突破发达国家对中国的技术封锁就必须使我国集成电路国产化。

表 1.1 中国高端集成电路国产占有率[1]

	设备	核心集成电路	2018 年国产占有率
计算机系统	服务器	MPU	0%
	个人计算机	MPU	0%
	工业应用	MCU	2%
通用电子系统	可编程逻辑设备	FPGA/EPLD	0%
	数字信号处理设备	DSP	0%
通信设备	移动通信终端	Application Processor	18%
		Communication Processor	22%
		Embedded MPU	0%
		Embedded DSP	0%
	核心网络设备	NPU	15%
内存设备	半导体存储器	DRAM	0%
		NAND FLASH	0%
		NOR FLASH	5%
显示及视频系统	高清电视/智能电视	Image Processor	5%
		Display Processor	5%
		Display Driver	0%

另外,在人工智能方面,现如今,人们依旧采用的是冯·诺依曼体系结构,其特点在于程序存储于存储器中,与运算控制单元相分离。但冯·诺依曼瓶颈正阻碍计算机的进一步发展。为了满足速度和容量的需求,现代计算系统通常采取高速缓存 (SRAM)、主存 (DRAM)、外部存储 (NAND flash) 的三级存储结构,如图 1.1 所示。越靠近运算单元的存储器其速度越快,但受功耗、散热、芯片面积的制约,其相应的容量也越小。静态随机存储器 (SRAM) 响应时间通常在纳秒级,DRAM 则一般为 100ns 量级,与非型闪存 (NAND flash) 更是高达 100ns 级,当数据在这三级存储间传输时,后级的响应时间及传输带宽都将拖累整体的性能,形成 "存储墙"。因此,随着处理器性能的不断提升,"存储墙" 问题已经成为制约计算系统性能的主要因素。由于 DRAM 和 NAND flash 本身物理特性的限制,单纯依靠改良现有的存储器很难突破 "存储墙"。因此,新型存储开始受到广泛关注,其特点在于同时具备 DRAM 的读写速率与寿命以及 NAND flash 的非易失

特性。这使得新型存储理论上可以简化存储架构，将当前的内存和外存合并为持久内存，从而有望消除或缩小内存与外存间的"存储墙"。因此，类脑计算芯片应运而生，其可模拟人类大脑信息处理方式，能以极低的功耗对信息进行异步、并行、低速和分布式处理，并具备自主感知、识别和学习等多种能力。同传统计算芯片相比，类脑计算芯片实现了两个突破：一是突破传统"执行程序"计算范式的局限，有望形成"自主认知"的新范式；二是突破传统计算机体系结构的局限，实现数据并行传送、分布式处理，能够以极低功耗实时处理海量数据。

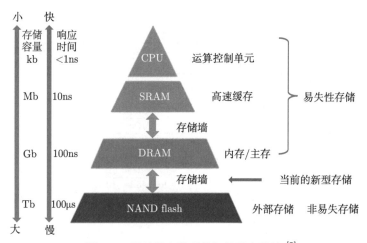

图 1.1　常见的存储系统架构及存储墙[2]

我国新型存储产业化能力及知识产权布局实际已经大幅落后。相变存储器和阻变存储器的专利从 2000 年左右开始逐渐增加，磁变存储器更是从 1990 年就开始有专利申请，英特尔公司在 2015 年发布 3D Xpoint，实际上已经经历了十余年的发展[2]。我国目前尚处于科学研究阶段，但是随着我国在 5G、人工智能等新兴领域的快速发展，新型存储必会有足够广阔的潜在市场。

综上，国内电子科学与技术的发展正面临严峻的考验：一方面，我国现有技术水平同国际水平还存在较大差距，核心技术掌握不足，创新能力不足，虽然我国也成为集成电路产业最大市场，但产业基础不足，最重要的核心技术仍未突破；另一方面，国际环境的压迫，我国芯片等领域的发展受到制约。对此，培养该方面的人才迫在眉睫。

本书旨在通过详细介绍当今半导体领域中忆阻器的研究现状，激光二极管、低维三极管、太阳能电池及传统的集成电路设计等相关领域研究的最新进展，以深入浅出的语言向普通大学生及研究生普及当今电子科学与技术相关领域的研究现状及前沿研究，也为专业的半导体研究领域的科研人员提供参考和借鉴。本书为普通本科生及研究生拓展视野、增加半导体相关知识面提供了一种重要的途径。

参 考 文 献

[1] 孙宇, 刘竞升, 罗军, 等. 中国集成电路产业发展现状及广东发展建议 [J]. 中国集成电路, 2019, 028(012):26-32.

[2] 黄阳棋, 朱邵歆. 这篇文章, 把新型存储说清楚了 [EB/OL]. 中国电子报 [2020-03-31]. http://www.cena.com.cn/semi/20200331/105722.html.

第 2 章 阻变存储器研究及性能提升

2.1 引　言

2.1.1 存储器的发展

如今的电子器件都有着强大的、速度很快的存储器，但是历史上的存储器并不是一直像今天这样高效。从软盘到闪存器件，从磁带到由磁力驱动的硬件器件，存储器的发展史需要我们了解和熟悉。

存储器的发展起源于几百年前一次不起眼的发明：穿孔卡片，如图 2.1(b) 所示。穿孔卡片是由 IBM 公司创始人之一的赫尔曼在 19 世纪发明的，它是存储和访问数据的一种原始形式，利用打孔与不打孔来表示数据信息。穿孔卡片曾被用于完成美国的人口普查报告，在 19 世纪末到 20 世纪初广泛应用于商业和相关文书。已输入信息的穿孔卡片也被当作早期数字计算机的计算机程序和数据的主要输入媒介[1]。

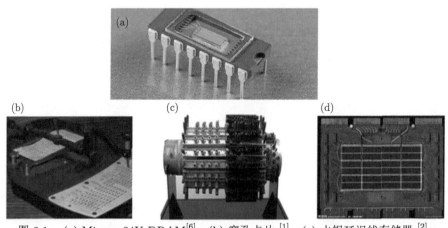

图 2.1　(a) Micron 64K DRAM[6]；(b) 穿孔卡片[1]；(c) 水银延迟线存储器[2]；
(d) 磁芯存储器[5]

水银延迟线 (mercury delay line) 存储器作为最早的一种计算机内存 (主存器)(图 2.1(c))，用于第二次世界大战期间军用雷达的开发[2]，后来，工程师莫齐利提出了利用水银延迟线作为最早的计算机上的内存。每一个数据位 (比特，bit)

2.1 引言

用机械波来表示,波峰表示为 1、波谷表示为 0。表示数据的机械波从汞柱的一端开始,一定厚度的熔融态金属汞通过振动膜片沿着纵向从一端传到另一端,在管的另一端,采用一个传感器感受传来的机械波,从而得到每一位的信息,并反馈到起点[3]。但是,这种存储器会因环境条件的影响而不准确。

20 世纪 50 年代初,磁带录音技术开始应用于计算机领域,后来出现了数据存储磁带机[4]。磁带是数据存储方面的一个巨大飞跃:它可以在半英寸 (1in=2.54cm) 长的氧化皮胶带上长时间存储大量数据,一个卷轴可以容纳相当于 1 万张穿孔卡片的数据,这个优势永远改变了计算机行业。磁芯存储器是美籍华裔王安于 1948 年发明的,如图 2.1(d) 所示,它使用微小的磁环,每个磁芯代表一位信息且通过线程来写入和读取信息。磁芯可以以两种不同的方式 (顺时针或逆时针) 磁化,磁芯的磁化方向代表着存储的数据为 1 或 0。合理的布线使得磁芯允许被设置为 1 或 0,并且通过向所选择的导线发送适当的电流脉冲来改变磁芯的磁化方向。从 20 世纪 50 年代初到 70 年初,磁芯存储器在计算机的主存储器领域占据着统治地位[5]。

1966 年,罗伯特·登纳德利用金属氧化物半导体晶体管制作存储器芯片,同年,登纳德在 IBM 公司的 Thomas J. Watson 研究中心成功研发出 DRAM。1970 年,英特尔公司向公众展现了首批量产的 DRAM 器件,随后它很快成为世界上最畅销的半导体芯片[6]。DRAM 在 20 世纪 70 年代后半期开始超越磁芯存储器,它可以应用于装有个人计算机 (PC) 的工作场所。DRAM 的应用使得个人有机会将大量信息存储在一个小芯片上,而不是存储在一张卡片或一卷磁带上,图 2.1(a) 为 Micron 64K DRAM。

从 20 世纪 70 年代开始,软盘是一种非常流行的解决个人计算机存储问题的方案。"软盘" 一词源自旧的 8in 磁盘,软盘是一条磁膜,上面覆盖着一层保护性塑料,这使得它们有很强的弯曲性。软盘以各种不同的形式出现,并且体积一直在逐步减小,到后期出现了 5.25in 和 3in 软盘,且后者成为了 1990 年最流行的数据存储形式之一。后期出现的软盘不再特别松散,并且比以前的磁盘更耐用,直到今天仍被人们当作一种存储方法[7]。

随着科技的飞速发展,世界信息总量呈爆炸式增长,人们对存储器的容量、体积、功耗、速度和使用寿命提出了更高的要求。随着研发的深入,半导体存储器因其容量大、体积小、功耗低、存取速度快、使用寿命长而逐渐进入了人们的视野。半导体存储器一般分为挥发性存储器和非挥发性存储器。挥发性存储器是需要外界供电来维持信息,而断电后会丢失所存储的信息。非挥发性存储器则不需要外界供电即可维持信息的保存[8]。在我们现在所处的数字时代,应用较广泛的非挥发性半导体存储器主要有只读存储器 (ROM,如机械硬盘驱动器 (HDD)) 和 flash 存储器 (如 U 盘、固态硬盘)。随着技术的发展,flash 存储器已经取得了很大的市场份额。尽管 flash 存储器具有读、写延迟较低、电可擦除且引入的块擦除机制使得擦除速度较快、可靠性较高等优点，但是其写操作电压较高、可擦写次数有限，并且制作工艺较为复杂，尤其在缩小存储器尺寸以提高数据存储密度的大趋势下，flash 存储器的确不够理想。例如，NAND flash 是当今最常用的存储器之一，但是当其达到纳米级以实现高密度时，其主要缺点是电流泄漏，这主要是由于其中集成了电容器，电流泄漏会导致信息的丢失[9]。为了克服 flash 存储器的种种缺点，现阶段主要有两种发展方向：一种是对 flash 存储器继续改进，如纳米晶存储器 (NCM)、电荷俘获存储器 (CTM) 等；另一种是发展全新的非易失性存储器，如铁电存储器 (FeRAM)、磁性随机存储器 (MRAM)、相变存储器 (PRAM)、阻变存储器 (RRAM) 等[10]。

2.1.2 阻变存储器的分类

阻变存储器 (resistive random access memory, RRAM) 是以材料的电阻在外加电场作用下可在高阻态和低阻态之间实现可逆转换为基础的一类前瞻性下一代非挥发性存储器，它具有简易的"三明治"结构。RRAM 最基本的存储单元由金属–绝缘体 (阻变层)–金属 (MIM) 结构构成，如图 2.2 所示，紧凑且简单，并且其工艺可以与互补金属氧化物半导体 (CMOS) 兼容[11]。它具有在 32nm 节点及以下取代现有主流 flash 存储器的潜力，成为目前新型存储器的一个重要研究方向。

图 2.2　阻变存储器的 MIM 结构

RRAM 作为典型的硅基存储器的替代品，存在以下几个很突出的特性[12]：结构简单[13]、微缩性好[14]、低功耗[15]、低成本[16]、强大的数据存储能力[16]、理想的多位数据存储单元、与 CMOS 技术的兼容性[17,18] 和与制造技术相兼容的特性[19,20]。除此之外，还可以通过获得中间电阻状态来应用于亚神经处理、滤波器、逻辑、算术和多位存储等领域。根据相关资料显示，2016 年 RRAM 市场已经达到了 470 亿美元[16]。RRAM 所提供的与硬盘和闪存不同的灵活性使得它很可能成为电子工业的未来。

2.1 引言

为了能清楚地给阻变存储器进行分类,我们要介绍一些相关的术语和一些特性参数。一般刚制备好的阻变存储器处于初始阻态(一般是高阻态),还没有阻变特性。为了使其可以正常工作,需要在两端施加一个比较大的电压脉冲,这个过程称为电铸激活(forming)。置位过程一般会使存储器从高阻态(HRS,二进制 0)转变到低阻态(LRS,二进制 1),但由于电铸激活之后阻变存储器内部缺陷较多,所以需要的电压相对较小。在复位过程中,在其两端施加电压使阻变存储器从低阻态转变到高阻态。阻变存储器的特性参数是评判阻变存储器性能好坏的标准[11],主要包括操作电压、运行速度、开关比、耐受性、保持特性与器件良率等。

写入电压又称操作电压,对阻变存储器进行置位/复位操作时均需要施加相应的操作电压激励来实现高低阻态的转变。在置位过程和复位过程中,过高的操作电压会增大器件的功耗,也会在阻变层内产生缺陷和损伤,所以操作电压对器件的稳定工作有着很重要的作用。

运行速度被定义为器件写入或擦除所需要的最短时间。阻变存储器的运行速度一般在几十纳秒到几百纳秒。

开关比是指器件高阻态电阻值与低阻态电阻值的比值。开关比的大小直接影响到判读数据的准确性。一般情况下阻变存储器的开关比要大于 10。

耐受性是指阻变存储器器件能够发生高低阻态转变的周期数,也称作疲劳特性。

保持特性是指存储在存储单元中的数据能够被保持多长时间。一般商业化的存储器产品都需要保证其保持时间能达到 10 年以上。

器件良率是指性能优良的器件占总生产器件的比率,该参数对任意产品都很重要。采用合适的电极材料、使用掺杂技术等方法均可提高阻变器件的器件良率。

阻变存储器一般有以下三种分类方式。

(1) 按极性划分。

单极性:在单极性的 RRAM 中,置位/复位操作是由相同极性的电压脉冲触发,如图 2.3(a) 所示。

双极性:在双极性的 RRAM 中,置位/复位操作由极性相反的电压脉冲触发,如图 2.3(b) 所示。

非极性:在非极性 RRAM 中,置位/复位操作可以由任何极性的电压脉冲触发。

(2) 按阻变层材料划分:氧化物阻变存储器、非氧化物无机材料阻变存储器、有机材料阻变存储器、二维材料阻变存储器等。

(3) 按阻变机制划分:电化学金属化导电细丝阻变存储器、价态变化存储器、相变阻变存储器、铁电阻变存储器等。

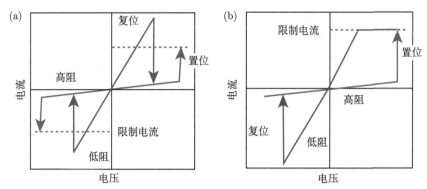

图 2.3 单极性 RRAM 器件 (a) 和双极性 RRAM 器件 (b) 的 I-V 特性曲线示意图 [21]

2.2 阻变存储器的材料

2.2.1 氧化物阻变存储器

人们已经注意到，名义上为绝缘体的氧化物可能会经历突然的电阻转换而转换为导电状态 [21]。20 世纪 60 年代第一次出现了关于这种电阻转变现象的报道 [22]，但当时的电阻转变特性未实现很大的应用价值，这导致人们逐渐丧失了对存储器的研究兴趣。到了 20 世纪 90 年代，人们再次掀起了对电阻转变现象的研究热潮，很多的二元金属氧化物都发现了电阻转换行为，其中大部分为过渡金属氧化物，还有很多的镧系金属氧化物。我们无法给出一个涵盖所有材料的综述，只能选择部分氧化物材料进行介绍，这几种氧化物材料为 HfO_x、AlO_x、NiO_x、TiO_x 和 TaO_x，因为它们在过去几年中引起了最多的关注。基于这几种材料的阻变存储器的性能将在下面进行讨论 [21]。

1. 铪氧化物 (HfO_x)

缺陷丰富的 HfO_x 是性能优异的 RRAM 材料。在早期的 HfO_x 存储器研究中，$TiN/HfO_x/Pt$ 结构 [23] 受到研究人员的广泛关注。为了使制备工艺与 CMOS 工艺更加兼容，人们在 TiN 与 HfO_x 之间加了一层金属覆盖层。Lee 等 [24] 通过沉积退火调整了带有 Ti 覆盖层的 HfO_x 的介电强度，该双极性 HfO_x RRAM 的运行速度小于 10ns、开关比大于 100、耐受性大于 10^6 个周期、高温条件下仍有较长的寿命、能够实现多位存储以及器件良率较高。Govoreanu [25] 报道了一个以 Hf 为覆盖金属层的 $TiN/Hf/HfO_x/TiN$ 的 RRAM 器件，器件的透射电子显微镜 (TEM) 照片如图 2.4(a) 所示，该器件展现出了非常优异的性能。HfO_{2-x} 有很大的带隙宽度 [26] 和电阻变化范围 [27]。根据近几年的研究进展，HfO_{2-x} RRAM 在高温和低温条件下的保持特性均得到了改进 [28,29]。Chand 等 [28] 在 Ti 顶电极

2.2 阻变存储器的材料

和 HfO_{2-x} 层之间加入 Al_2O_3 层并且进行了高温真空退火和金属后退火 (PMA) 处理，提升了器件的热稳定性和可靠性。另外，Fang 等[29]采用 $Pt/HfO_x/TiN$ 结构来提高 RRAM 器件在低温条件下的性能稳定性，相关性质可见图 2.4(b) 与 (c)。目前，较高的保持特性和稳定的多级电阻仍是研究人员努力研究的方向。Qi 团队[30]制造了 $W/HfO_{2-x}/Pt$ RRAM 结构，该结构表现出出色的保持特性、耐受性和受负扫描电压或正 CC(compliance current) 控制的可靠的多级电阻。为了在交叉开关阵列上实现大规模集成，该团队还制造了具有 $Ag/HfO_{2-x}/Ag$ 结构的选择器，选择器和存储器相连具有处理漏电流的巨大潜力，并能够简化两个器件 (选择器和阻变存储器) 的制造过程，这为将来的多值存储提供了一种新的方法和指南。到目前为止，HfO_x 是探索的最成熟的 RRAM 材料之一。

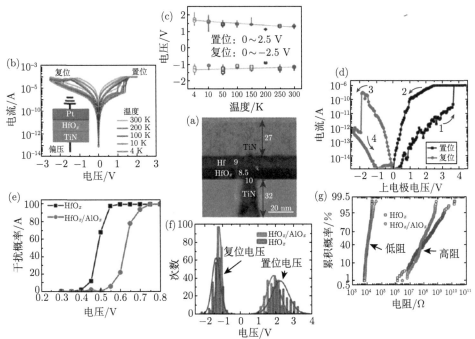

图 2.4 (a)$TiN/Hf/HfO_x/TiN$ RRAM 器件的 TEM 照片[25]；(b)$Pt/HfO_x/TiN$ RRAM 器件从 4~300 K 的 I-V 曲线[29]；(c) 器件在不同温度下的置位以及复位电压[29]；(d)$Al/AlO_x/Pt$ RRAM 器件的典型直流开关特性[32]；(e)AlO_x 层可以将器件的时间扰动电压提高到 0.5 V[34]；(f) 通过 200 个直流扫描周期获得的置位/复位电压的分布[35]；(g) 通过 200 个直流扫描周期获得的 HRS/LRS 电阻分布[35]

2. 铝氧化物 (AlO_x)

Hickmott 的研究首次报道了阳极 AlO_x 的负阻现象[22]，之后研究人员对 AlO_x 又展开了大量的研究。AlO_x 较大的带隙导致 AlO_x RRAM 器件有较低的复位电流,很多 AlO$_x$ RRAM 器件的典型转换电流可达到毫安量级或数百微安,而 Wu 等证实了在 Al/AlO$_x$/Pt 结构中复位电流可以下降到 1 μA 以下 (图 2.4(d))[31]。最近,Kim 团队通过向 AlO$_x$ 中掺氮实现了小于 100 nA 的复位电流[32]。较小的复位电流降低了 RRAM 器件的能耗;小的复位电流会导致器件低电阻状态时的电阻值较大 (可能位于兆欧姆量级),较大的 LRS 电阻可实现不依赖选择器件的超大规模存储阵列,因为大的 LRS 电阻可显著降低潜在漏电流[33]。除此之外,AlO$_x$ 也可以与其他 RRAM 材料共同使用以提升器件特性的均匀性。Chen 等[34]在 HfO$_x$ 电阻层下方使用了 AlO$_x$ 缓冲层,以增强读取状态下的抗干扰性 (图 2.4(e))。Yu 等[35]利用双层 HfO$_x$/AlO$_x$ 结构,获得了比纯 HfO$_x$ 层更好的开关均匀性,可见图 2.4(f) 与 (g)。

3. 镍氧化物 (NiO$_x$)

NiO$_x$ 是人们最早研究的可应用于 RRAM 的材料。NiO$_x$ RRAM 的复位电流一般比较大,但最近有研究团队将复位电流缩小到 10 μA 以内[36]。当使用铂、金、钨和镍等金属作为电极时,NiO$_x$ RRAM 经常表现出单极性电阻转换特性,该特性使得 NiO$_x$ RRAM 允许在写和读操作期间使用二极管而不是晶体管来作为存储阵列中的存储单元,即可形成 1D1R 结构[37]。但是对于单极性的 NiO$_x$ RRAM 来说,由于复位和置位具有相同的电压极性,所以复位和置位参数的窄分布对 NiO$_x$ RRAM 至关重要。NiO$_x$ RRAM 众所周知的问题之一是开关均匀性较差,必须克服这一问题才能降低故障的发生并且实现多级电阻。后来有研究表明,通过使用新型的纳米导电细丝约束结构可以改善 NiO$_x$ RRAM 的开关均匀性[38]。近年来,Kang 团队利用离子液体在正电压下对 NiO$_x$ 层进行了预处理,他们发现离子液体预处理过的 NiO$_x$ RRAM 具有均匀的工作电压和电阻,且在电阻转换期间有增大的 R_{off}/R_{on} 比[39],如图 2.5 所示。Tang 团队制作了 Pt/NiO$_x$/ITO 结构的 RRAM,且观察到该器件可再现的单极性电阻转换 (URS) 和双极性电阻转换 (BRS) 性质。根据 X 射线光电子能谱 (XPS) 的测试结果,在 NiO$_x$ 膜中存在氧空位,并且这种缺陷被证明与 URS 和 BRS 效应密切相关[40]。

4. 钛氧化物 (TiO$_x$)

TiO$_x$ 也是最早被研究的可用于 RRAM 的材料之一。多数研究主要集中在其阻变转换机制方面,对于 TiO$_x$ RRAM 器件本身性能的研究还是较少。Kwon 等[41]通过 TEM 直接观察到存在于 TiO$_x$ 忆阻器中的纳米丝,见图 2.6(a) 与 (b);Pt/TiO$_2$/Pt RRAM 具有单极性电阻转换特性[42],而 Pt/TiO$_x$/TiN RRAM 表现出双极性电阻转换行为,但是在 Pt/TiO$_x$ 的接触面仍可以观察到单极复位特性[43]。有很多研究团队将亚化学计量的 TiO$_{2-x}$ 层与化学计量的 TiO$_2$ 层故意堆叠在一起以形成双层结构,例如,Yang 等[44]制作出 50 nm×50 nm 的

2.2 阻变存储器的材料

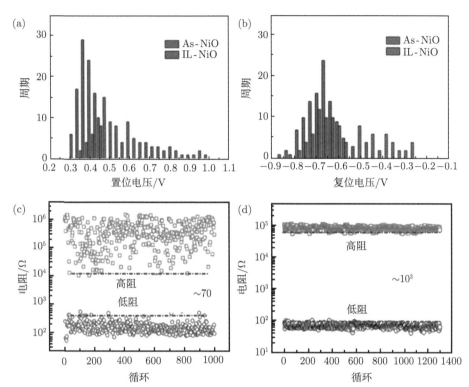

图 2.5 (a)100 次周期的置位电压分布图；(b)100 次扫描周期的复位电压分布图；(c) 未经过离子溶液处理的 NiO RRAM 器件的保持特性；(d) 经过离子溶液处理的 NiO RRAM 器件的保持特性。As-NiO 与 IL-NiO 分别指的是未经过与经过离子溶液预处理的 NiO RRAM 器件[39]

Pt/TiO$_2$/TiO$_{2-x}$/Pt 器件，其原子力显微镜 (AFM) 图像可见图 2.6(c)，该器件表现出双极性转换特性并且顶电极加负向偏压时发生了置位；Strukov 等[45] 认为这个器件的转换机制是带正电荷氧空位的迁移使得富氧/缺氧区域的交界前线向前或者向后推移。随后，Do 等[46] 研制了一个相似的 Pt/TiO$_2$/TiO$_{2-x}$/Pt 存储器，且认为双极性转换行为可能归于 TiO$_{2-x}$ 和 TiO$_2$ 层界面处的氧化还原反应，此氧化还原反应是由氧离子从富氧区域迁移到缺氧区域以及 TiO$_{2-x}$ 层中的部分氧陷阱而引起的。所以说，关于这种富氧/缺氧双层结构的转换机理尚未达成共识。近来，Zhidik 团队[47] 利用化学计量的 TiO$_2$ 和非化学计量的 TiO$_x$ 制备出不含贵金属和稀土元素的 Mo/TiO$_x$/TiO$_2$/Cu RRAM，并且证明了该 RRAM 只能在真空电铸激活过程之后才可操作和使用；在电铸激活过程中，该结构发生了不可逆的变化，导致在伏安特性中形成负微分电阻区。Ding 团队[48] 分别制备了 10 nm 与 30 nm 厚度的 TiO$_x$ 薄膜 TiN/TiO$_x$/HfO$_2$/Pt RRAM 器件；通过测试证明了厚的 TiO$_x$ 薄膜的 RRAM 有较大的开关比和较小的操作电压，该器件可在最小 20 ns 脉冲宽度下实现 10 倍的电阻变化且耐受性可达到 10^6 个脉冲循环。

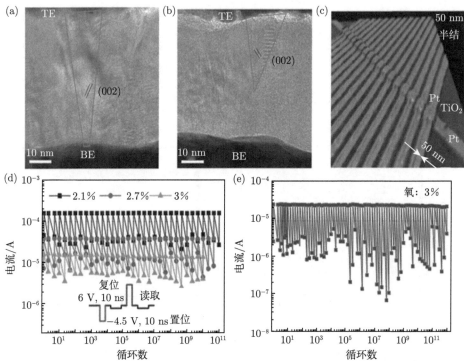

图 2.6 (a)Ti_4O_7 纳米丝的高分辨率 TEM 图像；(b) 未连接的锥形 Ti_4O_7 结构[41]；(c) 器件的 AFM 图像[44]；(d) 尺寸为 0.5μm×0.5 μm 的器件在沉积条件分别为氧分压 2.1%、2.7%与 3%时的耐受性数据；(e) 尺寸为 30μm×30 μm 的器件在沉积条件为氧分压 3%时的耐受性数据[51]

5. 钽氧化物 (TaO_x)

TaO_x RRAM 因为其较好的耐受性而广受关注。有研究团队用 XPS[49]和 EELS(电子能量损失谱)[50]揭示了 TaO_x 通常包含两相，其中一相更接近导电性较强的 TaO_2 相，另一相更接近绝缘性更强的 Ta_2O_5 相，电阻转换时氧迁移存在于两相之间。Lee 团队[51]利用 TaO_2/Ta_2O_5 双层结构研制了 RRAM 器件，该存储器有速度快、高耐受性和微缩性强等性能。在 TaO_x RRAM 器件耐受性方面也有很多瞩目的研究成果。Wei 等[50]报道了耐受性有 10^9 个周期的 TaO_x RRAM 器件。Yang 等[52]报道的 TaO_x RRAM 器件的耐受性达到了 10^{10} 个周期。而 Lee[51]发现 TaO_x RRAM 的耐受性可以达到 10^{12} 个周期，耐受性数据可见图 2.6(d) 与 (e)。除此之外，研究人员也在减少 TaO_x RRAM 器件的干扰和提高器件的参数均匀性等方面做了很多尝试。Li 团队[53] 提出了一种利用小宽度多脉冲进行复位脉冲操作编程的方法，该方法减少了复位过程中的能量消耗，极大地改善了 TaO_x RRAM 器件的读出干扰。另一研究团队[54] 研制的 ITO/TaO_x/TiN RRAM 器件可以在低工作电压下保持大于 10^4 个周期的较高的均匀性，置位和复位的平均电压分别为 0.036 V 和 −0.109 V，这些特性归因于氧化铟锡 (ITO) 电极和 TiON 层的富氧特性。TaO_x RRAM 器件有很好的嵌入式应用前景，并且有可能改变存储器的层次结构。

2.2.2 非氧化物无机材料阻变存储器

RRAM 在高密度存储器应用中引起了人们的广泛关注，然而还是有很多问题亟待解决，例如如何降低其操作电流并且提高器件的可靠性，为此人们对非氧化物无机材料阻变存储器展开了研究。我们在下文中将要介绍几种性能优异的非氧化物无机材料以及它们为 RRAM 带来的性能提升。

1. 硫族化合物

硫族化合物或硫属化物 (chalcogenide) 是指至少含有一种硫族元素 (氧族元素中除了氧以外的元素) 离子及一种电负性较小的元素的化合物。一般硫族元素是指硫、硒、碲、钋等元素，而电负性较小的元素一般是指砷、锗、磷、锑、铅、硼、铝、镓、铟、钛、钠等元素。一般来说，基于硫族化合物的信息存储器件可以分为两大类[55]。

一类是相变存储器，该类存储器的信息存储是通过将硫族化合物材料的纳米晶粒从无定形状态转变为晶体状态来实现的[56]，这种转换需要将热量施加到纳米级的晶粒上，可以通过光学或者电学方法实现。光学相变存储器在 20 世纪 90 年代得到发展并商业化，目前已经被人们广泛应用，例如可重写的 DVD 光盘。因为电荷存储器 (EPROM、flash 等) 的广泛发展，电学相变存储在最初阶段并没有得到广泛发展。直到近期电荷存储器的微缩问题日益突出，电学相变存储才重新获得人们的关注。含有 Sb-Te 的合金，Ge-Sb-Te(GST) 等多用作相变存储的功能层[57−59]。

另一类的存储器其存储机制则较为新颖，它被称为可编程金属化单元存储器。其机理是运用电化学的方法让固态电介质中的金属阳离子迁移来改变其电阻进而进行信息的存储[60,61]。可编程金属化单元 (programmable metallization cell, PMC) 最初是由 M. Kozicki 在 21 世纪初期提出[60]，目前很多研究团队都在这个方向展开研究。该类存储器的功能材料多为含有金属的硫族化合物，例如 Ag-Ge-Se[60,61]、Ag-Ge-S[62]、Cu-S[63]。可编程金属化单元存储器相较于相变存储器在功耗和微缩性上都更有优势。然而，在阻变存储器中使用的硫族化合物大多为二维材料，所以具体的研究成果我们将在二维材料部分进行介绍。

2. 氮化物

氮化物也是很有研究和应用价值的 RRAM 材料，以 AlN 和 Cu_xN 等材料为基础的 RRAM 器件有着较低的操作电压/电流和很好的 CMOS 工艺兼容性。氮化物薄膜在 RRAM 应用中有很多杰出的性能，例如较高的导热性、很好的绝缘性以及较高的介电常数[64,65]。Zhang 团队[66] 提出一种完全由金属氮化物构成的 TiN/AlN/TiN RRAM 器件，其操作电流小于 100 μA，150°C 环境下其保持时间可达 3×10^5 s，耐受性可达 10^5 个脉冲周期，如图 2.7(a)~(c) 所示。全金属氮化物 RRAM 器件在未来的非易失性存储器应用中具有很大的潜力。Kim 团队[67] 提出了 Ti/AlN/Pt 结构，其中 AlN 薄膜厚度为 130 nm 且通过射频溅射形成，该器件显示了更加优异的电阻转换性能，运行速度很快 (小于 10 ns)、操作电流较低 (小于 10 μA)、耐受性超过了 10^8 个周期，在 85°C 的环境下保持时间可超过 10 年。Du 团队[68] 研究了氧等离子体注入对氮化铜薄膜电阻转换性能的影响，电

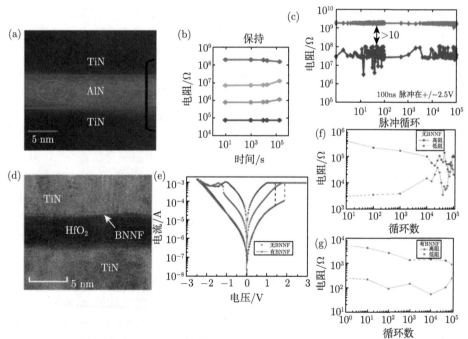

图 2.7 TiN/AlN/TiN RRAM 器件[66]：(a) 器件的 TEM 图像；(b) 器件在 150°C 条件下的保持特性 (读电压为 0.3 V)；(c)100 ns 宽度的脉冲下器件的耐受性 (置位电压为 2.5 V，复位电压为 −2.5 V，读电压为 3 V)。TiN/BNNF/HfO_2/TiN RRAM 器件[69]：(d) 器件的 TEM 图像；(e) 有 BNNF 与无 BNNF 的器件的电流电压特性之间的比较；(f) 无 BNNF 时器件的耐受性；(g) 有 BNNF 时器件的耐受性 (读电压均为 0.2 V)

学测量的线性拟合结果表明,形成氧化铜 (CuO) 有助于延长 Cu_xN RRAM 器件的耐受性,而 Cu_2O 的增加则会导致严重的性能下降。Chen 团队[69] 利用六方氮化硼提升了 RRAM 器件的性能,六方氮化硼纳米膜 (BNNF) 位于锡顶电极与 HfO_2 电阻转变层之间的界面上,从而形成了 $BNNF/HfO_2$ 双层结构,如图 2.7(d)~(g) 所示。$TiN/BNNF/HfO_2/TiN$ RRAM 器件有很多优异的性能,如优异的耐久性和均匀性。近年来,研究人员对氮化硅材料进行了多项研究工作,因其表现出了非常优越的耐高温、耐腐蚀、机械性能,以及具有理想的稳定性和耐压性等介电响应性能。Kim 等成功展示了具有优良存储器特性的 $Ag/Si_3N_4/Al$ 器件,其中 Si_3N_4 薄膜的厚度分布在 5~20 nm,该器件在低复位电流 (1 nA) 时复位操作有双极电阻转换特性,对于 10 nm 厚的 Si_3N_4 层,在 ±1.2 V 时开关比可达到 10^7 以上[64]。$Cu/Si_3N_4/p+Si$ RRAM 具有单极性和双极性的电阻转换性能。双极性电阻转换不需要电铸激活过程,且在耐受性、器件参数均匀性和开关比等方面更加优异[70]。利用氧掺杂的 ZrN_x 薄膜制备的透明 $Pt/ZrN_x/Ti$ RRAM 在室温下和 85℃ 的环境下均有大于 10^4 s 的保持时间,并且在室温下该器件在正偏压区电流比为 $5×10^3$,在负偏压区电流比为 $5×10^5$,该器件的性质意味着,通过控制溅射过程中的工作压力来沉积 ZrN_x 薄膜的透明存储器可能成为未来透明电子器件的一个里程碑[71]。这些结果表明,基于氮化物的 RRAM 可以用作有前途的高速存储器件。

2.2.3 有机材料阻变存储器

有机材料阻变存储器就是由夹在两个电极之间的有机材料构成的存储器,它至少要有两个由外部电路控制的稳定的电阻状态。有机材料的引入使它有很多优异的性能,例如简单的存储器结构、较低的制造费用和较高的灵活性。除此之外,它也满足高的开关比和耐受性、较好的保持性能以及较快的运行速度等重要的性能指标。

到目前为止,很多有机材料都展现出了在外加电场下的电阻转变性能,有机小分子物质就是其中之一。在过去二十多年里,有机小分子一直都是阻变存储器应用中的重要研究对象。它的本征阻变性能依赖于给-受体结构的协同作用,三苯胺 (TPA)、甲氧基等具有给电子能力的给体基团,而咪唑-[4,5-b] 吩嗪 (BIP)、1,8-邻苯二甲酰亚胺等具有吸电子特性的受体基团[72]。Wang 团队则报道了一系列具有不同 TPA/BIP 比值的具有电阻转换性能的小分子物质[73-75]。给-受体体系在外电场的作用下发生电荷转移,导致薄膜中的导电沟道导通,从而使器件的电阻状态从高阻态切换为低阻态。除此之外,还可通过单原子取代、端基取代、烷基链长度调整等方式对有机小分子的阻变特性进行调节。

除了有机小分子,有些聚合物材料在阻变存储器应用中也展现出了优秀的性能。聚合物材料具有高柔韧性，易于溶液加工，较有机小分子更适合应用于阻变存储器中。到目前为止，研究人员们已经在单组分电活性聚合物、聚合物共混物、聚合物-小分子混合物、聚合物-纳米颗粒或聚合物-无机化合物混合物等材料中观测到阻变存储特性[76]。Song 团队[77]提出了一种基于 MIM 三明治结构的阻变器件。其中聚(n-乙烯基咔唑)(PVK) 与金纳米颗粒 (GNP) 的混合物作为两个金属电极之间的活性层。通电之后，该器件可从低导电状态过渡到高导电状态。通过施加反向偏压，高导电状态可以恢复到低导电状态。在室温下，该器件的开关电流比高达 10^5，且在两种状态下的应力测试中均表现出良好的稳定性。Kim 团队[78]研究了亚微米级通孔器件结构中聚芴衍生物聚合物材料的电阻转换特性。亚微米级聚合物存储器件表现出出色的电阻转换性能，例如大的开关比 (10^4)，良好的耐受性和保持特性 (大于 10000 s)。除此之外，Cho 团队[79]使用耐热的聚酰亚胺和 PCBM 复合膜制作了高分子阻变存储器，该挥发性存储器在没有封装的情况下具有大于 1000 个写入/擦除周期以及 1 周的数据保持能力，对该器件的顶层和底层进行加热 (大于 300℃ 的热量且持续 1h)，其并未展现出性能上的衰退。郑州大学的 Cao 团队[80]利用聚乙烯胺 (PVAm) 来部分替代甲胺离子 (MA^+) 制造出了稳定且灵活的聚合物 OHP(organic-inorganic hybrid perovskite)RRAM 器件，与未经修改的 OHP 器件相比，经 PVAm 改善的 OHP RRAM 器件显示出显著提高的开关比、长期稳定性和灵活性。有机复合材料通常用于引发电阻转换，包含有机或无机纳米颗粒的聚合物共混体系主要构成了阻变存储器件的活性材料[81]。

2.2.4 二维材料阻变存储器

2004 年，曼彻斯特大学的 Geim 和 Novoselov 成功分离出单原子层的石墨材料——石墨烯，随后二维材料的概念被提出。二维材料指的是电子仅可以在两个维度的纳米尺度 (1~100 nm) 上自由运动的材料。二维材料超薄、柔韧且具有多层结构的特性使得它具有独特的电气、化学、机械和物理性能。石墨烯、石墨烯衍生物和二硫化钼是 RRAM 应用较为广泛的二维材料，最近其他的一些二维材料如 hBN、$MoSe_2$、WS_2 和 WSe_2 也被作为非挥发性 RRAM 器件的功能材料进行了研究[82]。金属或固有离子的迁移是这些器件导电的主要原因。基于二维材料的 RRAM 器件最突出的优点是开关速度快 (小于 10 ns)、功率损耗低 (10 pJ)、阈值电压低 (小于 1 V)、保持时间长 (大于 10 年)、耐受性高 (大于 10^8 个电压周期) 和扩展的机械鲁棒性 (500 次弯曲)。通过将它们与各种形式的金属纳米颗粒、有机聚合物或无机半导体混合，可以进一步增强它们的电阻转换性能，下面将简要介绍一些二维材料阻变存储器[82]。

1. 石墨烯

很多以石墨烯为基础的 RRAM 器件均有独特的机械、光学、电气以及热性能。Shin 团队[83]和 Doh 团队[84]被认为是将石墨烯应用为非挥发性存储器活性层的先驱者。其中，Shin 团队[83]研制的存储器具有双极性双稳态的开关特性，耐受性可达到 100 次电压扫描，但是其开关比较低 (仅为 13)。较低的开关比可能与石墨烯较高的导电性有关。Doh 团队[84]探究了石墨烯活性层的厚度对 RRAM 器件的影响，但是该器件大于 30V 的操作电压使其能耗较大，且其开关比仍较低。

由于前些年报道的二维材料具有鼓舞人心的特性，2012 年人们发表了更多石墨烯应用为 RRAM 器件活性层的文章。Wu 等[85]利用单层石墨烯片制备了一个开关比可达 10^6 的 RRAM 器件，如图 2.8(e) 与 (f) 所示，该器件与之前的基于石墨烯的 RRAM 器件的主要区别在于结构，它是平面结构而不是典型的夹心结构的两端子器件。He 团队[86]提出了另外一种有很多优良性能的基于石墨烯的 RRAM 器件；这个存储器也是平面结构的两端子器件，并且有总共五个电阻状态，其运行速度很快 (500 ns) 且器件良率很高 (98%)；阳极和阴极之间是纳米尺寸而不是微观尺寸使得该器件有很小的操作电压和能耗，弥补了之前 Wu 等[85]的微型平面存储器件的不足；此 RRAM 器件的缺点在于，在其电极上施加双电压扫描以引起电击穿之前，需要进行电铸激活步骤 ($V_{break} = 5.5$ V)；与典型的 RRAM 器件不同，该器件均未显示单极性或双极性电阻切换行为，而是表现出独立于电压极性的非极性电阻切换行为。Kim 等[87]认识到石墨烯单层极好的柔韧性，并在透明的聚萘二甲酸乙二醇酯 (PEN) 衬底上制造了 RRAM 器件；由于在其表面上存在高度透明的石墨烯层，因此仅观察到 PEN 基材的透明度损失了 8%；该器件具有 8.6 V 的存储窗口，保留 10 年；他们的器件在连续的压缩和拉伸应力循环下显示出高度稳定且可重复的双稳态电阻状态；该 RRAM 器件有可能被用作在弯曲下具有高耐久性的柔性透明电子存储器件。

Lai 等[88]研制了一种基于石墨烯纳米片的 RRAM 器件，功能材料为溶液可加工的聚乙烯醇与石墨烯纳米片的混合物，这是关于通过低温单步处理技术将石墨烯直接集成到 RRAM 器件的首个报告；该器件的操作电压相对较低 ($-1.3 \sim +1.4$ V)，电耐受性可达 10^7 个电压周期；该器件最突出的特性为通过极其简单且廉价的溶液处理方法实现了 RRAM 器件极好的电阻转换性能。Khurana 等[89]报道了一种具有三层结构的 RRAM 器件，中间是石墨烯片，这层石墨烯片被两层聚偏二氟乙烯 (PVDF) 夹在中间，底部电极是 ITO，顶部电极是 Pt；该器件有四个稳定的电阻状态，保持时间可达到 10^4 s，电压耐受性很好，其最突出的特性为具有多级电阻，可以实现高密度信息存储。Wu 等[90]将聚合物/SLG/聚合物/单层石墨烯 (single-layer graphene)/聚合物夹在两个金属电极之间研制了基于单层石墨烯的 RRAM 器件。它有很高的开关比 (10^4)，但是其结构复杂并且稳定的电阻态较少，使得信息存储的密度较小。Liu 等[91] 使用石墨烯制造出能耗低于千万分之一焦耳的 RRAM 器件，该器件在未来很可能应用于神经形态计算系统。Yalagala 等[92] 研发了一种基于 MgO-PVP-石墨烯的 RRAM 器件，该研究的显著特点是由于聚乙烯吡咯烷酮 (PVP) 和 MgO 的高水溶速率，这个器件在小于 3 s 的时间内即可溶于水，如图 2.8(a)~(d) 所示。

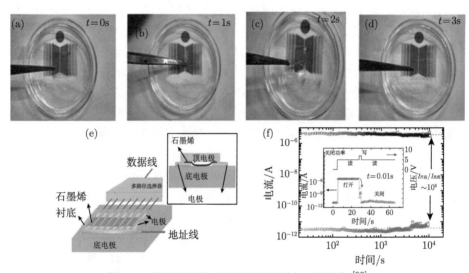

图 2.8　存储器活性层逐渐溶解的瞬态光学图像[92]
(a) 初始状态；(b)1 s 后浸入器件；(c)2 s 后部分溶解；(d)4 s 后完全溶解。Wu 等利用石墨烯片制备的 RRAM 器件[90]：(e) 器件的原理图；(f) 器件的开关比以及保持特性测试

2. MoS_2

Liu 等[93] 揭示了夹在还原氧化石墨烯和铝之间的 MoS_2-PVP 混合纳米复合材料的电阻转换效应，该效应是由混合有 PVP 的 MoS_2 薄片的载流子释放和捕获所致，从此 MoS_2 可实现非易失存储器的潜力受到研究人员的关注。除了 MoS_2 纳米薄片，Xu 等[94] 使用 MoS_2 纳米微球作为三明治结构中还原氧化石墨烯和 ITO 电极之间的功能层，该基于 MoS_2 纳米微球的 RRAM 器件有着更加优异的性能，其开关比可高达 10^4，且操作电压较低 (± 2 V)，保持时间可以达到 10^4 s。Bessonov 等[95] 提出了一种溶液处理的由 MoS_2 和 MoO_x 组成的双分子层结构，它表现出了双极性电阻转换行为并且实现了多级电阻，该器件开关比较高 (10^6)，操作电压为 ± 0.2 V；该 RRAM 器件的制备过程基于印刷技术，因此其制备工艺简单且耗资较低，性能优于大部分基于石墨烯或 MoS_2 的 RRAM 器件。Sangwan 等[96] 报道了基于 MoS_2 单层的双极性电阻转换现象，该器件最突出的特性为阈值电压可调，调整范围为 3.5~8 V，这种独特的性能为双极和互补开关器件的设计提供了新的思路。Wang 等[97] 提出了基于 MoS_2 纳米微球的双稳态光阻 RRAM

器件，对该器件的控制主要是通过纳米微球在光照和黑暗条件下的极化；可以通过改变白光的照明强度来实现多级电阻转换，然而其开关比较小是一大缺点 (仅为 10 左右)。Rehman 等[98] 还报道了基于片状 MoS_2 的纳米复合材料的双极电阻转换行为，其中片状 MoS_2 散布在聚乙烯醇 (PVA) 的有机聚合物中以形成杂化纳米复合材料。该器件最突出的特性为其保持时间可达到 10^5 s，即使达到 1500 个弯曲周期，其性能仍没有很明显的衰退。

Fan 等[99] 报道了 Au/MoS_2-PVK/ITO RRAM 器件的电阻转换行为，功能材料为 MoS_2 和 PVK 的杂化纳米复合材料，这是 MoS_2 纳米片与 PVK 共价键合的首个表现出可重写性能的 RRAM 器件。Wang 等[100] 研究了夹在两层聚甲基丙烯酸甲酯 (PMMA) 聚合物之间的 MoS_2 和 PMMA 杂化纳米复合材料的电阻转换行为，该研究的特别之处在于 MoS_2 首次以量子点的形式被应用于 RRAM 器件，而该器件有比基于其他量子点材料的 RRAM 器件更低的能耗。Li 等[101] 研究了基于数层 MoS_2 的 RRAM 器件的电阻转变行为，与众不同的是该器件是横向/平面结构而不是传统的三明治结构，该工作的主要目标就是为模拟人脑神经系统提供一种方法，并且成功地证明了该器件在模拟人脑方面的潜力。之后，Chen 等[102] 提出了具有两个三明治层的 RRAM 器件，MoS_2 起到分离和阻断层的作用，位于过渡金属氧化物 HfO_2 和 TiO_2 之间，该器件的成功之处在于成功地模拟了长程可塑性 (LTP) 和短程可塑性 (STP)。Kim 等[103] 报道了一项很有突破性的研究，他们利用 MoS_2 的电阻转换特性为下一代物联网和可重新配置的通信系统制造了极低功率的射频开关。Qiu 等[104] 提出了一种重要的 RRAM 器件结构，通过优化 MoS_2 的界面层，该器件可在双极性、整流性和互补性电阻开关之间切换。Das 等[105] 报道了基于 MoS_2 的多层多功能 RRAM 器件，它有多位存储能力且对可见光敏感。Feng 等[106] 报道了一种基于 MoS_2 的全打印的 RRAM 器件，该器件有挥发性和非挥发性存储的功能，开关比高达 10^7。而 Zhai 等[107] 提出的基于 MoS_2 的 RRAM 器件既可以用作存储器件，又可以用作红外探测器。

3. 其他二维材料

Yan 等[108] 在平面结构的器件中使用 $MoSe_2$ 连接 Ag 阳极和 Au 阴极并发现了电阻转换现象。之后，Han 团队[109] 研究了嵌入 TiO_2 中的 $MoSe_2$ 阵列的纳米岛中的阻变效应，该器件具有光控电阻转换效应，在光照和黑暗条件下都表现出了稳定的电阻转换行为。而后，有研究团队分别研究了基于 $MoSe_2$ 纳米岛[110]、$MoSe_2$ 掺杂的 Se 微丝[111] 及六角 $MoSe_2$ 纳米片[112] 的 RRAM 器件，然而都未能展示出突出的性能。Li 等[113] 报道了一项十分重要的研究，他们研究了温度对 $MoSe_2$ 电阻转换效应和磁特性的影响，$MoSe_2$ 的磁力与温度呈反比关系，而其电阻转换行为受温度的调节，这为未来高性能 $MoSe_2$ RRAM 器件的制备提供了新的思路。

WS$_2$ 是具有良好电学和机械性能的二维材料,在电学耐受性和机械鲁棒性等方面,基于 WS$_2$ 的 RRAM 器件相较于其他二维材料显示出了更加优异的性能。Das 团队[114]制备了基于 WS$_2$ 的 RRAM 器件,其中的 WS$_2$ 薄膜由化学气相沉积 (CVD) 方法制备,该器件的开关比高达 10^3,耐受性为 100 个电压周期。除此之外,Li 等[115]提出了一种通过气溶胶喷射印刷技术制造的完全印刷的 WSe$_2$ RRAM 器件,它既表现出挥发性又表现出非挥发性的单极性电阻转换行为。

2.3 阻变存储器的阻变机制

2.3.1 电化学金属化导电细丝

电化学金属化导电细丝主要依赖于电化学活性电极来形成基于金属阳离子的导电细丝[116]。导电细丝直径为纳米量级,它连接顶部电极和底部电极。当导电细丝连接顶部与底部电极之时,阻变存储器处于低电阻状态 (LRS);而当顶部与底部电极之间无导电细丝连接,也即两电极之间出现空隙之时,阻变存储器处于高电阻状态 (HRS)[117]。

电化学金属化导电细丝存储器的英文名称为 electrochemical metallization of conductive filament memory,其电阻转换现象主要基于金属离子的迁移以及随后发生的氧化还原反应[118]。从结构上来讲,电化学金属化导电细丝存储器主要包括三部分,第一部分是可氧化的顶部电极,第二部分是金属氧化物层,最后一部分是相对于顶部电极来说化学惰性较强的底部电极。活性顶部电极溶解,金属阳离子迁移并在惰性底部电极上发生还原反应以金属态沉积。如图 2.9 所示,顶部电极采用 Ag 而底部电极采用 Pd。对顶部电极 Ag 施加正偏压,氧化反应在此发生使得 Ag 失电子变为 Ag$^+$,随后 Ag$^+$ 进入介质层中。加在 Pd 电极上的负偏压能够吸引 Ag$^+$,Ag$^+$ 到达底部电极之后发生还原反应成为 Ag 原子。Ag 原子随着堆积时间增加而逐渐堆积成电化学金属化导电细丝,此时 RRAM 器件处于 LRS。当所施加的电压的极性反转时,形成的电化学金属化导电细丝几乎能够完全溶解,此时 RRAM 器件处于 HRS[119]。

2.3.2 价态变化存储器

价态变化存储器内部的导电细丝由氧空位缺陷组成,而不是金属原子。如图 2.10 所示,该阻变机制与氧空位的产生和随后氧离子的堆积有关,因此中间介质层的软击穿是很关键的一步。一旦软击穿发生,氧原子会在指向阳极界面方向的强电场作用下脱离晶格位置变成氧离子 (O^{2-}),同时氧空位会留在氧化物层

2.3 阻变存储器的阻变机制

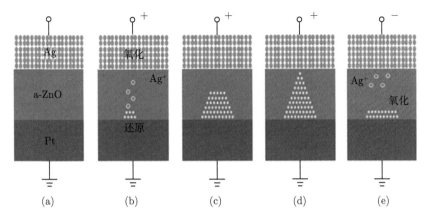

图 2.9　电化学金属化导电细丝 RRAM 的电阻转换机制示意图[116]
(a) 原始状态；(b) 与 (c)Ag 的氧化和 Ag^+ 阳离子向阴极的迁移及其还原；(d)Ag 原子堆积成电化学金属化导电细丝，电极之间形成通路使得器件处于 LRS；(e) 施加相反极性的电压会发生导电细丝的溶解

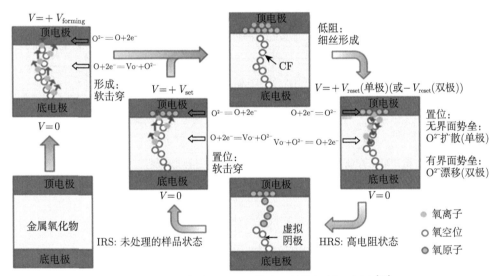

图 2.10　价态变化 RRAM 的电阻转换机制示意图[21]

中[116]。如果贵金属被用作阳极材料以形成界面氧化层，那么氧离子 (O^{2-}) 则会与阳极材料反应或以中性非晶格氧的形式排出，因此电极/氧化物界面起到了储氧层的作用[120]。堆积的氧空位形成氧空位导电细丝从而使得 RRAM 呈现出 LRS。通过复位过程我们可将器件由低电阻状态切换回高电阻状态。在复位过程中，氧离子 (O^{2-}) 从阳极界面迁移回氧化体并与氧空位结合或氧化导电细丝使其部分断裂，从而 RRAM 器件重新回到高阻态。对于表现出单极性电阻转换特性的 RRAM 而言，氧离子 (O^{2-}) 的扩散由焦耳热激活，在浓度梯度的辅助下扩散；而在双极性电阻转换特性的 RRAM 中，由于界面层扩散势垒的存在，氧离子 (O^{2-})

的扩散需在电场的辅助下进行,单纯的热扩散是不够的[116]。

2.3.3 相变阻变存储器

相变阻变存储器 (phase change resistive memory) 是一种纳米尺度的非易失性存储器,其典型的器件结构和基本工作机制示意图如图 2.11(a) 与 (b) 所示,导电性强的晶相和导电性较弱的非晶相之间的电阻比是人们使用 PCM 进行信息存储的关键[121]。PCM 的导电机制为:PCM 结构中的相变材料可在有序的晶相和无序的非晶相之间来回转变[122],相对应地,相变材料的电学和光学特性也会发生很大变化。一般来说,非晶相的电阻较高而光学反射率较低,晶相的电阻较低而光学反射率较高。非晶相是热不稳定的玻璃,其在室温下有较长的结晶时间。但是,当我们对非晶相加热使其温度足够高但在熔点之下时,它会迅速结晶。对晶相进行加热使其熔化并迅速冷却,那么晶相会转化为非晶相。在 PCM 中,使相变材料发生相变的热量由焦耳热产生。

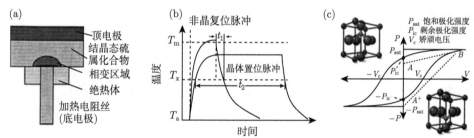

图 2.11　(a) 相变阻变存储器的典型器件结构;(b) 相变阻变存储器的基本工作机制示意图;(c) 基于 PZT 材料 FRAM 的存储原理示意图[126]

2.3.4 铁电阻变存储器

铁电材料 (如 PZT、BLZ(Bi(La, Zr, Ti)O_3)、SBT(SrBi$_2$Ta$_2$O$_9$) 等) 是一类晶体,其有较低的对称性,这导致该类晶体沿着一个或者多个晶体轴发生自发极化,铁电材料的极化-电场滞回曲线与磁滞回线高度类似。铁电晶体具有极化矢量,当对其施加外部电场时,极化矢量可以沿着两个相反的方向取向。这种相反的极化方向一般来自于带正电的金属离子和负氧离子在相反方向上的位移,该位移能够降低晶体的对称性,例如从立方晶系转换到四方晶系。如图 2.11(c) 所示,我们以 PZT 为例,对 PZT 施加正电压或者负电压时,立方晶格中心的 Ti^{4+} 或 Zr^{4+} 离子向上或向下移动,而 O^{2-} 会向相反的方向移动,PZT 晶体的极化方向会因外加电场的加入而重新定向。铁电材料在两个稳定极化状态之间切换时极化方向也会相应改变,不同的极化方向对应着大小不同的电流,这就是铁电阻变存储器 (ferroelectric random access memory, FRAM) 的物理机理[123,124]。

2.4 阻变存储器的性能提升

2.4.1 利用二维材料提升器件性能

Son 等[125] 提出了一种柔性双稳态存储器件，如图 2.12(a) 所示，该器件基于嵌入石墨烯的有机聚合物的三层结构。石墨烯层夹在两层 PMMA 之间，以提高电流稳定性并增强电荷存储能力。这项工作可以看作是石墨烯 RRAM 器件的一项突破，因为它具有非易失性存储器件的非凡特性。该器件的开关比可达 10^7，耐受性可达 1.5×10^5 个循环，保持时间能够达到 1×10^5 s。与之前提出的基于石墨烯的三端器件相比，该器件的工作电压 (±2 V) 较低，因此功耗更小。该器件即使被弯曲，当弯曲半径在 10 mm 的情况下其性能仍没有任何明显的下降，相关数据可见图 2.12(b) 与 (c)。

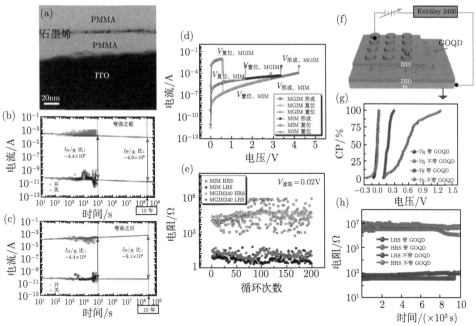

图 2.12 (a)PMMA/石墨烯/PMMA/ITO 层的 TEM 图像；(b) 器件弯曲前与 (c) 弯曲后在室温下的保持能力测试，弯曲半径为 10 mm[127]。插入石墨烯层的 MGIM 结构和传统 MIM 结构的对比[129]：(d) 初始电流电压特性对比；(e) 高低电阻值随电阻转换周期数的变化，在常温和标准大气压下测量。Ag/ZHO/GOQD/ZHO/Pt 器件[137]：(f) 结构原理图；(g)GOQD 层对器件置位电压和复位电压的影响；(h)GOQD 层对器件高低电阻态和保持特性的影响

Wu 等[126] 将石墨烯薄片嵌入两层高分子聚合物中，一层是聚苯乙烯，另一层是聚乙烯咔唑，与以前报道的器件相比，该器件的工作电压相对较高，在 ±6 V

的范围内,但是,开关比可达到 10^7。Ji 等 [127] 将多层石墨烯夹在两层有机聚酰亚胺聚合物之中,该 RRAM 器件的保持时间 (大于 10^6 s) 远高于一般的石墨烯基 RRAM 器件 (10^4 s)。Atab 等 [128] 发现,在基于 ZnO 的 RRAM 器件中掺入石墨烯纳米片能够显著提高器件的性能,石墨烯纳米片是电荷俘获材料,并且因为其有很高的电导率可以降低阈值电压从而使器件的能耗下降,同时也使得器件的保持时间有所提高。Lee 团队 [129] 发现在氧化物层中加入带有缺陷的石墨烯层可以有效改善氧化物层的电阻转换性能,因为石墨烯层能够减少导电细丝的随机断裂,同时器件的阈值电压也会有效减少进而使器件的功耗降低,可见图 2.12(d) 和 (e)。Lee 团队 [130] 在氧化钽中插入带有纳米孔的石墨烯薄片,该层石墨烯薄片可以起到阻止氧化还原反应和离子运输的作用,薄片上纳米孔的尺寸和 RRAM 器件的 LRS 与 HRS 有着直接的关系。因为石墨烯层的氧存储作用,有石墨烯层的 HfO_2 RRAM 器件比无该层的器件在低阻状态下有更强的保持能力 [131]。Sun 团队 [132] 通过在基于 ITO/MER/Al 的 RRAM 器件中加入石墨烯层的方法成功消除了负极反应。

2.4.2 利用量子点提升器件性能

Cristea 团队 [133] 利用石墨烯量子点 (GOQD) 显著提高了 RRAM 器件的电阻转换性能,该团队将石墨烯量子点插入 P3HT 的有机聚合物中。石墨烯量子点是电荷俘获材料,可以有效增强空穴输运能力。Qi 团队 [134] 研究了基于氧化石墨烯 (GO) 的存储器的导电机制,并且通过嵌入氧化碳量子点 (OCQD) 将该 RRAM 器件的易失性存储性质更改为稳定的双极性电阻切换;通过实验,该团队证实了由于 OCQD 的加入,LRS 的保持能力有了很明显的增强。Yan 教授团队 [135] 在 ZHO($Zr_{0.5}Hf_{0.5}O_2$) 薄膜中插入氧化石墨烯量子点,以解决 RRAM 器件中由导电细丝的成核和生长的随机性而导致的开关参数的不稳定,如图 2.12(f)~(h) 所示;通过与未掺入氧化石墨烯量子点的 RRAM 器件进行对比,显然氧化石墨烯量子点的掺入能显著提升器件的性能;Ag/ZHO/GOQD/ZHO/Pt RRAM 器件有更低的阈值电压,保持能力更强,开关速度更快且置位与复位电压分布更加均匀。Yan 团队 [136] 利用自组装硫化铅量子点 (PbS QD) 来提高 RRAM 电阻转换参数的均匀性,如图 2.13(c)~(e) 所示,自组装 PbS QD RRAM 器件电阻开关 (RS) 性能优于纯 Ga_2O_3 和随机分布的 PbS QD 的 RRAM 器件,例如,阈值电压较低、置位和复位电压均匀分布、保持能力强、时间响应快速和功耗低,并且成功模拟了生物突触功能和可塑性。Younis 团队 [137] 通过引入二氧化铈 (CeO_2) 量子点作为表面电荷陷阱,来提高半导体纳米棒 (NR) 阵列的电阻切换性能,与纯 ZnO 和 CeO_2 纳米结构相比,这种器件具有出色的电阻开关特性,具有更高的开关比、更好的均匀性和稳定性。Han 团队 [138] 研制了基于溶液处理的黑磷量子点 RRAM 器件，该器件具有可调特性，有多级数据存储能力和超高的开关比，且通过调整黑磷量子点层的厚度，该器件的开关比和置位电压可以实现可预测性的调整。

2.4.3 利用外围电路提升器件性能

一般而言，通常使用施加外部串联电阻等方式来限制流过 RRAM 器件的电流，以此来防止器件出现过度编程等情况。针对串联电阻在防止器件过度编程方面效率较低的问题，Gaba 团队[139]使用片上电阻代替了串联电阻，该串联电阻的应用达到了很好的效果，该团队制造了以多晶硅为底部电极 (多晶硅电极起到片上电阻的作用) 和以金属为底部电极的 RRAM 器件，通过测试可以发现，片上电阻的存在可以增强 RRAM 器件的耐受性；当器件从 HRS 过渡到 LRS 时，片上电阻可以防止存储的电流通过器件放电。Hsu 团队[140]公开了一种温度补偿的传感电路，该电路由温度敏感原件、温度补偿电路、阻变存储器阵列和读出放大器构成；温度敏感原件能发出响应给温度补偿电路，接受响应的补偿电路可生成温度相关信号，以补偿由温度变化导致的阻变存储器的电阻值变化；该外围电路能提高 RRAM 针对温度变化的可读性。

一般来说，RRAM 器件的置位操作速度较快，而复位操作的速度较慢且成功率较低[141]。因此，三星公司提出了恒定信号脉冲编程 (constant signal pulse programming, CSPP)[142]，该算法的核心思想是：一次复位，若高阻未达预设值则再执行一次复位操作，且二次复位的脉冲电压宽度和幅度与第一次相同；若二次复位失败则再执行第三次；然而，该算法的问题在于多次同样的复位操作的变阻成功率仍然很低。为解决此问题，复旦大学课题组提出斜坡式脉冲写电路[143]，该电路与 CSPP 的不同点在于，若高阻未达到预设值则下一次的复位电压幅值较前一个多 V_{step}，直到最终实现所需的 HRS；该外围电路成功提高了写操作的成功率，避免了 RRAM 器件 HRS 过低，同时提高了器件高阻值的集中度。限流问题也是 RRAM 器件的关键问题。限流，即是在电铸激活或置位时对晶体管的栅极加一定的电压从而控制电流的大小。通过运用限流电路能够准确地控制 RRAM 器件低阻值区间，有效改善器件低阻值分布与高低阻值窗口[145]，降低器件的功耗并且提高耐久性。

2.4.4 通过杂质掺杂提升器件性能

Gao 团队[144]应用第一原理计算指出，将 Al、La、Ga 等三价元素掺入 HfO_2、ZrO_2 等四价金属氧化物中可以很好地控制沿着掺杂位点的氧空位细丝的形成，进而有效改善 RRAM 器件的电阻转换行为，提高了器件性能参数的均匀性；同时该团队研制出 Al 掺杂的 HfO_2 RRAM 器件，并通过实验和理论计算证明了器件性能参数的均匀性和电阻转换行为均有所改善。由氧空位形成的局部导电细丝在

RRAM 器件的电阻转换行为中起到了很关键的作用,然而,导电细丝的随机形成对器件参数产生了很大的影响。因此,分别有研究团队在 HfO$_2$ 层中掺入 Cu[145] 和 Gd[146] 而有效抑制了氧空位的随机形成,进而使得器件的参数均匀性有较大提升。

一般来说,中间氧化物层的厚度对三明治结构金属氧化物 RRAM 器件的性能有很大影响,较厚的氧化层会使得器件的编程电压较高,而氧化物层较薄则器件的漏电流较大且窗口较小。Wang 团队[147]意图通过优化氧化物界面本身来解决此问题,他们将一层 Pd 掺杂的 MoS$_2$ 层添加在传统 ITO/HfO$_2$/ITO 结构的电极和功能层中,可以观测到器件的编程电压显著降低,器件的阻变窗口提高了 30 倍,并且器件参数的均匀性也有较大提升。Tan 团队[148]研究了 Au 掺杂对 Cu/HfO$_2$/Pt RRAM 器件的性能影响。经过测试,Cu/HfO$_2$:Au/Pt 器件显然有更强的性能,例如,置位电压变得更低,器件的参数均匀性有较大提高,并且开关比变得更大。阻变存储器可以模仿生物突触功能,被认为是克服冯·诺依曼架构瓶颈的关键器件之一,然而当前的阻变存储器难以连续地调节传导和完备地模拟生物突触的功能。Yan 教授团队[149]研制出 Ag 掺杂的 Ag/TiO$_2$:Ag/Pt 阻变存储器。如图 2.13(f)~(i) 所示,该 RRAM 器件在正脉冲和负脉冲序列下表现出了逐渐增强和抑制的传导,实现了全面的生物突触功能和可塑性,对于进一步创建人工神经系统至关重要。

2.4.5 特殊器件形貌提升器件性能

导电细丝型 HfO$_2$ RRAM 器件被认为是未来非易失性存储器最有希望的候选者之一。因为氧空位的分布和导电细丝的生长不受控制,该类 RRAM 器件的稳定性,尤其是高阻状态下的稳定性很难进一步提高。而 Niu 团队[150]使用与 CMOS 兼容的纳米尖端研制出具有高度稳定耐受性的 TiN/Ti/HfO$_2$/Si-tip RRAM 器件,可见图 2.13(a) 与 (b),纳米尖端底部电极能为电场和离子电流密度提供局部限制,因此该方法能有效约束氧空位的分布和纳米导电细丝的位置,从而有效提升 RRAM 器件的可靠性和稳定性。

基于低温非晶 TiO$_2$ 薄膜的交叉阵列型双极性 RRAM 器件是一种非常有前途的器件,然而在实际应用时会出现电迁移或电击穿等较为严重的问题。Jeong 团队[151]通过设计顶部和底部的接口域而显著提升了器件实用能力;通过在顶部界面区插入超薄金属层或在底部界面中加入薄的阻挡层,均可以实现更强的电阻转换性能和更优越的耐受性。

Lee 团队[51]一改传统的 Pt/Ta$_x$O$_y$/Pt 对称结构,制造了 Pt/Ta$_2$O$_{5-x}$/TaO$_{2-x}$/Pt 非对称无源 RRAM 器件。该器件有较低的开关电流及功耗,同时电压耐受性高达 10^{12} 个电压周期。除此之外,可以将两个该器件分别与本征肖特

2.4 阻变存储器的性能提升

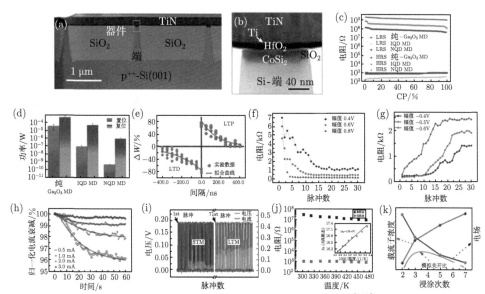

图 2.13 TiN/Ti/HfO$_2$/CoSi$_2$/Si-tip RRAM 器件的 STEM 图像[150]：(a)STEM 图像显示了三个 Si-tip；(b) 高倍率 STEM 图像以清楚显示器件结构和 Si-tip。纯 Ga$_2$O$_3$、随机分布的 PbS QD(IQD)、自组装 PbS QD(NQD) RRAM 器件[136]：(c) 超过 60 圈 I-V 扫描之后的电阻值分布；(d) 器件电阻切换功率的不同，误差线表示针对每个类别测得的最小值和最大值；(e)NQD RRAM 器件对生物突触 STDP 特性的模拟。器件电导对正负脉冲幅度的依赖性[151]：(f) 正脉冲序列幅度对电导调制的影响；(g) 负脉冲序列幅度对电导调制的影响。不同大小的限制电流和重复的刺激引发的短时程可塑性 (STP) 和长时程可塑性 (LTP) 的转变[149]；(h) 在置位操作后测定了 I-t 曲线，限制电流的大小分别为 0.5 mA、1 mA、2 mA 和 3 mA，突触的权重变化趋势与心理学中人类记忆的"遗忘曲线"类似；(i) 通过重复的脉冲刺激 (幅度为 0.2 V，持续时间 100 ns，周期 400 ns) 使得突触器件的电导变化，STM 过渡到 LTM。Au/CeO$_2$/Au/Si RRAM 器件[152]：(j) 器件的高低阻态电阻值对温度的依赖性，插图是 OFF 状态的阿伦乌斯 (Arrhenius) 图；(k) 载流子浓度，电场和开关比随 CeO$_2$ 膜厚度的变化 (彩图请扫封底二维码)

基势垒结合，来解决高密度纵横制阵列中的杂散泄漏电流路径问题，而不再像传统方法中使用分立晶体管或二极管。

Younis 团队[152]使用自组装的 CeO$_2$ 纳米立方体制备了 Au/CeO$_2$/Au/Si RRAM 器件。如图 2.13(j) 与 (k) 所示，纳米立方体的独特几何形状使得器件有出色的电阻转换行为。沿着每个纳米立方体形成了直的和可延伸的导电细丝，从而使得器件的开关比在较小的数值区间内分布；再者，由于纳米立方体之间存在着大量的界面，这些界面显著增加了器件在 HRS 的值，使得器件的开关比较大 (约为 10^4)，在 480 K 的环境温度下仍有很高的稳定性。

2.5 小　　结

本章首先简述了存储器的发展史，介绍了阻变存储器的基本概念、工作原理和特性参数，对阻变存储器进行了分类。接着，以材料为分类依据分别介绍了氧化物阻变存储器、非氧化物无机材料阻变存储器、有机材料阻变存储器以及二维材料阻变存储器的相关研究成果和发展状况。随后，以阻变存储器的阻变机制为分类依据分别介绍了电化学金属化导电细丝、价态变化存储器、相变阻变存储器和铁电存储器的研究和发展情况。最后，我们总结了利用二维材料、量子点、外围电路、杂质掺杂以及特殊器件形貌提升阻变存储器器件性能的方法与相关的研究。

参 考 文 献

[1] Truesdell L E. The Development of Punch Card Tabulation in the Bureau of the Census[M]. Washington: US Government Printing Office, 1965.

[2] 苏儒. 主要的存储技术进展年表 [J]. 无线电工程,1987,(05):100-101.

[3] 柯研. 国外电子计算机系统发展部分情况 [J]. 电子技术应用,1975,(01):99-112.

[4] Exabyte 公司. 磁带机——计算机系统备份设备 [J]. 上海微型计算机,1997,(32):45.

[5] 佚名. 磁芯存储器长期领先 [J]. 磁性材料及器件, 1971,(3):14-18.

[6] 黄阳棋,刘超,乔路. 英特尔 50 年存储发展史对我国的启示 [J]. 河南科技,2020,39(28):25-27.

[7] Amankwah-Amoah J .Competing technologies, competing forces: The rise and fall of the floppy disk, 1971‐2010[J].Technological Forecasting and Social Change, 2016, 107(jun.):121-129.

[8] Chua L. Memristor-The missing circuit element[J]. IEEE Trans Circuit Theory, 1971, 18(5): 507-519.

[9] Aritome S. Scaling Challenge of Nand Flash Memory Cells[M]. New York: Wiley-IEEE Press, 2015.

[10] 祁菁,任天爽,汪博. 阻变存储器的研究概述与展望 [J]. 基地物理, 2014, 1(5): 18-22.

[11] 孙一慧. 氧化物阻变存储器的性能调控及功能化基础 [D]. 北京: 北京科技大学, 2017.

[12] Rehman M M，Gul J Z，Rehman H M M U，et al. Decade of 2D materials based RRAM devices: a review[J]. Science and Technology of Advanced Materials, 2020, 21(1): 147-186.

[13] Siddiqui G U, Rehman M M, Choi K H, et al.Enhanced resistive switching in all-printed, hybrid and flexible memory device based on perovskite $ZnSnO_3$ via PVOH polymer[J]. Polymer, 2016, 100:102-110.

[14] Siddiqui G U, Rehman M M, Yang Y J, et al. A two-dimensional hexagonal boron nitride/polymer nanocomposite for flexible resistive switching devices[J]. Journal of Materials Chemistry C, 2017, 5(4):862-871.

[15] Rehman M M, Uddin G, Gui J Z, et al. Resistive switching in all-printed, flexible and hybrid MoS$_2$-PVA nanocomposite based memristive device fabricated by reverse offset[J]. Scientific Reports, 2016, 4(6): 36195.

[16] Hui F, Grustan-Gutierrez E, Long S, et al. Graphene and related materials for resistive random access memories[J]. Advanced Electronic Materials, 2017, 3(8): 1600195.

[17] Sarkar B, Lee B, Misra V, et al. Understanding the gradual reset in Pt/Al$_2$O$_3$/Ni RRAM for synaptic applications[J]. Semiconductor Science & Technology, 2015, 30(10): 105014.

[18] Breuer T, Siemon A, Linn E, et al. A HfO$_2$-based complementary switching crossbar adder[J]. Advanced Electronic Materials, 2015, 1(10): 1500138.

[19] Rehman M M, Yang B S, Yang Y J, et al. Effect of device structure on the resistive switching characteristics of organic polymers fabricated through all printed technology[J]. Current Applied Physics: the Official Journal of the Korean Physical Society, 2017, 17(4):533-540.

[20] Jin Y L, Jin K J, Ge C, et al. Resistive switching phenomena in complex oxide heterostructures[J]. World Scientific Publishing Company, 2013, 27(29): 1330021.

[21] Wong H S P . Metal-oxide RRAM[J]. Proceedings of the IEEE, 2012, 100(6):1951-1970.

[22] Hickmott T W. Low-frequency negative resistance in thin anodic oxide films[J]. Journal of Applied Physics, 1962, 33(9):2669-2682.

[23] Goux L, Czarnecki P, Chen Y Y, et al. Evidences of oxygen-mediated resistive-switching mechanism in TiNHfO$_2$ Pt cells[J]. Applied Physics Letters, 2010, 97(24): 1-3.

[24] Lee H Y, Chen P S, Wu T Y, et al. Low power and high speed bipolar switching with a thin reactive Ti buffer layer in robust HfO$_2$ based RRAM[C]// 2008 International. Electron Devices Meeting. IEEE, 2009.

[25] Govoreanu B. 10×10nm^2 Hf/HfO$_x$ crossbar resistive RAM with excellent performance, reliability and low-energy operation[C]// 2011 International Electron Devices Meeting. IEEE, 2012.

[26] Chung K B, Cho M H, Hwang U, et al. Effects of postnitridation annealing on band gap and band offsets of nitrided Hf-silicate films[J]. Applied Physics Letters, 2008, 92(2): 2912.

[27] Yoon J H, Kim K M, Song S J, et al. Pt/Ta$_2$O$_5$/HfO$_{2x}$/Ti resistive switching memory competing with multilevel NAND flash[J]. Advanced Materials, 2015, 27(25):3811-3816.

[28] Chand U, Huang K C, Huang C Y, et al. Investigation of thermal stability and reliability of HfO$_2$ based resistive random access memory devices with cross-bar structure[J]. Journal of Applied Physics, 2015, 117(18):1-9.

[29] Fang R, Chen W, Gao L, et al. Low-temperature characteristics of HfO[J]. IEEE Electron Device Letters, 2015, 36(6):567-569.

[30] Qi M, Guo C, Zeng M. Oxygen vacancy kinetics mechanism of the negative forming-free process and multilevel resistance based on hafnium oxide RRAM[J]. Journal of Nanomaterials, 2019, 2019:1-9.

[31] Wu Y, Lee B, Wong H. Al$_2$O$_3$ -Based RRAM using atomic layer deposition (ALD) with 1-μA reset current[J]. Electron Device Letters, IEEE, 2010, 31(12): 1449-1451.

[32] Kim W, Park S I, Zhang Z P, et al. Forming-free nitrogen-doped AlO$_x$ RRAM with sub-μA programming current[J]. 2011 Symposium on VLSI Technology-Digest of Technical Papers, 2011:22-23.

[33] Liang J, Wong H S P. Cross-point memory array without cell selectors—device characteristics and data storage pattern dependencies[J]. IEEE Transactions on Electron Devices, 2010, 57(10):2531-2538.

[34] Chen Y S, Lee H Y, Chen P S, et al. Highly scalable hafnium oxide memory with improvements of resistive distribution and read disturb immunity[C]// Electron Devices Meeting (IEDM), 2009 IEEE International. IEEE, 2010.

[35] Yu S, Yi W, Yang C, et al. Characterization of switching parameters and multilevel capability in HfO$_x$/AlO$_x$ bi-layer RRAM devices[C]// International Symposium on VLSI Technology Systems & Applications Proceedings, 2011:1-2.

[36] Nardi F, Ielmini D, Cagli C, et al. Sub-10 μA reset in NiO-based resistive switching memory (RRAM) cells[C]// Memory Workshop. IEEE, 2010.

[37] Lee M J, Park Y, Kang B S, et al. 2-stack 1D-1R cross-point structure with oxide diodes as switch elements for high density resistance RAM applications[C]// IEEE International Electron Devices Meeting. IEEE, 2008.

[38] Lee B, Wong H. NiO resistance change memory with a novel structure for 3D integration and improved confinement of conduction path[C]// VLSI Technology, 2009 Symposium on. 2009.

[39] Kang X, Guo J, Gao Y, et al. NiO-based resistive memory devices with highly improved uniformity boosted by ionic liquid pre-treatment[J]. Applied Surface Science, 2019, 480(JUN.30):57-62.

[40] Tang H H, Whang T J, Su Y K. Interpretations for coexistence of unipolar and bipolar resistive switching behaviors in Pt/NiO/ITO structure[J]. Japanese Journal of Applied Physics, 2019, 58(SD):SDDE14.

[41] Kwon D H, Kim K, Jang J, et al. Atomic structure of conducting nanofilaments in TiO$_2$ resistive switching memory[J].Nature Nanotech, 2010, 5: 148-153.

[42] Chae S C, Lee J S, Choi W S, et al. Multilevel unipolar resistance switching in TiO$_2$ thin films[J]. Applied Physics Letters, 2009, 95(9):139.

[43] Fujimoto M, Koyama H, Hosoi Y, et al. High-speed resistive switching of TiO$_2$/TiN nano-crystalline thin film[J]. Japanese Journal of Applied Physics, 2006, 45(11): 20-29.

[44] Yang J J, Pickett M, Li X M, et al. Memristive switching mechanism for metal/oxide/metal nanodevices[J]. Nature Nanotechnology, 2008, (7):3.

[45] Strukov D B, Snider G S, Stewart D R, et al. The missing memristor found[J]. Nature, 2008, 453(7191): 80-83.

[46] Do Y H, Kwak J S, Bae Y C, et al. Hysteretic bipolar resistive switching characteristics in TiO$_2$/TiO$_{2x}$ multilayer homojunctions[J]. Applied Physics Letters, 2009, 95(9):833.

[47] Zhidik E V, Troyan P E, Sakharov Y V, et al. Study and production of thin-filmmemristors based on TiO_2-TiO_x layers[J]. IOP Conference Series: Materials Science and Engineering, 2019, 498(1):012022.

[48] Ding X, Liu L, Feng Y, et al. Effect of TiO_x film thickness on resistive switching behavior of TiN/TiO_x/HfO_2/Pt RRAM device[C]. 2019 Silicon Nanoelectronics Workshop (SNW), 2019: 1-2.

[49] Wei Z, Kanzawa Y, Arita K, et al. Highly reliable TaO_x ReRAM and direct evidence of redox reaction mechanism[C]// IEEE International Electron Devices Meeting. IEEE, 2008.

[50] Wei Z, Takagi T, Kanzawa Y, et al. Demonstration of high-density ReRAM ensuring 10-year retention at 85°C based on a newly developed reliability model[C]//IEEE International Electron Devices Meeting. IEEE, 2011.

[51] Lee M J, Lee C B, Lee D, et al. A fast, high-endurance and scalable non-volatile memory device made from asymmetric Ta_2O_{5-x}/TaO_{2-x} bilayer structures[J]. Nature Materials, 2011, 10(8): 625-630.

[52] Yang J J, Zhang M X, Strachan J P, et al. High switching endurance in TaO_x memristive devices[J]. Applied Physics Letters, 2010, 97(23): 232102.

[53] Li X, Yu J, Xu X, et al. Improvement of read disturb on TaO_x-based RRAM cells with optimized pulse programming method[C]//2019 IEEE International Conference on Integrated Circuits, Technologies and Applications (ICTA). IEEE, 2019: 176-177.

[54] Li C, Wang F, Zhang J, et al. Improved uniformity of TaO_x-based resistive random access memory with ultralow operating voltage by electrodes engineering[J]. ECS Journal of Solid State Science and Technology, 2020, 9(4): 041005.

[55] Wang F, Wu X. Non-volatile memory devices based on chalcogenide materials[C]//2009 Sixth International Conference on Information Technology: New Generations. IEEE, 2009: 5-9.

[56] Ovshinsky S R. Reversible electrical switching phenomena in disordered structures[J]. Physical Review Letters, 1968, 21(20): 1450.

[57] Ohta T. Phase-change optical memory promotes the DVD optical disk[J]. Journal of Optoelectronics and Advanced Materials, 2001, 3(3): 609-626.

[58] Siegel J, Schropp A, Solis J, et al. Rewritable phase-change optical recording in $Ge_2Sb_2Te_5$ films induced by picosecond laser pulses[J]. Applied Physics Letters, 2004, 84(13): 2250-2252.

[59] Naito M, Ishimaru M, Hirotsu Y, et al. Electron microscopy study on amorphous Ge-Sb-Te thin film for phase change optical recording[J]. Japanese Journal of Applied Physics, 2003, 42(10A): L1158.

[60] Kozicki M N, Yun M, Yang S J, et al. Nanoscale effects in devices based on chalcogenide solid solutions[J]. Superlattices and Microstructures, 2000, 27(5-6): 485-488.

[61] Kozicki M N, Gopalan C, Balakrishnan M, et al. Nonvolatile memory based on solid electrolytes[C]//Proceedings 2004 IEEE Computational Systems Bioinformatics Con-

ference. IEEE, 2004: 10-17.

[62] Kozicki M N, Balakrishnan M, Gopalan C, et al. Programmable metallization cell memory based on Ag-Ge-S and Cu-Ge-S solid electrolytes[C]//2005 Symposium Non-Volatile Memory Technology. IEEE, 2005.

[63] Sakamoto T, Sunamura H, Kawaura H, et al. Nanometer-scale switches using copper sulfide[J]. Applied Physics Letters, 2003, 82(18): 3032-3034.

[64] Kim H D, An H M, Kim K C, et al. Large resistive-switching phenomena observed in Ag/Si$_3$N$_4$/Al memory cells[J]. Semiconductor Science and Technology, 2010, 25(6): 065002.

[65] Kim H D, An H M, Kim T G. Improved reliability of Au/Si$_3$N$_4$/Ti resistive switching memory cells due to a hydrogen postannealing treatment[J]. 2011, 109(1): 016105-016105-3.

[66] Zhang Z, Gao B, Fang Z, et al. All-metal-nitride RRAM devices[J]. IEEE Electron Device Letters, 2014, 36(1): 29-31.

[67] Kim H D, An H M, Lee E B, et al. Stable bipolar resistive switching characteristics and resistive switching mechanisms observed in aluminum nitride-based ReRAM devices[J]. IEEE Transactions on Electron Devices, 2011, 58(10): 3566-3573.

[68] Du X, Zhou Q, Yan Z, et al. The effects of oxygen plasma implantation on bipolar resistive-switching properties of copper nitride thin films[J]. Thin Solid Films, 2017, 625: 100-105.

[69] Chen Y S, Wang Z, Zhang Z, et al. Enhancement of HfO$_2$ based RRAM performance through hexagonal boron nitride interface layer[C]//2018 Non-Volatile Memory Technology Symposium (NVMTS). IEEE, 2018: 1-3.

[70] Kim S, Kim M H, Cho S, et al. Bias polarity dependent resistive switching behaviors in silicon nitride-based memory cell[J]. IEICE Transactions on Electronics, 2016, 99(5): 547-550.

[71] Kim H D, Yun M J, Kim K H, et al. Oxygen-doped zirconium nitride based transparent resistive random access memory devices fabricated by radio frequency sputtering method[J]. Journal of Alloys and Compounds, 2016, 675: 183-186.

[72] Gao S, Yi X, Shang J, et al. Organic and hybrid resistive switching materials and devices[J]. Chemical Society Reviews, 2019, 48(6): 1531-1565.

[73] Wang C, Wang J, Li P Z, et al. Synthesis, characterization, and non-volatile memory device application of an N-substituted heteroacene[J]. Chemistry—An Asian Journal, 2014, 9(3): 779-783.

[74] Wang C, Hu B, Wang J, et al. Rewritable multilevel memory performance of a tetraaza-tetracene donor-acceptor derivative with good endurance[J]. Chemistry—An Asian Journal, 2015, 10(1): 116-119.

[75] Wang C, Yamashita M, Hu B, et al. Synthesis, characterization, and memory performance of two phenazine/triphenylamine-based organic small molecules through donor-acceptor design[J]. Asian Journal of Organic Chemistry, 2015, 4(7): 646-651.

[76] 陈威林, 高双, 伊晓辉, 等. 有机和杂化阻变材料与器件 [J]. 功能高分子学报, 2019, 32(4): 434-447.

[77] Song Y, Ling Q D, Lim S L, et al. Electrically bistable thin-film device based on PVK and GNPs polymer material[J]. IEEE Electron Device Letters, 2007, 28(2): 107-110.

[78] Kim T W, Oh S H, Choi H, et al. Reliable organic nonvolatile memory device using a polyfluorene-derivative single-layer film[J]. IEEE Electron Device Letters, 2008, 29(8): 852-855.

[79] Cho B O, Yasue T, Yoon H, et al. Thermally robust multi-layer non-volatile polymer resistive memory[C]//2006 International Electron Devices Meeting. IEEE, 2006: 1-4.

[80] Cao X, Han Y, Zhou J, et al. Enhanced switching ratio and long-term stability of flexible RRAM by anchoring polyvinylammonium on perovskite grains[J]. ACS Applied Materials & Interfaces, 2019, 11(39): 35914-35923.

[81] Lee T, Chen Y. Organic resistive nonvolatile memory materials[J]. MRS Bulletin, 2012, 37(2): 144-149.

[82] Rehman M M, Rehman H M M U, Gul J Z, et al. Decade of 2D-materials-based RRAM devices: a review[J]. Science and Technology of Advanced Materials, 2020, 21(1): 147-186.

[83] Shin Y J, Kwon J H, Kalon G, et al. Ambipolar bistable switching effect of graphene[J]. Applied Physics Letters, 2010, 97(26): 262105.

[84] Doh Y J, Yi G C. Nonvolatile memory devices based on few-layer graphene films[J]. Nanotechnology, 2010, 21(10): 105204.

[85] Wu C, Li F, Zhang Y, et al. Recoverable electrical transition in a single graphene sheet for application in nonvolatile memories[J]. Applied Physics Letters, 2012, 100(4): 042105.

[86] He C, Shi Z, Zhang L, et al. Multilevel resistive switching in planar graphene/SiO_2 nanogap structures[J]. ACS Nano, 2012, 6(5): 4214-4221.

[87] Kim S M, Song E B, Lee S, et al. Transparent and flexible graphene charge-trap memory[J]. ACS Nano, 2012, 6(9): 7879-7884.

[88] Lai Y C, Wang D Y, Huang I S, et al. Low operation voltage macromolecular composite memory assisted by graphene nanoflakes[J]. Journal of Materials Chemistry C, 2013, 1(3): 552-559.

[89] Khurana G, Misra P, Katiyar R S. Multilevel resistive memory switching in graphene sandwiched organic polymer heterostructure[J]. Carbon, 2014, 76: 341-347.

[90] Wu C, Li F, Guo T. Efficient tristable resistive memory based on single layer graphene/insulating polymer multi-stacking layer[J]. Applied Physics Letters, 2014, 104(18): 183105.

[91] Liu B, Liu Z, Chiu I S, et al. Programmable synaptic metaplasticity and below femtojoule spiking energy realized in graphene-based neuromorphic memristor[J]. ACS Applied Materials & Interfaces, 2018, 10(24): 20237-20243.

[92] Yalagala B, Khandelwal S, Deepika J, et al. Wirelessly destructible MgO-PVP-graphene

composite based flexible transient memristor for security applications[J]. Materials Science in Semiconductor Processing, 2019, 104: 104673.

[93] Liu J, Zeng Z, Cao X, et al. Preparation of MoS_2-polyvinylpyrrolidone nanocomposites for flexible nonvolatile rewritable memory devices with reduced graphene oxide electrodes[J]. Small, 2012, 8(22): 3517-3522.

[94] Xu X Y, Yin Z Y, Xu C X, et al. Resistive switching memories in MoS_2 nanosphere assemblies[J]. Applied Physics Letters, 2014, 104(3): 033504.

[95] Bessonov A A, Kirikova M N, Petukhov D I, et al. Layered memristive and memcapacitive switches for printable electronics[J]. Nature Materials, 2015, 14(2): 199-204.

[96] Sangwan V K, Jariwala D, Kim I S, et al. Gate-tunable memristive phenomena mediated by grain boundaries in single-layer MoS_2[J]. Nature Nanotechnology, 2015, 10(5): 403-406.

[97] Wang W, Panin G N, Fu X, et al. MoS_2 memristor with photoresistive switching[J]. Scientific Reports, 2016, 6(1): 1-11.

[98] Rehman M M, Siddiqui G U, Gul J Z, et al. Resistive switching in all-printed, flexible and hybrid MoS_2-PVA nanocomposite based memristive device fabricated by reverse offset[J]. Scientific Reports, 2016, 6(1): 1-10.

[99] Fan F, Zhang B, Cao Y, et al. Solution-processable poly (N-vinylcarbazole)-covalently grafted MoS_2 nanosheets for nonvolatile rewritable memory devices[J]. Nanoscale, 2017, 9(7): 2449-2456.

[100] Wang D, Ji F, Chen X, et al. Quantum conductance in MoS_2 quantum dots-based nonvolatile resistive memory device[J]. Applied Physics Letters, 2017, 110(9): 093501.

[101] Li D, Wu B, Zhu X, et al. MoS_2 memristors exhibiting variable switching characteristics toward biorealistic synaptic emulation[J]. ACS Nano, 2018, 12(9): 9240-9252.

[102] Chen P A, Ge R J, Lee J W, et al. An RRAM with a 2D material embedded double switching layer for neuromorphic computing[C]//2018 IEEE 13th Nanotechnology Materials and Devices Conference (NMDC). IEEE, 2018: 1-4.

[103] Kim M, Ge R, Wu X, et al. Zero-static power radio-frequency switches based on MoS_2 atomristors[J]. Nature Communications, 2018, 9(1): 1-7.

[104] Qiu J T, Samanta S, Dutta M, et al. Controlling resistive switching by using an optimized MoS_2 interfacial layer and the role of top electrodes on ascorbic acid sensing in TaO_x-based RRAM[J]. Langmuir, 2019, 35(11): 3897-3906.

[105] Das U, Bhattacharjee S, Sarkar P K, et al. A multi-level bipolar memristive device based on visible light sensing MoS_2 thin film[J]. Materials Research Express, 2019, 6(7): 075037.

[106] Feng X, Li Y, Wang L, et al. First demonstration of a fully-printed MoS_2 RRAM on flexible substrate with ultra-low switching voltage and its application as electronic synapse[C]//2019 Symposium on VLSI Technology. IEEE, 2019: T88-T89.

[107] Zhai Y, Yang X, Wang F, et al. Infrared-sensitive memory based on direct-grown MoS_2–upconversion-nanoparticle heterostructure[J]. Advanced Materials, 2018, 30(49):

1803563.

[108] Yan Y, Sun B, Ma D. Resistive switching memory characteristics of single MoSe$_2$ nanorods[J]. Chemical Physics Letters, 2015, 638: 103-107.

[109] Han P, Sun B, Cheng S, et al. Preparation of MoSe$_2$ nano-islands array embedded in a TiO$_2$ matrix for photo-regulated resistive switching memory[J]. Journal of Alloys and Compounds, 2016, 664: 619-625.

[110] Zhang X, Qiao H, Nian X, et al. Resistive switching memory behaviours of MoSe$_2$ nano-islands array[J]. Journal of Materials Science: Materials in Electronics, 2016, 27(7): 7609-7613.

[111] Zhou G, Sun B, Yao Y, et al. Investigation of the behaviour of electronic resistive switching memory based on MoSe$_2$-doped ultralong Se microwires[J]. Applied Physics Letters, 2016, 109(14): 143904.

[112] Han P, Sun B, Li J, et al. Ag filament induced nonvolatile resistive switching memory behaviour in hexagonal MoSe$_2$ nanosheets[J]. Journal of Colloid and Interface Science, 2017, 505: 148-153.

[113] Li P, Sun B, Zhang X, et al. Effect of temperature on the magnetism and memristive memory behavior of MoSc$_2$ nanosheets[J]. Materials Letters, 2017, 202: 13-16.

[114] Das U, Mahato B, Sarkar P K, et al. Bipolar resistive switching behaviour of WS$_2$ thin films grown by chemical vapour deposition[C]//AIP Conference Proceedings. AIP Publishing LLC, 2019, 2115(1): 030274.

[115] Li Y, Sivan M, Niu J X, et al. Aerosol jet printed WSe$_2$ based RRAM on kapton suitable for flexible monolithic memory integration[C]//2019 IEEE International Conference on Flexible and Printable Sensors and Systems (FLEPS). IEEE, 2019: 1-3.

[116] Zahoor F, Azni Zulkifli T Z, Khanday F A. Resistive random access memory (RRAM): an overview of materials, switching mechanism, performance, multilevel cell (MLC) storage, modeling, and applications[J]. Nanoscale Research Letters, 2020, 15(1): 1-26.

[117] Kumar D, Aluguri R, Chand U, et al. Metal oxide resistive switching memory: materials, properties and switching mechanisms[J]. Ceramics International, 2017, 43: S547-S556.

[118] Kozicki M N, Barnaby H J. Conductive bridging random access memory—materials, devices and applications[J]. Semiconductor Science and Technology, 2016, 31(11): 113001.

[119] Huang Y, Shen Z, Wu Y, et al. Amorphous ZnO based resistive random access memory[J]. RSC Advances, 2016, 6(22): 17867-17872.

[120] Chand U, Huang C Y, Jieng J H, et al. Suppression of endurance degradation by utilizing oxygen plasma treatment in HfO$_2$ resistive switching memory[J]. Applied Physics Letters, 2015, 106(15): 153502.

[121] Le Gallo M, Sebastian A. An overview of phase-change memory device physics[J]. Journal of Physics D: Applied Physics, 2020, 53(21): 213002.

[122] Ovshinsky S R. An introduction to ovonic research[J]. Journal of Non-Crystalline Solids, 1970, 2: 99-106.

[123] Setter N, Damjanovic D, Eng L, et al. Ferroelectric thin films: Review of materials, properties, and applications[J]. Journal of Applied Physics, 2006, 100(5): 051606.

[124] 孙鹏霄. 阻变存储器阻变机理及物理模型研究 [D]. 兰州: 兰州大学,2015.

[125] Son D I, Kim T W, Shim J H, et al. Flexible organic bistable devices based on graphene embedded in an insulating poly (methyl methacrylate) polymer layer[J]. Nano Letters, 2010, 10(7): 2441-2447.

[126] Wu C, Li F, Guo T, et al. Controlling memory effects of three-layer structured hybrid bistable devices based on graphene sheets sandwiched between two laminated polymer layers[J]. Organic Electronics, 2012, 13(1): 178-183.

[127] Ji Y, Choe M, Cho B, et al. Organic nonvolatile memory devices with charge trapping multilayer graphene film[J]. Nanotechnology, 2012, 23(10): 105202.

[128] El-Atab N, Cimen F, Alkis S, et al. Enhanced memory effect with embedded graphene nanoplatelets in ZnO charge trapping layer[J]. Applied Physics Letters, 2014, 105(3): 033102.

[129] Lee K, Hwang I, Lee S, et al. Enhancement of resistive switching under confined current path distribution enabled by insertion of atomically thin defective monolayer graphene[J]. Scientific Reports, 2015, 5(1): 1-8.

[130] Lee J, Du C, Sun K, et al. Tuning ionic transport in memristive devices by graphene with engineered nanopores[J]. ACS Nano, 2016, 10(3): 3571-3579.

[131] Mannequin C, Delamoreanu A, Latu-Romain L, et al. Graphene-HfO_2-based resistive RAM memories[J]. Microelectronic Engineering, 2016, 161: 82-86.

[132] Sun Y, Wen D, Sun F. Eliminating negative-set behavior by adding a graphene blocking layer in resistive switching memory devices based on epoxy resin[J]. Applied Physics Express, 2019, 12(7): 074006.

[133] Obreja A C, Cristea D, Mihalache I, et al. Charge transport and memristive properties of graphene quantum dots embedded in poly (3-hexylthiophene) matrix[J]. Applied Physics Letters, 2014, 105: 083303.

[134] Qi M, Bai L, Xu H, et al. Oxidized carbon quantum dot-graphene oxide nanocomposites for improving data retention of resistive switching memory[J]. Journal of Materials Chemistry C, 2018, 6(8): 2026-2033.

[135] Yan X, Zhang L, Yang Y, et al. Highly improved performance in $Zr_{0.5}Hf_{0.5}O_2$ films inserted with graphene oxide quantum dots layer for resistive switching non-volatile memory[J]. Journal of Materials Chemistry C, 2017, 5(42): 11046-11052.

[136] Yan X, Pei Y, Chen H, et al. Self-assembled networked PbS distribution quantum dots for resistive switching and artificial synapse performance boost of memristors[J]. Advanced Materials, 2019, 31(7): 1805284.

[137] Younis A, Chu D, Lin X, et al. High-performance nanocomposite based memristor with controlled quantum dots as charge traps[J]. ACS Applied Materials & Interfaces, 2013, 5(6): 2249-2254.

[138] Han S T, Hu L, Wang X, et al. Black phosphorus quantum dots with tunable memory

properties and multilevel resistive switching characteristics[J]. Advanced Science, 2017, 4(8): 1600435.

[139] Gaba S, Choi S, Sheridan P, et al. Improvement of RRAM device performance through on-chip resistors[J]. MRS Online Proceedings Library (OPL), 2012, 1430: mrss 12-1430-e09-09.

[140] Hsu S T. Temperature compensated RRAM circuit: U.S. Patent 6868025[P]. 2005-3-15.

[141] 焦斌, 邓宁, 陈培毅. 阻变存储器外围电路关键技术研究进展 [J]. 固体电子学研究与进展,2013,33(4):363-370.

[142] Lee S R, Kim Y B, Chang M, et al. Multi-level switching of triple-layered TaO_x RRAM with excellent reliability for storage class memory[C]//2012 Symposium on VLSI Technology (VLSIT). IEEE, 2012: 71-72.

[143] 金钢, 吴雨欣, 张佶, 等. 基于 0.13μm 标准逻辑工艺的 1Mb 阻变存储器设计与实现 [J]. 固体电子学研究与进展, 2011, 31(2): 174-179.

[144] Gao B, Zhang H W, Yu S, et al. Oxide-based RRAM: Uniformity improvement using a new material-oriented methodology[C]//2009 Symposium on VLSI Technology. IEEE, 2009: 30-31.

[145] Wang Y, Liu Q, Long S, et al. Investigation of resistive switching in Cu-doped HfO_2 thin film for multilevel non-volatile memory applications[J]. Nanotechnology, 2009, 21(4): 045202.

[146] Zhang H, Liu L, Gao B, et al. Gd-doping effect on performance of HfO_2 based resistive switching memory devices using implantation approach[J]. Applied Physics Letters, 2011, 98(4): 042105.

[147] Wang X F, Tian H, Zhao H M, et al. Interface engineering with MoS_2-Pd nanoparticles hybrid structure for a low voltage resistive switching memory[J]. Small, 2018, 14(2): 1702525.

[148] Tan T, Guo T, Chen X, et al. Impacts of Au-doping on the performance of Cu/HfO_2/Pt RRAM devices[J]. Applied Surface Science, 2014, 317: 982-985.

[149] Yan X, Zhao J, Liu S, et al. Memristor with Ag-cluster-doped TiO_2 films as artificial synapse for neuroinspired computing[J]. Advanced Functional Materials, 2018, 28(1): 1705320.

[150] Niu G, Calka P, der Maur M A, et al. Geometric conductive filament confinement by nanotips for resistive switching of HfO_2-RRAM devices with high performance[J]. Sci. Rep, 2016, 6(1): 1-9.

[151] Jeong H Y, Lee J Y, Choi S Y. Interface-engineered amorphous TiO_2-based resistive memory devices[J]. Advanced Functional Materials, 2010, 20(22): 3912-3917.

[152] Younis A, Chu D, Mihail I, et al. Interface-engineered resistive switching: CeO_2 nanocubes as high-performance memory cells[J]. ACS Applied Materials & Interfaces, 2013, 5(19): 9429-9434.

第 3 章 忆阻器的应用

3.1 引　言

近年来，正如戈登·摩尔 (Gordon Moore) 预测的那样，微处理器芯片上的晶体管数量每两年左右翻一番，CMOS 晶体管的尺寸不断缩小，基于 CMOS 晶体管的数字计算机的性能近年来显著提高[1]，自 1945 年冯·诺依曼体系结构建立以来，基础计算机一直无法摆脱这种体系结构[2]。此外，随着物联网 (IoT) 和云计算的出现，数据量呈指数增长。未来几年，计算机的能效和处理速度将成为制约发展的重要因素。因此，传统数字计算机体系结构的缺点是一个日益严重的问题。在器件层面，随着晶体管的尺寸接近物理极限，泄漏电流逐渐成为一个问题。在体系结构层面，在冯·诺依曼体系结构中，当数据输入或输出时，中央处理器 (CPU) 将处于空闲状态。由于 CPU 速度和内存容量的增长速度远大于双方之间的数据流量，瓶颈问题变得越来越严重 (冯·诺依曼瓶颈)[3]。两个单元之间的性能不匹配导致相当长的延迟 (内存墙)[4]，这将迫使处理器设计使用具有高度并行性的异构内核[5]。图形处理单元 (GPU) 具有多核和快速内存访问结构，是提高计算速度和能源效率的成功尝试之一[6,7]。但是延迟问题仍然存在，线路的延迟变得比闸门的延迟更重要。长线路不仅存在传输延迟问题，还存在能耗问题[8]。内存计算作为冯·诺依曼体系结构的替代品重新出现[9]。内存计算最早提出于 20 世纪 60 年代，它可以具有简单的结构和更快的计算速度，而无须将数据移出结构，如图 3.1 所示。然而，它没有得到足够的重视，可能是因为在过去几十年里晶体管的快速发展推动了计算能力的令人满意的提高。利用自旋、相变或离子输运等物理现象的各种新兴电子器件[10-13]在理论和实践上都越来越成熟。基于内存计算概念的新型计算系统建立在先进的 CMOS 器件和纳米技术之上，为能源消耗和速度问题提供了极具吸引力的解决方案。随着摩尔定律逐渐达到物理极限，传统的冯·诺依曼架构正面临挑战[14]。特别地，基于忆阻器的神经网络展现出极大的潜力。

电阻切换只需要少量的能量，因为忆阻开关器件可以制作得非常小，它可以具有非常高的堆叠密度 (小于 2 nm)[15]，并且开关速度可以非常快 (小于 0.1 ns)[16]，加快响应时间，因为当器件关闭时没有信息丢失。此外，大多数忆阻器电导率的变化是离子运动的结果，例如金属离子[17]、氧气真空[18]和界面状态[19]中捕获

3.1 引言

的载流子的迁移，这与生物突触和神经元中观察到的现象非常相似，所以它可以被称为类似大脑的计算。特别是，相变忆阻器可以以典型的神经方式工作，对脉冲有独特的响应[20]。尖峰神经网络 (SNN) 作为第三代神经网络模型，相较人工神经网络而言，更节能，而且其体积小得多，能耗低[21]。类似于甚至小于生物神经系统。近年来，许多 SNN 被开发出来，通过突触可塑性、时空识别、长、短期记忆等来模拟生物神经系统的实际功能[22,23]。忆阻器和基于忆阻器的 SNN 必然与当今和未来的计算需求相关，产生了广泛的应用，例如机器人技术、机器视觉、语音识别、触觉感知、医学和认知处理等 (图 3.2)。其中，忆阻器最有前途的应用

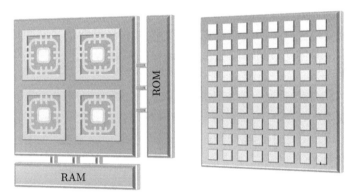

图 3.1　一种传统结构，其中 CPU 与内存 (左侧) 和内存处理单元 (右侧) 分离，用于内存计算

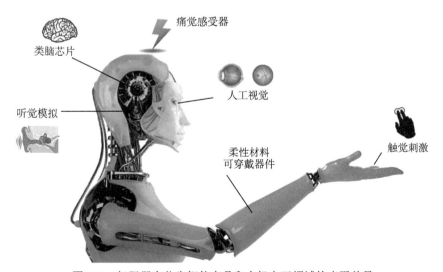

图 3.2　忆阻器在仿生智能产品和人机交互领域的应用前景

之一是人工突触。在这里，我们回顾了现有的忆阻器材料 (传统氧化物、二维材料等)，并详细解释了它们的阻变机制 (金属离子的导电丝或通道以及氧空位等)，它们与神经生理学的关联揭示了类脑神经计算的基础。同时，我们也关注了忆阻器的先进应用、基于忆阻器的神经网络技术，以及忆阻器和基于忆阻器的神经网络的未来前景和挑战。

3.2 忆阻器在神经仿生中的应用

3.2.1 生物突触和人工突触

1872 年，首次在人类身上发现了神经元细胞。虽然人们对单个神经元的细胞水平上的生物电现象有了明确的认识，但对生物神经网络的了解却很少。随着科学技术的发展，即使是最简单的昆虫 (如蚂蚁或蟑螂)，人们仍然不了解其神经系统的具体运作。

突触是中枢神经系统中信息传播的基础，如图 3.3 所示。突触间的信号传递是由神经递质完成的。在突触前部，当动作电位出现时，Ca^{2+} 被允许进入突触前神经元，促进突触小泡与质膜融合，从而释放神经递质，如图 3.3(b) 所示。神经递质与突触后细胞膜受体结合，引起突触后电位的变化，完成动作电位的传递。

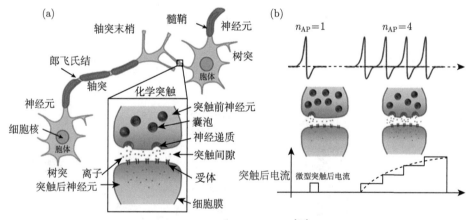

图 3.3　生物突触的描述 [24]

(a) 动作电位传递路径和放大的突触视图；(b) 突触后电流随着突触前电位的增加而增加

人工突触被设计用来模拟生物记忆系统的结构和记忆功能 [25−30]。大脑实现记忆和学习能力的基本和决定性因素是突触可塑性 [31−33]。因此，人工突触的制备和随后的主要活动取决于突触可塑性的模拟，如长时程增强 (LTP)、长时程

3.2 忆阻器在神经仿生中的应用

抑制 (LTD)、短时程增强 (STP)、短时程抑制 (STD) 和尖峰时间依赖性可塑性 (STDP)、成对脉冲抑制 (PPD) 和成对脉冲促进 (PPF)[34,35]。在心理学的研究中，记忆行为分为短期记忆 (STM) 和长期记忆 (LTM)，它们被认为是突触可塑性的结果。具体而言，PPF 的意义在于：在生物突触中，突触后电流 (PSC) 引起的第二个动作电位峰高于前一个，且增强程度与脉冲间隔有关。如图 3.4(b) 所示，人工突触在更高频率 (5000 Hz) 脉冲刺激下增强。

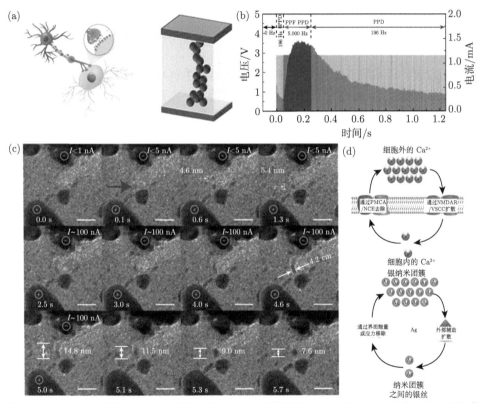

图 3.4 (a) 生物突触可以用 RS 器件来表示；(b) 模拟突触 PPF 和 PPD 行为，以及银氧化物忆阻器对 PPF 行为的跟踪；(c) 原位 TEM 下 Ag-CF 的生长和断裂过程；(d) Ca^{2+} 和 Ag 动力学之间类比的示意图 [50]

其中，STDP 描述了神经元之间连接的强度，完善了 1949 年提出的传统希伯来突触可塑性模型 [36]。这种知识的结合导致了 STDP 学习系统潜在人工架构的提出 [37-43]。STDP 最初被认为是一个计算机学习算法家族 [44,45]，机器智能和计算神经科学界正在使用它 [46,47]。同时，它的生物学和生理学基础在过去十年中已经相当完善 [48,49]。在本节中，我们介绍了基于不同机制的忆阻器的神经形态计算。

3.2.2 导电丝忆阻器人工突触

忆阻器导电丝 (CF) 的生长和断裂行为与生物突触中突触重量的变化非常相似。因此，基于 CF 的忆阻器是人工突触的候选者[51]。基于 CF 机制的忆阻器对突触前和突触后电位变化的模拟如图 3.4(a) 所示。到目前为止，许多基于 CF 的忆阻器已经被用作人工突触，模拟关键的突触学习规则。对于非挥发性 RS 机制电化学金属化存储器 (electrochemical metallization memory，ECM) 和价态变化存储器 (valence change memory，VCM)，由于 Ca^{2+} 扩散动力学的限制，通常使用调制脉冲来实现 STDP，这表明可以通过调整突触前和突触后峰值之间的时间间隔来调节和控制突触重量。

2017 年，杨和同事在氧化物基挥发性忆阻器中展示了银纳米团簇 (图 3.4(c))，其中银原子的扩散动力学类似于生物突触中的 Ca^{2+} 迁移[50]。忆阻器中类 Ca^{2+} 动力学的模拟应该是为了追求更复杂的人工突触，因为 Ca^{2+} 动力学 (包括 Ca^{2+} 积累和挤出) 是实现突触后电位变化的关键步骤。

在电场的帮助下，银扩散到银纳米团簇之间的间隙区域，类似于 Ca^{2+} 的流入过程。当去除电刺激时，界面能和可能的机械应力将银纳米颗粒从间隙区域桥接，从而复制 Ca^{2+} 挤出过程。因此，扩散忆阻器的银动力学是生物突触的功能模拟 (图 3.4(d))[50]。此外，在生物突触中也发现，长时间或过多的高频脉冲刺激最终会导致从促进到抑制的转变，这只是由相同频率的刺激脉冲数量增加引起的[52]。经实验验证，PPD 和 PPF 的行为如图 3.4(b) 所示，类似于突触的 STP，暗示了自主计算的潜力[53,54]。

3.2.3 相变忆阻器人工突触

PCM 是神经启发计算技术的主要候选之一，技术成熟度高。如图 3.5(a) 所示，展示了生物突触的基本概念以及使用 PCM 进行模拟的方法[55]。与 CF 机制类似，通过晶相的生长 (图 3.5(b))，相变材料的电阻水平可以进行连续过渡，以模拟生物突触的方式实现突触可塑性。成功的商业化的相变材料在模拟生物突触的功能和可塑性方面显示出巨大的潜力。图 3.5(c)、(d) 所示的人工突触很好地模拟了生物突触，实现了 STDP[56,57]。

3.2.4 光电忆阻器人工突触

近年来，基于光电电阻随机存取存储器 (optoelectronic resistive random access memory，ORRAM) 的人工光电突触已经成为一个热门话题，并被许多团队研究，在模拟生物视觉系统方面具有巨大潜力，甚至可以超越可见光区域。虽然这项技术尚未成熟，但它将降低集成电路的复杂性和功耗，并有助于机器视觉中神经网络的发展。

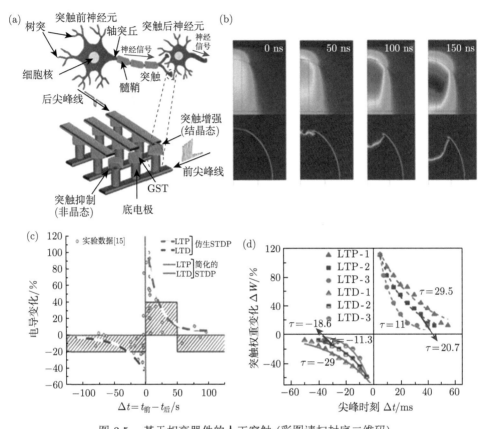

图 3.5 基于相变器件的人工突触 (彩图请扫封底二维码)

(a) 在交叉杆阵列结构中，PCM 突触对应于生物突触；(b) 随着温度升高，结晶相随着时间的增加而增长 (突触重量变化)[55]；(c) 生物 STDP 和简化的相变突触 STDP[56]；(d) 对于不同间隔的预脉冲振幅，在 10～30 ms 范围内的峰值时间延迟用于增强和 −10～−30 ms 范围内的峰值时间延迟用于抑制[57]

2017 年，Tan 等发现，ITO/CeO$_{2-x}$/AlO$_y$/Al 的能带曲率和固有导电结构受到光照脉冲和电刺激脉冲时可以被显著调制 (图 3.6(a)、(b))，在器件中产生持久且可调节的光响应。所有电阻状态以及原始状态都可以在循环操作中进行光学编程、维护、电读取和擦除 (图 3.6(c))。因此，该器件可以将入射光束的波长 (或频率) 信息转换为电信号，并同时存储它们[58]。2019 年，Wang 等报道了 MoS$_2$/苝-3,4,9,10-四羧基二酐 (PTCDA) 杂化异质结 ORRAM(图 3.6(d)) 的光调制特性，包括电调制和光调制、有效的栅极可调谐性以及显著的 STP 和 LTP。脉冲宽度为 400 ms、间隔为 100 ms 的光下刺激的 PPF 行为如图 3.6(e) 所示[59]。

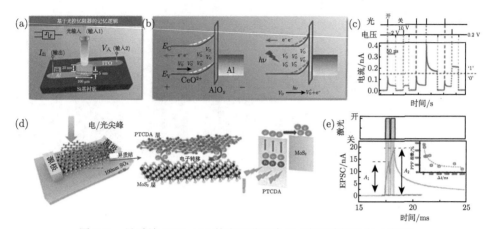

图 3.6 异质结 ORRAM 的光调节示意图 (彩图请扫封底二维码)

(a) ITO/CeO$_{2-x}$/AlO$_y$/Al 结构光调节原理图; (b) 光照对 ITO/CeO$_{2-x}$/AlO$_y$/Al 异质结薄膜肖特基势垒的影响; (c) ITO/CeO$_{2-x}$/AlO$_y$/Al 的光响应和电可擦除性, 读取电压为 0.2 V[58]; (d) MoS$_2$/PTCDA 结构的 3D 示意图, 以及光脉冲下的电荷转移; (e) MoS$_2$/PTCDA 异质结在双光脉冲刺激下的 PPF 行为 (V_{ds} 为 0.1 V, 绿色区域代表激光脉冲辐照)[59]

有趣的是，几乎所有的异质结都表现出突触行为。Gao 等将脉冲光刺激应用于 ITO/Nb:SrTiO$_3$ 异质结，获得的光响应结果显示了典型的神经行为[60]。如图 3.7(a) 所示，PPF 测量结果显示为一对 $\Delta t = 0.5$ s 的光脉冲。图 3.7(b) 中的 PPF 曲线也显示了相当好的性能，$\Delta t = 20$ s 时 PPF 仍高于 115%。

同时，该器件类似人脑的 "重复学习" 现象也得到了证实。如图 3.7(c) 所示，在持续应用 100 个光脉冲后，人工光电突触达到了高学习水平 (高突触重量)，而且在黑暗中持续 100s 后 (遗忘过程)，重新应用光脉冲 (重新学习)，可以发现，只有 7 个光脉冲需要达到第一次学习的高突触重量水平，这远低于第一次学习的 35 个脉冲。此外，在接下来的 100 个突触遗忘中，神经元的突触重量比第一次遗忘时减少得更少，这很好地模拟了人脑学习遗忘的过程。这种行为可以描述如下：一个人在失去知识之前重新学习所需的时间更少，而在重新学习之后记忆更深刻[61]。

2018 年，Wang 等报道了基于无机钙钛矿量子点的人工光电突触。在图 3.7(d) 中，在光子脉冲的累积调制期间，易失性记忆被转换为非易失性记忆，从而刺激记忆以维持从 STM 到大脑可塑性的 LTM。图 3.7(e)、(f) 总结了由两个相同光脉冲模拟的 PPF 效应，这表明平均 PPF 在不同强度和波长下逐渐增加，与光强度和波长正相关。图 3.7(g) 显示了在不同刺激下可靠的增强抑制功能，模拟了大脑的记忆印象。图 3.7(h) 显示了光子脉冲在不同时间间隔的应用，将突触功能从 PPF 调节到 PPD，这模拟了生物神经网络中的易化-抑制转换[62]。因此，异质结可以合理地视为一种人工光电突触[63-67]。

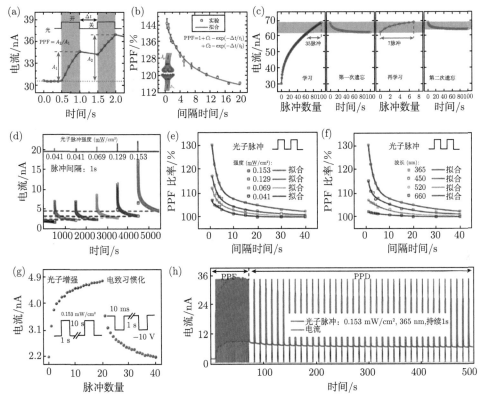

图 3.7 异质结人工光电突触在脉冲光刺激下的光响应特性 (彩图请扫封底二维码)
(a) 作用在 ITO/Nb:SrTiO$_3$ 异质结上的蓝光脉冲 ($\Delta t = 0.5$ s), 光强为 30 mW·cm^{-2}; (b) PPF 由两个光脉冲之间的间隔表示; (c) 生物突触重复学习的现象[60]; (d) 对于固定波长和不同光强的单脉冲, CsPbBr$_3$ 钙钛矿量子点人工光子突触产生的兴奋性突触后电流; PPF 效应是用 (e) 固定波长、不同光强度和 (f) 固定光强度、不同波长的两个光脉冲模拟的; (g) 突触器件对光脉冲序列和负电脉冲序列的反应; (h) 器件中 PPD 跟随 PPF 的实验演示, 红线表示器件电流, 蓝线表示光子脉冲[62]

3.3 忆阻器神经网络

在由忆阻器构成的大型交叉阵列构成的神经网络中, 忆阻器可以直接利用物理定律, 通过应用学习算法来执行高效的大规模内存计算。虽然最近实现的忆阻器神经网络 (memristor neural network, MNN) 的能量效率和计算速度令人满意, 但大规模忆阻器交叉阵列的实验实现仍处于初级阶段。

3.3.1 人工神经网络

人工神经网络 (ANN) 是一项机器学习技术, 它是利用模拟生物神经网络来进行人工智能 (AI)。人类大脑中的神经网络十分密集且复杂。根据科学估算, 在成人大脑中有约 1000 亿个神经元, 而每个神经元大约有 10000 个连接[68]。神经网络将会应用到大数据、云计算、人工智能等领域。企业越来越多地采用人工智能来创建新的应用程序，推动了人工智能优化芯片的发展。这些应用包括连接设备、自动驾驶车辆、设备上的个人界面、语音交互和增强现实 (AR)。

最近，基于忆阻器的人工神经元已经被开发出来，但其生物现实动力学有限，并且没有与集成网络中的人工突触直接交互[69]。目前大多数研究仅限于小尺寸 (少于 10^{24} 个忆阻器)、二进制器件状态或有限的可重构性。当然，科学家们已经开发了一个用 PCM 阵列实现的大型神经网络 (165000 个突触)[70]，但这项工作受到顺序接口的限制[71]。人工神经网络由大量相互连接的神经元组成[72]。图 3.8(a) 是一个简单的人工神经网络拓扑图。每个神经元 (圆圈) 代表一个特定的输出函数，称为刺激函数。每两个神经元之间的连接表示通过该连接的信号的权重值，这等效于人工神经网络的记忆。网络的输出随网络连接方式、权值和激励函数而变化。此外，MNN 的硬件实现依赖于忆阻器交叉杆 (图 3.8(b))，这可以通过与周围神经元电路匹配来实现[73]。与忆阻器交叉杆密切相关的一个核心计算操作是向量矩阵乘法 (VMM)，它可以在单个模拟计算步骤中，利用欧姆定律和基尔霍夫电流定律自然地在密集忆阻器交叉杆几何结构中实现求和。

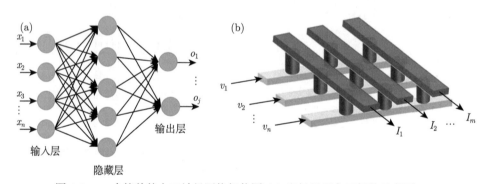

图 3.8　一个简单的人工神经网络拓扑图 (a) 和忆阻器交叉杆的示意图 (b)

学习作为神经网络研究的主要部分，神经网络的适应性也是借助于学习来完成的。随着环境的改变，可以调节权重以改善系统的行为。根据学习环境的不同，神经网络可以分为监督式学习和无监督式学习。

(1) 在监督式学习中，将样本数据集输入网络，将网络输出与期望输出进行比较，训练后，每个神经元收敛到一个权值。

当输入任务时，神经网络计算最优解 (可以理解为，我们为神经网络提供特定的训练，使其具备完成相应任务的能力)。

(2) 在无监督式学习中，样本数据是输入的，但没有给出预期的输出，数据之间的关系是根据聚类或某个模型得到的。

3.3 忆阻器神经网络

因此,无监督学习更像是自我学习,让神经网络学会自己做事情。

感知器是人工神经网络的起源算法,属于监督学习。2015 年,Prezioso 等构建了一个集成的无晶体管忆阻器交叉杆神经网络。如图 3.9(a)、(b) 所示,对于 3×3 的二值图像,$V_1 \sim V_9$ 是输入信号,V_{10} 是恒定的偏压。$W_{i,j}(i=1,2,3, j=1,\cdots,10)$ 是神经网络中突触的可调节权重。

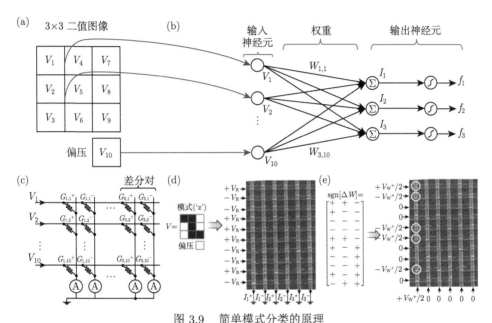

图 3.9 简单模式分类的原理
(a) 输入图像;(b) 用于 3×3 二值图像分类的单层感知器;(c) 单层感知器用 10×6 忆阻器横杆实现;(d) 特定模式 "z" 的分类培训示例;(e) 特定列 (图中的第一列) 的重量调整示例 [74]

这是一个单层感知器,有 10 个输入和 3 个输出,完全连接 $10\times 3=30$ 个突触权重,每个权重对应于两个忆阻器电导值之间的差值。

$$W_{i,j} = G_{i,j}^+ - G_{i,j}^-$$

因此,交叉杆中的忆阻器总数为 $30\times 2=60$(图 3.9(c)),I_i 可以使用向量矩阵乘法计算:

$$I_i = \sum_{j=1}^{10} W_{i,j}V_j$$

因此,如图 3.9(d) 所示,交叉杆中的输入信号 V_R 根据像素的颜色确定 ($+V_R$ 表示深色,$-V_R$ 表示浅色)。对于特定的误差矩阵,重量调整示例如图 3.9(e) 所示。在所示的步骤中，只有重量应该增加的突触 (在左边的表格中用 "+" 标记) 被调整，也就是说，忆阻器的电导 $G_{1,1}^+$、$G_{1,2}^+$、$G_{1,5}^+$、$G_{1,6}^+$ 和 $G_{1,9}^+$ 被增加。

3.3.2 深度神经网络

深度神经网络 (DNN) 又称深度学习，是一种基于数据表示学习的机器学习方法，这也是一种能够模仿人脑神经构造的机器学习方式。深度学习的定义主要来自于人工神经网络的研究。DNN 是一个大的类别，可以分为卷积神经网络 (CNN)、递归神经网络 (RNN) 和生成性对抗网络 (GAN)，这些网络都在近些年崛起，它们有自己的专业领域，例如，CNN 在图像处理和图像识别方面具有卓越的能力，RNN 在处理声音和时间序列数据方面具有卓越的能力，而 GAN 在生成图像方面具有卓越的能力。其中，基于忆阻器的 GAN 仍处于模拟阶段[75,76]。

深度学习的 "深度" 指的是隐藏层的数量，因此选择问题越复杂，需要的隐藏层就越多。例如，AlphaGo 的策略网络的隐藏层是 13 层，每层中的神经元数量是 192[77]。本节介绍了使用忆阻器实现的两种 DNN。

Hubel 和 Wiesel 在 20 世纪 60 年代，深入研究了猫的大脑皮层神经元的局部敏感性和方向选择性，并发现特殊的网络结构可以减少反馈神经网络的复杂度，于是提出了一种 CNN[78]。1980 年，Fukushima 提出了 CNN 的重要实现——新识别机[79]。在 CNN 中，卷积层中的神经元只与前一层中的一些神经元节点连接，它们之间共用同层中几个神经元之间连接的权重 w 和偏移量 b，这缩减了所需训练参数的数量。近年来，CNN 因其避免了复杂的图像预处理，在模式识别和分类等领域已经展现出了极其广泛的应用范围。

最近，Yao 等构建了一个由八个 128×16 忆阻器交叉阵列 (图 3.10(b)) 组成的五层 CNN(图 3.10(a))，以实现高性能的数字识别；该阵列显示出显著可重复的多级电导状态 (图 3.10(c))；这是第一个基于忆阻器交叉杆的内存计算的硬件实现，提出了一种适合忆阻器特性的混合训练算法和一种新的卷积空间并行结构；与当代最先进的 GPU 相比，其电源效率和性能密度都得到了提高。实践证明，DNN 可以利用忆阻器解决 CMOS 的物理极限问题[80]。

20 世纪 80 年代，约翰·霍普菲尔德 (J. Hopfield) 提出了一种新的神经网络——混合神经网络 (HNN)[81]，它既能解决模式识别的问题，还能给出一类组合优化问题的近似解。HNN 是一种从输出到输入具有反馈连接的 RNN，它将物理学的相关思想 (动力学) 引入到神经网络的结构中，从而形成 HNN。每个神经元与所有其他神经元相连，也被称为一个完全互联的网络 (图 3.11(a))。如果 HNN 是一个收敛稳定的网络，则这个反馈和迭代计算过程的变化会越来越小。一旦达到稳定的平衡状态，Hopfield 网络将输出一个稳定的常量值[82]，然而，可能没有平衡点，表现出超混沌行为[83]。2017 年，Kumar 等使用基于 Nb_2 的莫特 (Mott)

3.3 忆阻器神经网络

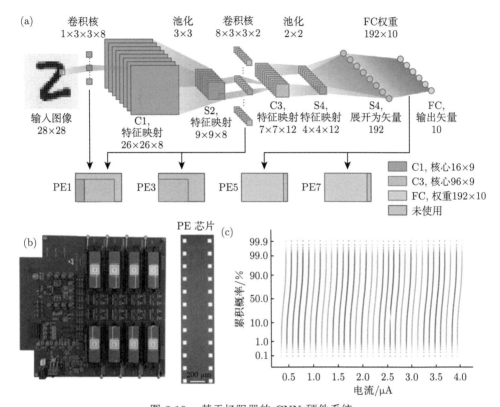

图 3.10 基于忆阻器的 CNN 硬件系统

(a) 用于修改的国家标准技术研究院 (MNIST) 图像识别的五层忆阻器 CNN 的结构, 具有交替卷积 (C1, C3) 和子采样 (S2, S4) 层, 输入为 28×28 灰度 (8 位) 数字图像; (b) 左边是 (a) 的硬件实现——集成电路板 (printed circuit board, PCB) 子系统, 8 核, 右边是由 128×16 忆阻器阵列构成的计算核心的放大图; (c) 关于 32 个独立电导状态的 10^{24} 个单元的累积概率分布[80]

忆阻器构建了一个内在耦合振荡器, 驱动系统显示混沌行为, 进一步证明了混沌振子的集成可以改善硬件实现 HNN[84]。

自 HNN 引入以来, 其在各个应用领域都受到了广泛关注[85,86]。它具有解决旅行推销员问题和地点分配问题的众所周知的能力[87,88], 但传统的 HNN 是通过构建金属氧化物半导体 (MOS) 晶体管作为电子突触来实现的[89]。1987 年, 贝尔实验室成功开发了第一个基于 HNN 的神经网络芯片[90], 为 HNN 带来了新的生命。随着忆阻器的出现, 基于忆阻器的 HNN 应运而生。2015 年, Duan 等提出了一种基于忆阻器的 "小世界"HNN 模型, 该模型更接近人脑神经网络, 在数字识别方面表现出优于原始规则神经的性能等[91]。图 3.11(b) 显示了具有突触权重的 "小世界"HNN。

HNN 还提供了模拟人类记忆的模型。2015 年, Hu 等提出了一种基于忆阻器

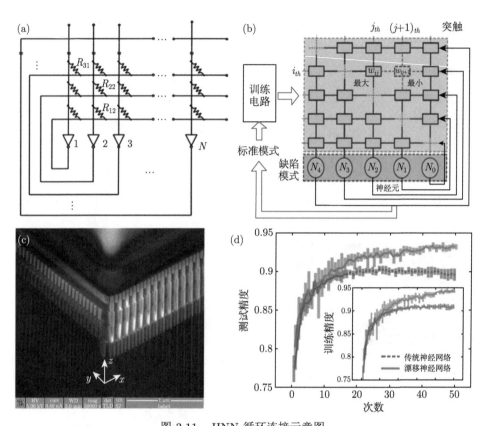

图 3.11　HNN 循环连接示意图

(a) N 个模拟放大器的示意图，这些放大器通过 N_2 电阻器 R 网络互连，将放大器 i 的输出连接到具有突触权重 j 的放大器；(b) HNN 的输入 [91]；(c) 使用的 PCM(39 nm, 1 Gbit) 阵列的 SEM 角视图；(d) 传统神经网络和漂移神经网络的分类精度的比较 [94]

的突触，由四个忆阻器桥电路和三个晶体管组成 [92]。2018 年，Sun 等改进了忆阻器桥电路，使其能够在不使用晶体管的情况下实现神经细胞的突触操作，他们用该器件构建了一个 3 位忆阻器 HNN，展示了联想记忆的能力 [93]。

最近 Lim 等 [94] 建立了一个 39nm、1Gbit 的相变存储器忆阻器阵列 (图 3.11(c))，并量化了独特的电阻漂移效应；在此基础上，提出了利用电阻漂移改善基于相变存储器的忆阻器网络训练的自发稀疏学习 (SSL) 方案；在训练过程中，SSL 将漂移效应视为基于自发一致性的蒸馏过程，在 HRS 下不断增强阵列权重，除非基于梯度的方法将其切换到低电阻状态；实验表明，SSL 不仅有助于网络的收敛，而且在手写数字分类中具有更好的性能和稀疏可控性 (图 3.11(d))，无需额外计算；这项工作促进了记忆器件固有特性的学习算法，为神经形态计算芯片的发展开辟了新的方向。此外，2017 年，Sheridan 等报道了首次使用 32×32 忆阻器交叉开关，使用生物启发式方法实现稀疏编码算法的实验[95]。2019 年，Wang 等首次实现了现场训练的大规模氧化还原忆阻器卷积神经网络 (convolutional neural networks，CNN) 和循环神经网络 (recurrent neural networks，RNN) 的模式和视频分类[96]。

3.3.3 脉冲神经网络

真正的生物神经元通过电脉冲"尖峰"相互通信，这影响了 Maass 撰写的颇具影响力的论文——《尖峰神经元网络：第三代神经网络模型》[97]。尖峰神经元网络 (SNN) 可以实现更高层次的生物神经模拟，并将时间的概念融入其操作中。虽然现代神经网络在许多领域取得突破，但它们不能模仿生物大脑神经元的工作机制。SNN 从根本上不同于当前主流的神经网络和机器学习方法。建立 SNN 有三个必要条件：外部刺激编码、建立神经元模型和开发学习规则。

到目前为止，已经提出了各种编码，但它们都是完全时态代码的特例。在全时态编码中，编码依赖于所有峰值的精确定时[98]。二进制编码是全有或全无编码：由于其简单性，它很有吸引力，但它完全忽略了计时特性和尖峰的多重性。速率编码是尖峰时间特性的另一种抽象，因为只有间隔中尖峰的速率被用作传输信息的度量。潜伏期编码使用的是棘波的计时，而不是棘波的多重性，与此密切相关的是顺序编码，即围绕神经元发出第一个峰值的顺序进行编码[99]。预测性棘波编码提供了一种方法，分别在神经元的体细胞和突触上执行模拟–数字转换和数字–模拟转换[100−102]。概率尖峰编码与有效使用尖峰神经元进行推理有关，通常在计算神经科学的背景下考虑[103]。

人脑中的真实神经元表现出多种尖峰行为，为了模拟真实神经元的行为，人们提出了几种具有 STDP 学习规则生物学特性的尖峰神经元模型。这些模型通常表示为各种复杂的动态系统，其中最著名的是霍奇金–赫胥黎 (H-H) 模型[104]，伊兹克维奇 (Izhikevich) 模型是霍奇金–赫胥黎模型的简化[105]，其他模型，如泄漏积分火灾 (LIF) 模型、二次积分和发射模型，以及其他更复杂的模型，代表神经科学现实主义和计算复杂性之间的不同权衡。其中，LIF 模型是应用最广泛的模型之一。动态系统模型的另一种替代方法是所谓的尖峰响应模型 (SRM)[106]，其中膜电位不需要通过微分方程模拟，但膜电位被用作积分核的总和。在这种观点中，神经元 (理解为包括其传入突触) 相当于一个数学运算符，因为它将输入中的峰值序列映射到输出中的峰值序列[107]。

以 LIF 模型为例。如图 3.12(a) 所示，ANN 简单地使用权重连续向前计算，找到中间层的相应值，然后使用反向传播 (BP) 来减少损失函数的值。但 SNN 发送的是神经尖峰 (在随机时间点发生的离散事件)，而不是连续值。神经元的膜电位在正常时间内不起作用，只要脉冲被发送过来，超过膜电位的阈值，兴奋的神经元就会向前传输脉冲 (图 3.12(b)),从而减少能量消耗并加快计算速度,膜电位在传输后立即降回原始水平,等待下一个脉冲。

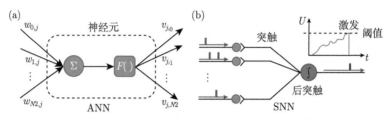

图 3.12 传统的人工神经元模型 (a) 和 LIF 尖峰神经元模型 (b)

此外,SNN 通常连接稀疏,并利用特殊的网络拓扑。传统的人工神经网络需要逐层计算,计算量要大得多。图 3.13 显示了两种不同的尖峰神经元模型。其中,如图 3.13(a) 所示,Yi 等设计了一个 H-H 莫特尖峰神经元,通过改变被动 R 和 C 元件的值来实现各种神经元动力学,而不改变 VO_2 器件的参数[108]。如图 3.13(b) 所示,Zhang 等设计了一个 LIF-莫特尖峰神经元,构建了高性能硬件 SNN,实现了高精度的 MNIST 数字识别[109]。尽管 SNN 比人工神经网络优越,但 SNN 的挑战来自几个方面。

首先,当大多数误差函数考虑实际值和时间连续性时,为了协调尖峰的不连续性,有几种解决方案及其缺点。

(1) 线性化尖峰过程,但仅适用于学习步骤较少的场合,如 SpikeProp[110]。

(2) 抹去尖峰,进行更连续的计算,如恢复算法[111]。

(3) 使用更复杂的不间断尖峰神经元模型或使用具有潜在平滑瞬态点火强度的随机神经元,但网络复杂性显著增加[112,113]。

(4) 预测脉冲编码方法实际上是近似连续的信号,其他方法依赖于提取的脉冲神经元信息[114,115]。

其次,脉冲的编码和解码是 SNN 的第二个挑战。

(1) 神经科学本身尚未被探索。

(2) SNN 中的速率编码或脉冲时间编码的标签也令人困惑。原则上,尖峰神经元取代了 S 型函数神经元,通信变得稀疏,只需要在链路被激活时进行计算,这与传统的人工神经网络不同,人工神经网络是在每个时间步执行的,但编码和解码是个绊脚石。

(3) 脉冲神经计算与生物脉冲神经元之间的合理关联也是一个挑战。由于 SNN 的计算与生物神经网络的计算非常相似,因此可以合理地预期,基于计算科学的计算驱动方法与基于数据的神经科学之间存在良好的交互作用。神经科学告诉我们 SNN 模型，有效的 SNN 可以阐明 STDP 等现象背后的计算原理和神经编码[116]。

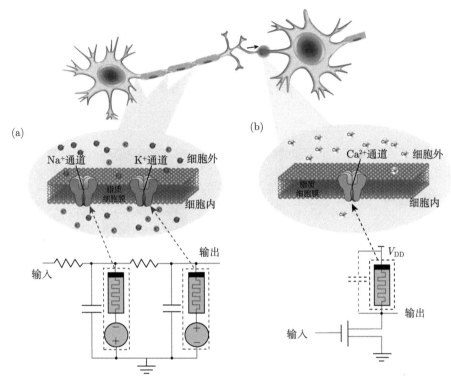

图 3.13　两种莫特峰神经元

(a)H-H 神经元由两个基于 VO_2 的忆阻器构建，模拟电压门控 Na^+ 和 K^+ 穿过细胞膜的机制，使细胞体和轴突节点的动作电位产生和重复[108]；(b)LIF 神经元是使用基于 NbO_2 的忆阻器构建的，该忆阻器模拟突触前电压门控 Ca^{2+} 的流入和流出，以实现突触间的信息传递[109]

3.4　忆阻器的应用

忆阻器作为电子突触和神经元的仿生特性，激发了神经形态计算中新信息技术的出现。信息技术的创新不仅依赖于复杂的算法和编程，还依赖于新材料和高性能器件来实现人机交互。如果存在能够将外部刺激传递到内部神经系统的电子受体，则忆阻器的应用可以扩展到人工神经。忆阻器实现了高速三维存储，将广泛应用于集成电路学科，并将极大地推动人工智能的研究向实用方向发展。

3.4.1　忆阻器在生物学上的应用

天然材料经过数百万年的发展，具有近乎完美的结构和功能，其中电子转移是最基本、最微妙的生物过程之一[117]。图 3.14(a)~(d) 显示了用作忆阻器 RS 层的几种生物材料, 包括木质素 (图 3.14(a))、蛋清 (图 3.14(b))、家蚕的血淋巴 (图 3.14(c))、丝素蛋白 (图 3.14(d))。丝素蛋白是一种具有代表性的生物分子, 已被广泛研究[118,119]。除了蛋白质, 糖 (尤其是多糖) 也被广泛用作开发新型绿色存储器件的基础[120]。最近, Fu 等报道了一种扩散型忆阻器, 由从硫还原的杆菌中获得的蛋白质纳米线制备成功, 在 40~100 mV 的生物电压下工作。由这些忆阻器构建的人工突触不仅表现出与生物神经元的时间整合, 并且还证明了直接处理生物传感信号的潜力[121]。生物材料的应用和仿生器件的发展得益于与生物学的交叉。

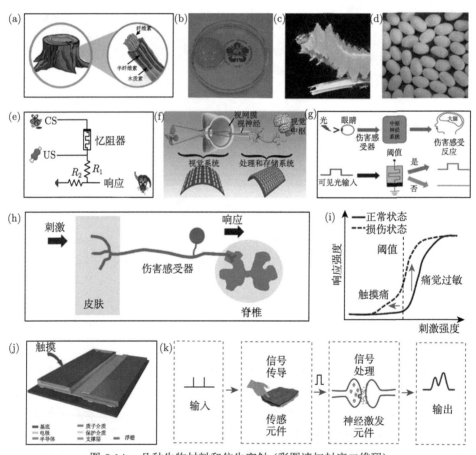

图 3.14 几种生物材料和仿生突触 (彩图请扫封底二维码)
(a) 木质素, 植物的主要成分之一[122]; (b) 鸡蛋和提取的蛋清[123]; (c) 家蚕的血淋巴是从家蚕身上获得的[124]; (d) 丝素蛋白[118]; (e) 经典条件反射的消失和恢复[128]; (f) 基于忆阻器图像传感器的视觉记忆器件, 模拟人类的视觉记忆[64]; (g) 人眼的工作机理示意图及本光电探测器阈值强度依赖性光响应特性[129]; (h) 典型伤害感受器神经系统的示意图; (i) 伤害感受器在正常和受损状态下的典型刺激与反应关系[130]; (j) 触觉传感元件示意图; (k) 人工触觉感知系统原理[131]

人体有一个复杂的神经系统。不同类型的神经元表现出不同的功能。忆阻器的多样性使模拟这些功能变得容易。巴甫洛夫条件反射是生物大脑中联想学习的经典案例。巴甫洛夫实验的忆阻器模拟已经被广泛研究[125-127]。如图 3.14(e) 所示，Li 等提出了一种记忆性神经形态回路，可以在一系列条件刺激 (CS) 和非条件刺激 (US) 下产生经典的巴甫洛夫条件反射[128]。2018 年，Chen 等提出了一种简单的体系结构，它使用光刻胶构建灵活的人工视觉记忆系统，这为模仿人类视觉记忆带来可能，如图 3.14(f) 所示，基于 RS 存储器件和电阻图像传感器的视觉存储器件可以模拟人类视觉存储[63]，仿真示意图如图 3.14（g）所示。2018 年，Kim 等[130] 利用 Pt/Ti/HfO$_2$/TiN 结构的阈值开关特性设计了一种固态伤害感受器 (图 3.14(h)、(i))，该结构显示了生物伤害感受器在从几毫秒到几十秒内的所有阈值，以及具有放松、镇痛和痛觉过敏的必要功能。2017 年，Zang 等设计了一种基于双有机晶体管的触觉感知元件，其信号传导和信号处理能力可以实现动态触觉传感，如图 3.14(j) 和 (k) 所示[131]，该系统具有显著的感知特性，在柔性、低成本仿生智能产品和人机界面元件等方面具有广阔的应用前景。此外，在听觉方面，Park 等[132] 在 2013 年首次利用脑电图实验证明了 MNN 上的听觉记忆。它可以应用于脑机接口的开发，以恢复瘫痪患者的语言能力。2019 年，Danesh 等[133] 使用 MNN 以并行模式模拟进行同步语音推理和学习，能量效率约为 1.6×10^{17} flops·W^{-1}，比 Summit 超级计算机高出约 7 个数量级。

3.4.2 忆阻器在模式识别中的应用

光电突触出现后，科学家们开始尝试将易失性光电探测器与非易失性记忆器结合起来，以实现图像的感知和存储[134-136]。2018 年，Chen 等开发了一个集感知和存储于一体的系统。次年，Zhou 等报告了一种基于 MoO$_x$ 的 ORRAM，其阵列是概念验证，可以简化神经形态视觉系统电路，并有助于边缘计算和物联网应用的开发。它为简化神经形态视觉系统电路提供了潜力，并有助于边缘计算和物联网应用的发展。如图 3.15(a) 所示，阵列的可调突触可塑性使我们能够进行图像的第一阶段处理，例如图像对比度增强和降噪等，大大提高了后续图像处理的效率和准确性。近年来，随着记忆型光电突触的发展，基于光电忆阻器的神经网络开始出现。这些网络应用于直接收集和处理视觉信息。2019 年，Feldmann 等[138] 首次利用 PCM 实现 SNN 的全光、集成、可扩展的神经形态框架，这种光子神经突触学网络保证了能够获得光学系统固有的高速和高带宽，这种光子神经突触网络有望实现光学系统固有的高速和高带宽，从而实现光学通信和视觉数据的直接处理。同年，Zhai 等利用近红外突触模拟单层感知器进行监督学习。最近，Mennel 等[139] 证明了图像传感器本身可以构成一个可以同时感知和处理光学图像的神经网络。一个图像传感器阵列，该阵列可以识别 3×3 像素图像，其中每个像素由三个 WSe$_2$ 光电二极管/亚像素组成。该器件实现了 50 ns 内的超快速模式识别，通过将光电流转换成电压，该器件在 50 ns 内实现了超高速模式识别。通过将光电流转换为电压，然后将其馈送到忆阻器交叉阵列中，就可以实现深度学习网络的模拟。随后，Wang 等利用 WSe$_2$/h-BN/Al$_2$O$_3$ 异质结构构建了视网膜形态学视觉传感器，实现了对输入图像的预处理，并作为 CNN 实现图像识别 [140]。

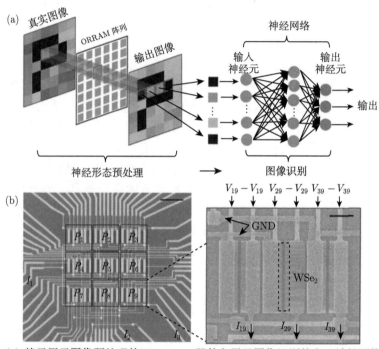

图 3.15 (a) 基于用于图像预处理的 ORRAM 器件和用于图像识别的人工神经网络的人工神经形态视觉系统的图示 [137]；(b) 光电二极管阵列的显微镜图像，包括 3×3 像素 (比例尺，15 μm) 和其中一个像素 (比例尺，3 μm) 的 SEM 图像 [139]

光电突触的人工神经网络实现的另一个有趣功能是对彩色和彩色混合模式进行模式识别 [13]。图 3.16 为光学神经突触器件的示意图，该器件将光学传感器件和突触器件串联在 h-BN/WSe$_2$ 异质结构上；将光学传感器件暴露在红光 (655 nm)、绿光 (532 nm) 和蓝光 (405 nm) 下，并保持突触前漏极偏置不变，通过向突触器件施加电压脉冲来测量突触后电流；光刺激降低了光学传感器件的电阻，同时增加了被困在突触器件电荷捕获层的载流子密度。光刺激降低了光学传感器件的电阻，并增加被困在突触器件电荷捕获层的载流子密度，这进一步调整了光学神经突触器件的突触动力学特性。在 600 次的兴奋和抑制脉冲下，发现光学神经突触在 0.3% 阈值 ΔG 和 1.5(增强)/1.5(抑制) 的非线性下达到 599 个有效电导状态，

3.4 忆阻器的应用

脉冲幅度为 ±0.3 V。对于需要 64~256 个状态的深层神经网络，光学神经突触器件实现了良好的线性和足够的状态数。

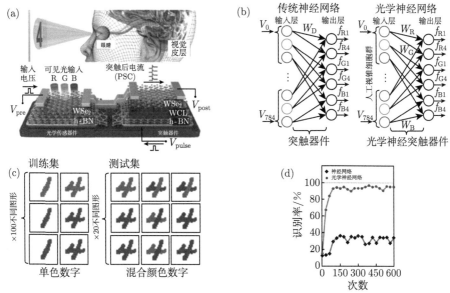

图 3.16 (a) 人类视神经系统和光学神经突触系统的示意图；(b) 用于识别 28×28 RGB 彩色图像的光学神经网络图示；(c) 分别由单色和彩色混合数字模式图像组成的训练和测试数据集示例；(d) 识别率是训练次数的函数

Strukov 团队开发了一种 64×64(4K) 无源模拟梯度忆阻器交叉杆阵列，通过优化开关层 Al_2O_3 和 TiO_x 以及集成技术，该 4K 无源忆阻器系统可以通过矩阵乘法实现 64×10 矢量，平均相对电导输入精度为 1%，从而通过现场训练的单层感知器模拟 MNIST 图像分类。4K 忆阻器阵列，每个交叉点由 Ti/Al/TiN/Al_2O_3/TiO_x/Ti/Al/TiN 忆阻器组成 (图 3.17(a))。15 个电导水平可以从 3~45 μS 的范围内获得，相对误差为 1%(图 3.17(b))。数字识别的实验结果表明，数字 "7" 的分类相对训练误差为 1%(图 3.17(c))[141]。

人工神经网络的能力取决于神经元中连接的数量。在 3D 中创建计算电路可以满足大规模连接和高效通信的需求，这也有利于语音识别、模式分类和目标检测等复杂任务。基于 3.5 nm HfO_2 开关层的忆阻器是为集成 3D 计算电路而开发的 (图 3.17(d))。利用 3D 忆阻器结构，由于忆阻器内核，利用 3D 忆阻器结构在收集一万张测试图像的过程中，软件 (98.11%) 和硬件 (98.10%) 之间的 MNIST 精度推断差异非常小。内核还用于处理输入视频帧 (图 3.17(e))，其边缘特征由 3D 忆阻器硬件精确提取 (图 3.17(f))[142]。当然，这些只是忆阻器和 MNN 应用的冰山一角。MNN 在实现行走步态分类[143]、CIFAR-10 分类[144,145] 等功能方面的应用也应运而生。未来忆阻器的应用将更加丰富多彩。

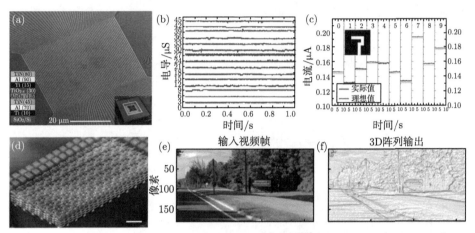

图 3.17 数字识别和视频处理的实现 (彩图请扫封底二维码)

(a)64×64 忆阻器交叉杆阵列,插图显示了器件结构和封装芯片;(b)15 个电导水平的间隔为 3~45 μS,相对误差为 1%;(c) 模式分类的实验结果表明,模式 "7" 的相对调谐误差为 1%[141];(d) 八层 3D 忆阻器的 SEM 图像,标尺为 2 μm;(e) 输入视频帧;(f)3D 忆阻器硬件检测结果 [142]

3.4.3 忆阻器在电路中的应用

作为一种新型的电路元件,忆阻器最直观的应用是在电路中的应用,其特点是利用忆阻器的特性来提高电路性能和简化电路。早在 2008 年,Itoh 等就证明了忆阻器对于振荡电路非常有用 [147]。2018 年,Zhang 等设计了一种带有忆阻器的约瑟夫森结电路,通过对输出序列的分析发现,适当的参数设置可以产生明显的混沌和周期性状态,研究了基于忆阻耦合约瑟夫森结电路的混沌加密 [146]。忆阻器耦合电路的方案图如图 3.18(a) 所示。2018 年,Sun 等仅使用少数忆阻器构成网络,并用极其简单的电路实现布尔逻辑运算。除了图 3.18(c)、(d) 所示的 NAND 和 NOR 操作外,他们还实现了异或操作和一位全加器操作 [148],大大减少了逻辑门所需的电路元件数量。

Minati 等实验证明了忆阻器适用于构建自治混沌电路 (图 3.19(a)),吸引子在几个平面上的投影 (图 3.19(e)) 显示了它的螺旋式相位相干结构,与 Rössler 系统等范例相似,这促使考虑忆阻器在其他混沌体系结构中使用,包括基于多个忆阻器的混沌体系结构,特别是在更广泛的非线性动力学背景下 [149]。例如,最近的理论结果表明,在通过忆阻器耦合的动力系统网络中,可以实现自适应一致性和同步。另一方面,需要承认的是,此处观察到的低维动力学有这样的特点:其特征在很大程度上与一些此前广为人知且结构相当简单的电路 (如单管振荡器) 易于产生的相位相干吸引子相重叠,这些电路在结构上也相当简单。忆阻器的出现可能为电路行业注入新的活力,应用前景十分广阔。

3.4 忆阻器的应用

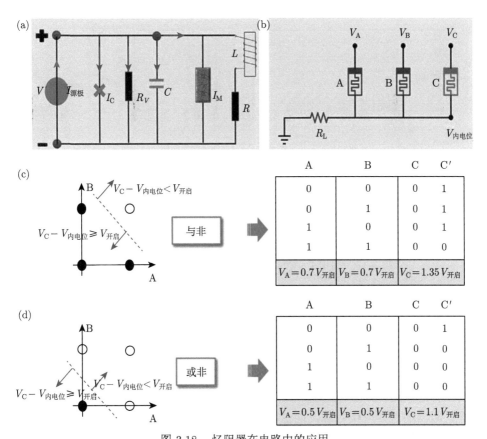

图 3.18 忆阻器在电路中的应用
(a) 忆阻器耦合电路的方案图 [146]；(b) 实现逻辑门功能的忆阻器阵列示意图；(c)NAND 操作和 (d)NOR 操作通过改变施加电压 V_A、V_B 和 V_C 来实现 [148]

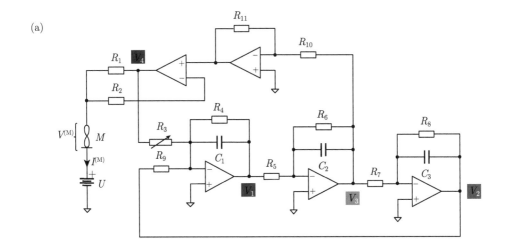

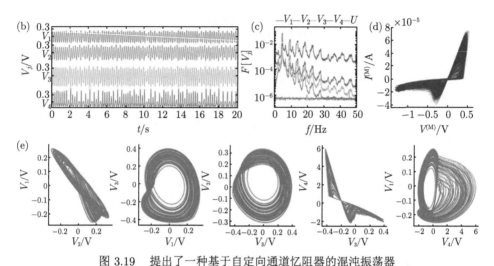

图 3.19 提出了一种基于自定向通道忆阻器的混沌振荡器
(a) 电路原理图；(b) 节点电压的时间序列 $V_1 \sim V_4$；(c) 相应的频谱，其中设置为恒定电压的源 U 额外显示为代表记录的噪声基线；(d) 收缩磁滞回线的电流–电压图；(e) 吸引子在多个平面上的投影 [149]

3.5 小　　结

尽管基于忆阻器的 DNN 比 CMOS 表现出更好的性能，但 DNN 在广义上属于人工神经网络，并没有逃脱第二代神经网络的范畴。因此，DNN(尤其是 GAN) 只是忆阻器的近期应用目标，而 SNN 则是忆阻器的长期应用目标。

尖峰训练增强了处理时空数据 (或真实世界的感官数据) 的能力。空间指的是神经元，时间指的是尖峰训练。这使我们能够自然地处理时间数据，而不需要 RNN 增加额外的复杂性。已经证明，尖峰神经元已经成为了更强大的计算单元。

然而，SNN 的广泛应用仍然面临许多挑战。主要是因为神经元尖峰行为的离散性和稀疏性使得尖峰训练不可微，我们无法在不丢失准确时间信息的情况下使用梯度下降来训练 SNN。虽然我们有无监督学习方法 (如 STDP 和 Hebb)，但其只能用于解决特定任务。为了将 SNN 应用到解决更具挑战性的现实任务 (如回归分析/推理/时态模式识别等) 中去，我们需要开发一种有效的监督学习方法。因为考虑到这些网络的生物现实性，涉及决定人类大脑如何学习，所以这是一项艰巨的任务。

另一个问题是，基于普通硬件模拟 SNN 需要大量的计算能力，因为它需要模拟微分方程。由于脉冲神经网络的复杂性，研究人员尚未找到有效的学习算法。该算法需要符合生物神经网络的特性 (生长和死亡)。

总之，在未来的工作中，为了适应 SNN 的应用，一方面，我们迫切需要优秀的功能材料。基于这种材料的忆阻器需要具有广泛的电阻值、低电阻值波动和漂移、线性和对称重量变化，以及用于推断的高绝对电阻值，并且需要高耐用性以便于重复编程和训练。同时，为了促进高密度单片集成，必须同时考虑成本和工艺难度。另一方面，我们需要制定一个通用的 SNN 算法，将脉冲神经计算与生物脉冲神经元合理关联，其开发需要整合不同学科（工程学、生物学、物理学等）和不同层次（算法、架构、系统、电路、器件、材料等）。令人欣慰的是，许多团队正在开发 SNN 监督学习规则，我们对 SNN 的未来持乐观态度。

如今，基于忆阻器的 SNN 还远不是大多数任务的实用工具。虽然 SNN 在实时图像和音频处理方面有一些应用，但它仍处于实验阶段。大多数基于忆阻器的 SNN 工作要么是理论上的，要么在简单的完全连接的第二代网络中显示性能。但应该相信，基于忆阻器的 SNN 和生物神经学可以有更强大的相互作用。在生物学上看似合理的 SNN 为实现脑-机接口带来了希望，在未来，它可能成为人脑的一部分，或者代替人脑来实现信息存储和思考。随着各种神经器件的发展，例如柔性皮下植入器件、人工生物感觉系统，真正意义上的"机器人"似乎就在眼前。

参考文献

[1] Moore G E. Cramming more components onto integrated circuits[J]. Proceedings of the IEEE, 2002, 86(1): 82-85.

[2] von Neumann J. First draft of a report on the EDVAC[J]. IEEE Annals of the History of Computing, 1993, 15(4): 27-75.

[3] Backus J. Can programming be liberated from the von Neumann style? A functional style and its algebra of programs[J]. Communications of the ACM, 1978, 21(8): 613-641.

[4] Wulf W A, McKee S A. Hitting the memory wall: Implications of the obvious[J]. ACM SIGARCH Computer Architecture News, 1995, 23(1): 20-24.

[5] Borkar S, Chien A A. The future of microprocessors[J]. Communications of the ACM, 2011, 54(5): 67-77.

[6] Afif M, Said Y, Atri M. Computer vision algorithms acceleration using graphic processors NVIDIA CUDA[J]. Cluster Computing, 2020, 23(4): 3335-3347.

[7] Owens J D, Houston M, Luebke D, et al. GPU computing[J]. Proceedings of the IEEE, 2008, 96(5): 879-899.

[8] Jouppi N P, Young C, Patil N, et al. In-datacenter performance analysis of a tensor processing unit[C]//Proceedings of the 44th Annual International Symposium on Computer Architecture, 2017: 1-12.

[9] Kautz W H. Cellular logic-in-memory arrays[J]. IEEE Transactions on Computers, 1969, 100(8): 719-727.

[10] Wolf S A, Awschalom D D, Buhrman R A, et al. Spintronics: a spin-based electronics vision for the future[J]. Science, 2001, 294(5546): 1488-1495.

[11] Stathopoulos S, Michalas L, Khiat A, et al. An electrical characterisation methodology for benchmarking memristive device technologies[J]. Scientific Reports, 2019, 9(1): 1-10.

[12] Locatelli N, Cros V, Grollier J. Spin-torque building blocks[J]. Nature Materials, 2014, 13(1): 11-20.
[13] Mikolajick T, Dehm C, Hartner W, et al. FeRAM technology for high density applications[J]. Microelectronics Reliability, 2001, 41(7): 947-950.
[14] Waldrop M M. The chips are down for Moore's law[J]. Nature News, 2016, 530(7589): 144.
[15] Pi S, Li C, Jiang H, et al. Memristor crossbar arrays with 6-nm half-pitch and 2-nm critical dimension[J]. Nature Nanotechnology, 2019, 14(1): 35-39.
[16] Choi B J, Torrezan A C, Strachan J P, et al. High-speed and low-energy nitride memristors[J]. Advanced Functional Materials, 2016, 26(29): 5290-5296.
[17] Nian Y B, Strozier J, Wu N J, et al. Evidence for an oxygen diffusion model for the electric pulse induced resistance change effect in transition-metal oxides[J]. Physical Review Letters, 2007, 98(14): 146403.
[18] Yang J J, Pickett M D, Li X, et al. Memristive switching mechanism for metal/oxide/metal nanodevices[J]. Nature Nanotechnology, 2008, 3(7): 429-433.
[19] Hur J H, Lee M J, Lee C B, et al. Modeling for bipolar resistive memory switching in transition-metal oxides[J]. Physical Review B, 2010, 82(15): 155321.
[20] Wuttig M, Yamada N. Phase-change materials for rewriteable data storage[J]. Nature Materials, 2007, 6(11): 824-832.
[21] Taherkhani A, Belatreche A, Li Y, et al. A review of learning in biologically plausible spiking neural networks[J]. Neural Networks, 2020, 122: 253-272.
[22] Wang X, Song T, Gong F, et al. On the computational power of spiking neural P systems with self-organization[J]. Scientific Reports, 2016, 6(1): 1-16.
[23] Poon C S, Zhou K. Neuromorphic silicon neurons and large-scale neural networks: challenges and opportunities[J]. Frontiers in Neuroscience, 2011, 5: 108.
[24] Arnold A J, Razavieh A, Nasr J R, et al. Mimicking neurotransmitter release in chemical synapses via hysteresis engineering in MoS_2 transistors[J]. ACS Nano, 2017, 11(3): 3110-3118.
[25] Tian H, Guo Q, Xie Y, et al. Anisotropic black phosphorus synaptic device for neuromorphic applications[J]. Advanced Materials, 2016, 28(25): 4991-4997.
[26] van de Burgt Y, Lubberman E, Fuller E J, et al. A non-volatile organic electrochemical device as a low-voltage artificial synapse for neuromorphic computing[J]. Nature Materials, 2017, 16(4): 414-418.
[27] Liu Y H, Zhu L Q, Feng P, et al. Freestanding artificial synapses based on laterally proton-coupled transistors on chitosan membranes[J]. Advanced Materials, 2015, 27(37): 5599-5604.
[28] Kim S, Choi B, Lim M, et al. Pattern recognition using carbon nanotube synaptic transistors with an adjustable weight update protocol[J]. ACS Nano, 2017, 11(3): 2814-2822.
[29] Zhu L Q, Wan C J, Guo L Q, et al. Artificial synapse network on inorganic proton

conductor for neuromorphic systems[J]. Nature Communications, 2014, 5(1): 1-7.

[30] Zhu L Q, Wan C J, Gao P Q, et al. Flexible proton-gated oxide synaptic transistors on Si membrane[J]. ACS Applied Materials & Interfaces, 2016, 8(33): 21770-21775.

[31] Martin S J, Grimwood P D, Morris R G M. Synaptic plasticity and memory: an evaluation of the hypothesis[J]. Annual Review of Neuroscience, 2000, 23(1): 649-711.

[32] Bliss T V P, Collingridge G L. A synaptic model of memory: long-term potentiation in the hippocampus[J]. Nature, 1993, 361(6407): 31-39.

[33] Bannerman D M, Sprengel R, Sanderson D J, et al. Hippocampal synaptic plasticity, spatial memory and anxiety[J]. Nature Reviews Neuroscience, 2014, 15(3): 181-192.

[34] Indiveri G, Linares-Barranco B, Legenstein R, et al. Integration of nanoscale memristor synapses in neuromorphic computing architectures[J]. Nanotechnology, 2013, 24(38): 384010.

[35] Yang J J, Strukov D B, Stewart D R. Memristive devices for computing[J]. Nature Nanotechnology, 2013, 8(1): 13-24.

[36] Hebb D O. The Organization of Behavior[M]. New York: Psychology Press, 1949.

[37] Pérez-Carrasco J A, Zamarreño-Ramos C, Serrano-Gotarredona T, et al. On neuromorphic spiking architectures for asynchronous STDP memristive systems[C]//Proceedings of 2010 IEEE International Symposium on Circuits and Systems. IEEE, 2010: 1659-1662.

[38] Linares-Barranco B, Serrano-Gotarredona T. Memristance can explain spike-time-dependent-plasticity in neural synapses[J]. Nature Precedings, 2009: 1-1.

[39] Zamarreño-Ramos C, Camuñas-Mesa L A, Pérez-Carrasco J A, et al. On spike-timing-dependent-plasticity, memristive devices, and building a self-learning visual cortex[J]. Frontiers in Neuroscience, 2011, 5: 26.

[40] Serrano-Gotarredona T, Masquelier T, Prodromakis T, et al. STDP and STDP variations with memristors for spiking neuromorphic learning systems[J]. Frontiers in Neuroscience, 2013, 7: 2.

[41] Bichler O, Querlioz D, Thorpe S J, et al. Extraction of temporally correlated features from dynamic vision sensors with spike-timing-dependent plasticity[J]. Neural Networks, 2012, 32: 339-348.

[42] Yu S, Wu Y, Jeyasingh R, et al. An electronic synapse device based on metal oxide resistive switching memory for neuromorphic computation[J]. IEEE Transactions on Electron Devices, 2011, 58(8): 2729-2737.

[43] Bichler O, Suri M, Querlioz D, et al. Visual pattern extraction using energy-efficient "2-PCM synapse" neuromorphic architecture[J]. IEEE Transactions on Electron Devices, 2012, 59(8): 2206-2214.

[44] Gerstner W, Ritz R, van Hemmen J L. Why spikes? Hebbian learning and retrieval of time-resolved excitation patterns[J]. Biological Cybernetics, 1993, 69(5): 503-515.

[45] Sjöström J, Gerstner W. Spike-timing dependent plasticity[J]. Scholarpedia, 2010, 5: 1362.

[46] Delorme A, Perrinet L, Thorpe S J. Networks of integrate-and-fire neurons using rank order coding B: Spike timing dependent plasticity and emergence of orientation selectivity[J]. Neurocomputing, 2001, 38: 539-545.

[47] Masquelier T, Guyonneau R, Thorpe S J. Competitive STDP-based spike pattern learning[J]. Neural Computation, 2009, 21(5): 1259-1276.

[48] Carlson K D, Richert M, Dutt N, et al. Biologically plausible models of homeostasis and STDP: stability and learning in spiking neural networks[C]//The 2013 international joint conference on neural networks (IJCNN). IEEE, 2013: 1-8.

[49] Bi G, Poo M. Synaptic modification by correlated activity: Hebb's postulate revisited[J]. Annual Review of Neuroscience, 2001, 24(1): 139-166.

[50] Wang Z, Joshi S, Savel'ev S E, et al. Memristors with diffusive dynamics as synaptic emulators for neuromorphic computing[J]. Nature Materials, 2017, 16(1): 101-108.

[51] Lv Z, Zhou Y, Han S T, et al. From biomaterial-based data storage to bio-inspired artificial synapse[J]. Materials Today, 2018, 21: 537-552.

[52] Zucker R S, Regehr W G. Short-term synaptic plasticity[J]. Annual Review of Physiology, 2002, 64: 355-405.

[53] Feng L, Molnár P, Nadler J V. Short-term frequency-dependent plasticity at recurrent mossy fiber synapses of the epileptic brain[J]. Journal of Neuroscience, 2003, 23(12): 5381-5390.

[54] Mulkey R M, Herron C E, Malenka R C. An essential role for protein phosphatases in hippocampal long-term depression[J]. Science, 1993, 261: 1051-1055.

[55] Kuzum D, Jeyasingh R G D, Lee B, et al. Nanoelectronic programmable synapses based on phase change materials for brain-inspired computing[J]. Nano letters, 2012, 12(5): 2179-2186.

[56] Cassinerio M, Ciocchini N, Ielmini D. Logic computation in phase change materials by threshold and memory switching[J]. Advanced Materials, 2013, 25(41): 5975-5980.

[57] Suri M, Bichler O, Querlioz D, et al. Phase change memory as synapse for ultra-dense neuromorphic systems: Application to complex visual pattern extraction[C]//2011 International Electron Devices Meeting. IEEE, 2011: 4.4.1-4.4.4.

[58] Tan H, Liu G, Yang H, et al. Light-gated memristor with integrated logic and memory functions[J]. ACS Nano, 2017, 11(11): 11298-11305.

[59] Wang S, Chen C, Yu Z, et al. A MoS_2/PTCDA hybrid heterojunction synapse with efficient photoelectric dual modulation and versatility[J]. Advanced Materials, 2019, 31(3): 1806227.

[60] Gao S, Liu G, Yang H, et al. An oxide Schottky junction artificial optoelectronic synapse[J]. ACS Nano, 2019, 13(2): 2634-2642.

[61] Liu G, Wang C, Zhang W, et al. Organic biomimicking memristor for information storage and processing applications[J]. Advanced Electronic Materials, 2016, 2(2): 1500298.

[62] Wang Y, Lv Z, Chen J, et al. Photonic synapses based on inorganic perovskite quantum dots for neuromorphic computing[J]. Advanced Materials, 2018, 30(38): 1802883.

[63] Chen S, Lou Z, Chen D, et al. An artificial flexible visual memory system based on an UV-motivated Memristor[J]. Advanced Materials, 2018, 30(7): 1705400.

[64] Lee M, Lee W, Choi S, et al. Brain-inspired photonic neuromorphic devices using photodynamic amorphous oxide semiconductors and their persistent photoconductivity[J]. Advanced Materials, 2017, 29(28): 1700951.

[65] Atkinson R C, Shiffrin R M. Human memory: A proposed system and its control processes[J]. Psychology of Learning and Motivation, 1968, 2: 89-195.

[66] Wang Z Q, Xu H Y, Li X H, et al. Synaptic learning and memory functions achieved using oxygen ion migration/diffusion in an amorphous InGaZnO memristor[J]. Advanced Functional Materials, 2012, 22(13): 2759-2765.

[67] Qian C, Oh S, Choi Y, et al. Solar-stimulated optoelectronic synapse based on organic heterojunction with linearly potentiated synaptic weight for neuromorphic computing[J]. Nano Energy, 2019, 66: 104095.

[68] Bishop C M. Neural Networks for Pattern Recognition[M]. Oxford: Oxford University Press, 1995.

[69] Wang Z, Joshi S, Savel'ev S, et al. Fully memristive neural networks for pattern classification with unsupervised learning[J]. Nature Electronics, 2018, 1(2): 137-145.

[70] Burr G W, Shelby R M, Sidler S, et al. Experimental demonstration and tolerancing of a large-scale neural network (165000 synapses) using phase-change memory as the synaptic weight element[J]. IEEE Transactions on Electron Devices, 2015, 62(11): 3498-3507.

[71] Hu M, Graves C E, Li C, et al. Memristor-based analog computation and neural network classification with a dot product engine[J]. Advanced Materials, 2018, 30(9): 1705914.

[72] Beale H D, Demuth H B, Hagan M T. Neural Network Design[M]. Boston: PWS Publishing Company, 1996.

[73] Xia Q, Yang J J. Memristive crossbar arrays for brain-inspired computing[J]. Nature Materials, 2019, 18(4): 309-323.

[74] Prezioso M, Merrikh-Bayat F, Hoskins B D, et al. Training and operation of an integrated neuromorphic network based on metal-oxide memristors[J]. Nature, 2015, 521(7550): 61-64.

[75] Liu F, Liu C. A memristor based unsupervised neuromorphic system towards fast and energy-efficient GAN[J]. 2018. DOI:10.48550/arXiv.1806.01775.

[76] Dong Z, Fang Y, Huang L, et al. A compact memristor-based GAN architecture with a case study on single image super-resolution[C]//2019 Chinese Control And Decision Conference (CCDC). IEEE, 2019: 3069-3074.

[77] Silver D, Huang A, Maddison C J, et al. Mastering the game of Go with deep neural networks and tree search[J]. Nature, 2016, 529(7587): 484-489.

[78] Hubel D H, Wiesel T N. Receptive fields, binocular interaction and functional architecture in the cat's visual cortex[J]. The Journal of physiology, 1962, 160(1): 106.

[79] Fukushima K. Neocognitron: A self-organizing neural network model for a mechanism

of pattern recognition unaffected by shift in position[J]. Biological Cybernetics, 1980, 36(4):193-202.

[80] Yao P, Wu H, Gao B, et al. Fully hardware-implemented memristor convolutional neural network[J]. Nature, 2020, 577(7792): 641-646.

[81] Hopfield J J. Neural networks and physical systems with emergent collective computational abilities[J]. Proceedings of the National Academy of Sciences, 1982, 79(8): 2554-2558.

[82] Zhang J, Jin X. Global stability analysis in delayed Hopfield neural network models[J]. Neural Networks, 2000, 13(7): 745-753.

[83] Pham V T, Jafari S, Vaidyanathan S, et al. A novel memristive neural network with hidden attractors and its circuitry implementation[J]. Science China Technological Sciences, 2016, 59(3): 358-363.

[84] Kumar S, Strachan J P, Williams R S. Chaotic dynamics in nanoscale NbO_2 Mott memristors for analogue computing[J]. Nature, 2017, 548(7667): 318-321.

[85] Hopfield J J. Neurons with graded response have collective computational properties like those of two state neurons[J]. Proceedings of the National Academy of Sciences, 1984, 81(10): 3088-3092.

[86] Hopfield J J, Tank D W. "Neural" computation of decisions in optimization problems[J]. Biological Cybernetics, 1985, 52(3): 141-152.

[87] Hopfield J J, Tank D W. Computing with neural circuits: A model[J]. Science, 1986, 233(4764): 625-633.

[88] Lázaro O, Girma D. A Hopfield neural-network-based dynamic channel allocation with handoff channel reservation control[J]. IEEE Transactions on Vehicular Technology, 2000, 49(5): 1578-1587.

[89] Verleysen M, Sirletti B, Vandemeulebroecke A, et al. A high-storage capacity content-addressable memory and its learning algorithm[J]. IEEE Transactions on Circuits and Systems, 1989, 36(5): 762-766.

[90] Howard R E, Schwartz D B, Denker J S, et al. An associative memory based on an electronic neural network architecture[J]. IEEE Transactions on Electron Devices, 1987, 34(7): 1553-1556.

[91] Duan S, Dong Z, Hu X, et al. Small-world Hopfield neural networks with weight salience priority and memristor synapses for digit recognition[J]. Neural Computing and Applications, 2016, 27(4): 837-844.

[92] Hu S G, Liu Y, Liu Z, et al. Associative memory realized by a reconfigurable memristive Hopfield neural network[J]. Nature Communications, 2015, 6(1): 1-8.

[93] Sun Z, Ambrosi E, Bricalli A, et al. Logic computing with stateful neural networks of resistive switches[J]. Advanced Materials, 2018, 30(38): 1802554.

[94] Lim D H, Wu S, Zhao R, et al. Spontaneous sparse learning for PCM-based memristor neural networks[J]. Nature Communications, 2021, 12(1): 319.

[95] Sheridan P M, Cai F, Du C, et al. Sparse coding with memristor networks[J]. Nature

Nanotechnology, 2017, 12(8): 784-789.

[96] Wang Z, Li C, Lin P, et al. In situ training of feed-forward and recurrent convolutional memristor networks[J]. Nature Machine Intelligence, 2019, 1(9): 434-442.

[97] Maass W. Networks of spiking neurons: the third generation of neural network models[J]. Neural Networks, 1997, 10(9): 1659-1671.

[98] Izhikevich E M. Polychronization: computation with spikes[J]. Neural Computation, 2006, 18(2): 245-282.

[99] Thorpe S, Gautrais J. Rank Order Coding[M]// Bower J M. Computational Neuroscience: Trends in Research. Boston: Springer, 1998: 113-118.

[100] Marre O, Amodei D, Deshmukh N, et al. Recording of a large and complete population in the retina[J]. Journal of Neuroscience, 2012, 32(43): 1485973.

[101] Rombouts J O, Van Ooyen A, Roelfsema P R, et al. Biologically plausible multidimensional reinforcement learning in neural networks[C]. Artificial Neural Networks and Machine Learning-ICANN 2012: 22nd International Conference on Artificial Neural Networks, Lausanne, Switzerland, September 11-14, 2012, Proceedings, Part I 22. Springer Berlin Heidelberg, 2012: 443-450.

[102] Bohte S M. Error-backpropagation in networks of fractionally predictive spiking neurons[C]//International Conference on Artificial Neural Networks. Berlin, Heidelberg: Springer, 2011: 60-68.

[103] Ma W J, Beck J M, Latham P E, et al. Bayesian inference with probabilistic population codes[J]. Nat. Neurosci., 2006, 9: 1432-1438.

[104] Hodgkin A L, Huxley A F. Propagation of electrical signals along giant nerve fibres[J]. Proceedings of the Royal Society of London. Series B-Biological Sciences, 1952, 140(899): 177-183.

[105] Izhikevich E M. Which model to use for cortical spiking neurons?[J]. IEEE Transactions on Neural Networks, 2004, 15(5): 1063-1070.

[106] Gerstner W, Kistler W M. Spiking Neuron Models: Single Neurons, Populations, Plasticity[M]. Cambridge: Cambridge University Press, 2002.

[107] Grüning A, Bohte S M. Spiking neural networks: Principles and challenges[C]//ESANN, 2014.

[108] Yi W, Tsang K K, Lam S K, et al. Biological plausibility and stochasticity in scalable VO_2 active memristor neurons[J]. Nature Communications, 2018, 9(1): 1-10.

[109] Zhang X, Wang Z, Song W, et al. Experimental demonstration of conversion-based SNNs with 1T1R mott neurons for neuromorphic inference[C]//2019 IEEE International Electron Devices Meeting (IEDM). IEEE, 2019: 6.7. 1-6.7. 4.

[110] Widrow B, Hoff M E. Adaptive switching circuits[R]. Stanford: Univ. Ca. Stanford Electronics Labs, 1960.

[111] Urbanczik R, Senn W. A gradient learning rule for the tempotron[J]. Neural Computation, 2009, 21(2): 340-352.

[112] Wade J J, McDaid L J, Santos J A, et al. SWAT: A spiking neural network training

[113] Natschläger T, Ruf B. Spatial and temporal pattern analysis via spiking neurons[J]. Network Computation in Neural Systems, 1998, 9(3): 319.

[114] Bohte S M, Kok J N, La Poutre H. Error-backpropagation in temporally encoded networks of spiking neurons[J]. Neurocomputing, 2002, 48(1): 17-37.

[115] Florian R V. The chronotron: A neuron that learns to fire temporally precise spike patterns[J]. PLoS One, 2012, 7(8): e40233.

[116] Urbanczik R, Senn W. Reinforcement learning in populations of spiking neurons[J]. Nature Neuroscience, 2009, 12(3): 250-252.

[117] Garg A, Onuchic J N, Ambegaokar V. Effect of friction on electron transfer in biomolecules[J]. The Journal of Chemical Physics, 1985, 83(9): 4491-4503.

[118] Wang H, Du Y, Li Y, et al. Configurable resistive switching between memory and threshold characteristics for protein-based devices[J]. Advanced Functional Materials, 2015, 25(25): 3825-3831.

[119] Wang H, Zhu B, Ma X, et al. Physically transient resistive switching memory based on silk protein[J]. Small, 2016, 12(20): 2715-2719.

[120] Raeis-Hosseini N, Lee J S. Resistive switching memory based on bioinspired natural solid polymer electrolytes[J]. ACS Nano, 2015, 9(1): 419-426.

[121] Fu T, Liu X, Gao H, et al. Bioinspired bio-voltage memristors[J]. Nature Communications, 2020, 11(1): 1-10.

[122] Park Y, Lee J S. Artificial synapses with short-and long-term memory for spiking neural networks based on renewable materials[J]. ACS Nano, 2017, 11(9): 8962-8969.

[123] Chen Y C, Yu H C, Huang C Y, et al. Nonvolatile bio-memristor fabricated with egg albumen film[J]. Scientific Reports, 2015, 5(1): 1-12.

[124] Wang L, Wen D. Nonvolatile bio-memristor based on silkworm hemolymph proteins[J]. Scientific Reports, 2017, 7(1): 1-8.

[125] Wu C, Kim T W, Guo T, et al. Mimicking classical conditioning based on a single flexible memristor[J]. Advanced Materials, 2017, 29(10): 1602890.

[126] Tan Z H, Yin X B, Yang R, et al. Pavlovian conditioning demonstrated with neuromorphic memristive devices[J]. Scientific Reports, 2017, 7(1): 1-10.

[127] John R A, Tiwari N, Yaoyi C, et al. Ultralow power dual-gated subthreshold oxide neuristors: an enabler for higher order neuronal temporal correlations[J]. ACS Nano, 2018, 12(11): 11263-11273.

[128] Li Y, Xu L, Zhong Y P, et al. Memristors: associative learning with temporal contiguity in a memristive circuit for large-scale neuromorphic networks[J]. Advanced Electronic Materials, 2015, 1(8): 1500125.

[129] Kumar M, Kim H S, Kim J. A highly transparent artificial photonic nociceptor[J]. Advanced Materials, 2019, 31(19): 1900021.

[130] Kim Y, Kwon Y J, Kwon D E, et al. Nociceptive memristor[J]. Advanced Materials, 2018, 30(8): 1704320.

[131] Zang Y, Shen H, Huang D, et al. A dual-organic-transistor-based tactile-perception system with signal-processing functionality[J]. Advanced Materials, 2017, 29(18): 1606088.

[132] Park S, Sheri A, Kim J, et al. Neuromorphic speech systems using advanced ReRAM-based synapse[C]//2013 IEEE International Electron Devices Meeting. IEEE, 2013: 25.6.1-25.6.4.

[133] Danesh C D, Shaffer C M, Nathan D, et al. Synaptic resistors for concurrent inference and learning with high energy efficiency[J]. Advanced Materials, 2019, 31(18): 1808032.

[134] Wang H, Zhao Q, Ni Z, et al. A ferroelectric/electrochemical modulated organic synapse for ultraflexible, artificial visual-perception system[J]. Advanced Materials, 2018, 30(46): 1803961.

[135] Nau S, Wolf C, Sax S, et al. Organic non-volatile resistive photo-switches for flexible image detector arrays[J]. Advanced Materials, 2015, 27(6): 1048-1052.

[136] Wang H, Liu H, Zhao Q, et al. A retina-like dual band organic photosensor array for filter-free near-infrared-to-memory operations[J]. Advanced Materials, 2017, 29(32): 1701772.

[137] Zhou F, Zhou Z, Chen J, et al. Optoelectronic resistive random access memory for neuromorphic vision sensors[J]. Nature Nanotechnology, 2019, 14(8): 776-782.

[138] Feldmann J, Youngblood N, Wright C D, et al. All-optical spiking neurosynaptic networks with self-learning capabilities[J]. Nature, 2019, 569(7755): 208-214.

[139] Mennel L, Symonowicz J, Wachter S, et al. Ultrafast machine vision with 2D material neural network image sensors[J]. Nature, 2020, 579(7797): 62-66.

[140] Wang C Y, Liang S J, Wang S, et al. Gate-tunable van der Waals heterostructure for reconfigurable neural network vision sensor[J]. Science Advances, 2020, 6(26): eaba6173.

[141] Jang B C, Kim S, Yang S Y, et al. Polymer analog memristive synapse with atomic-scale conductive filament for flexible neuromorphic computing system[J]. Nano Letters, 2019, 19(2): 839-849.

[142] Li W V, Li J J. An accurate and robust imputation method scImpute for single-cell RNA-seq data[J]. Nature Communications, 2018, 9(1): 1-9.

[143] Li C, Wang Z, Rao M, et al. Long short-term memory networks in memristor crossbar arrays[J]. Nature Machine Intelligence, 2019, 1(1): 49-57.

[144] Yan B, Yang Q, Chen W H, et al. RRAM-based spiking nonvolatile computing-in-memory processing engine with precision-configurable in situ nonlinear activation[C]//2019 Symposium on VLSI Technology. IEEE, 2019: T86-T87.

[145] Xue C X, Chen W H, Liu J S, et al. 24.1 A 1Mb multibit ReRAM computing-in-memory macro with 14.6 ns parallel MAC computing time for CNN based AI edge processors[C]//2019 IEEE International Solid-State Circuits Conference-(ISSCC). IEEE, 2019: 388-390.

[146] Zhang G, Ma J, Alsaedi A, et al. Dynamical behavior and application in Josephson

Junction coupled by memristor[J]. Applied Mathematics and Computation, 2018, 321: 290-299.

[147] Itoh M, Chua L O. Memristor oscillators[J]. International Journal of Bifurcation and Chaos, 2008, 18(11): 3183-3206.

[148] Sun Z, Ambrosi E, Bricalli A, et al. Logic computing with stateful neural networks of resistive switches[J]. Advanced Materials, 2018, 30(38): 1802554.

[149] Minati L, Gambuzza L V, Thio W J, et al. A chaotic circuit based on a physical memristor[J]. Chaos, Solitons & Fractals, 2020, 138: 109990.

第 4 章　基于新型二维材料的忆阻器

4.1　引　　言

在大数据时代,数据呈爆炸式增长。在未来的十多年里,数字字节将增长 300 倍[1]。巨大的数据量对下一代非易失性存储器 (NVM) 技术的存储容量和处理速率的要求更高。目前,许多新型的 NVM (如阻变随机存取存储器 (RRAM)[2-9]、磁性随机存取存储器 (MRAM)[10-13]、相变存储器 (PCM)[14-18] 等) 展现出可以同时永久存储和快速处理大量数据的突出优点。其中,忆阻器通过在不同的电阻态之间切换来存储和输出信息,忆阻器的可扩展纳米尺寸使高密度和大规模集成成为可能[19-23]。1971 年,Chua 基于对称性理论,预言了除电阻、电容和电感之外的第四种基本无源电路元件,并定义为忆阻器[24]。到 2008 年,惠普实验室在 *Nature* 杂志上报告了第一个工作的忆阻器件[25]。此后,忆阻器进入了快速发展阶段。2008 年,Yang 等制备了一种 TiO_2 基忆阻器件[26],该器件具有快速的双极非易失性开关行为,通过实验证明了在氧化物体系中电阻转换的一般模型。随后在 2010 年,Jo 等制备了纳米级硅基忆阻器件[27],证明了由互补金属氧化物半导体 (CMOS) 神经元和忆阻器突触组成的混合系统可以模拟重要的突触功能。需要指出的是,在不断发展的过程中,研究者们对真正的忆阻特性的认定也存在一些争议[28],这使得忆阻器的研究更加全面。简而言之,用忆阻器代替所有的阻变存储器件并不是一个精确的定义[28]。不过,这不是本书的重点,不做过多讨论。在本书中,我们把依赖于电阻变化的器件称为忆阻器。

此外,忆阻器具有亚纳秒级的开关速度[29,30]、低运行能耗[31]和长的写入/擦除耐久性[32]。重要的是,忆阻器的电阻态依赖于其过去的状态,因此可以用来模拟 (生物) 大脑中的突触连接,这在神经形态计算中引起了越来越多的关注[33,34]。而忆阻器有望解决传统计算机体系结构的关键问题[35]。人脑在许多复杂的学习过程中消耗的能量 (约 20 W) 远低于最先进的计算机 (约 1 MW),而且处理效率更好[36,37]。因此,在神经计算领域利用硬件计算模拟大脑,实现大脑功能,可能会对下一代人工智能产生深远影响[38-40]。迄今为止,Yan 等已经使用忆阻器实现了一系列突触功能,如长时程增强 (LTP) 和尖峰时间依赖性可塑性 (STDP)、长时程抑制 (LTD) 和双脉冲易化 (PPF)[41-45]。这促进了忆阻器在脑神经仿生学领域的进一步发展。

在器件结构和功能层方面，忆阻器多为上、下电极和功能层的三层结构，结构简单，与当今的 CMOS 工艺兼容[46-48]。忆阻器通过改变施加电压时功能层的电阻来存储信息[49,50]。以往忆阻器对功能层材料的研究主要是传统的氧化物材料，如 TiO_2[42,51]、钽氧化物[52,53]、HfO_x[54]、WO_x[55]、ZHO/IGZO[43] 等，氧化物基忆阻器具有许多优点，性能相对稳定，研究比较成熟。然而，基于传统三维氧化物的忆阻器很难减小尺寸，随着新材料的发展，当薄膜从三维尺寸减小到二维尺寸时，它将表现出其他有趣的特性，这些特性对于器件尺寸的减小是值得研究的。因此，以二维材料为代表的新材料是一个热门的研究方向，也可能是未来的研究热点。

随着石墨烯成功剥离，二维材料因其原子厚度[56-58]，产生优异的电学、光学[59-61]、热学和力学性能[62-67]，受到越来越多的关注。二维材料作为功能材料和结构材料已广泛应用于晶体管、电池、光伏发电、水处理等各个应用领域[68-81]。它们有望加入甚至替代传统的半导体材料，以制成尺寸更小、功耗更低、效率更高的纳米电子器件[82-85]。所研究的二维材料包括从导体、半导体到绝缘体的各种材料，如石墨烯[86-89]、BN[90,91]、拓扑绝缘体[92-94]、过渡金属二硫化物 (TMD，如 MoS_2 和 WS_2)[95-97]、二维钙钛矿[98,99]、InSe[100-103]、SnS[104,105]、TiS_2[106]、黑磷 (BP)[107-117]、MXenes[118-120]、锑[121-124]、铋[125-129]、硼烯[130]、碲烯[131,132-136]、硒纳米片[137,138]、二维金属[139] 等。重要的是，二维材料由于其优异的性能而在忆阻器方面得到了广泛的应用[86-88,90,91,95-99,107]。

石墨烯由于其优异的导电性、超高的结构和化学稳定性以及大的载流子迁移率，在忆阻器功能层的应用中显示出无与伦比的优势[86-88,140-142]。此外，一些研究人员使用石墨烯作为电极来制备忆阻器[143,144]，该忆阻器表现出非易失性特性，并改善了器件性能。一些研究人员在底部电极中加入单层石墨烯，以防止导电细丝 (CF) 穿透到底部电极中，基于单层石墨烯的器件显示出令人满意的电阻开关 (RS) 性能，包括高耐久性、快速和长保持率[145,146]。除石墨烯外，过渡金属二硫化物 (transition-metal dichalcogenide) 作为忆阻器的功能层材料也得到了广泛的研究[41,145,147]。例如，Bessonov 的 Skolkovo 报告称，使用层状 MoS_2 制备的忆阻器具有 0.1~0.2 V 的低运行电压和 10^2~$10^8 \Omega$ 的可调电阻[145]。随后，Yan 等用二维层状 WS_2 纳米片制作了一种人工电子器件。该器件功耗低，成功模拟生物突触功能[41]。此外，对于电阻开关器件，物理开关机制一直是研究的重点，比较流行的开关机理有细丝、原子空位和电子的捕获与释放。不同的机理导致不同的器件性能。图 4.1 全面概述了基于二维材料的忆阻器及其开关机理的研究进展，这些研究迅速推进了忆阻器的发展。基于二维材料忆阻器的快速发展，本章对二维材料在忆阻器中的应用及其物理开关机理进行了综述，并分析了基于二维材料忆阻器所面临的挑战和前景。

图 4.1　基于二维材料的忆阻器及其开关机理的研究进展

① 转载自参考文献 [86]；② 转载自参考文献 [144]；③ 转载自参考文献 [148]；④ 转载自参考文献 [149]；⑤ 转载自参考文献 [145]；⑥ 转载自参考文献 [150]；⑦ 转载自参考文献 [41]；⑧ 转载自参考文献 [96]；⑨ 转载自参考文献 [96]；⑩ 转载自参考文献 [90]；⑪ 转载自参考文献 [45]；⑫ 转载自参考文献 [151]；⑬ 转载自参考文献 [152]；⑭ 转载自参考文献 [91]；⑮ 转载自参考文献 [99]；⑯ 转载自参考文献 [107]

4.2　基于二维材料的忆阻器

4.2.1　石墨烯和衍生物

1. 作为功能层工作

横向 RRAM 由标准平面几何形状的金属–功能层–金属三层结构组成，类似于混合晶体管结构[41-43]。石墨烯等衍生物由于其低制备成本、高机械柔性和高光学透明度等优点，已被研究为横向 RRAM 的功能层介质[153,154]。2015 年，Porro 等报道了一种具有典型双极开关特性的 Ag/GO/ITO 器件[155]；图 4.2(a) 给出了该器件的光学照片；图 4.2(b) 是该器件典型的 I-V 曲线，红色实线是 I-V 曲线的第一个周期，点线和虚线分别是 I-V 曲线的第五个周期和第十个周期，箭头表示电压的扫描方向，可以看出，该器件具有稳定的双极行为。此外，Yi 等用 GO 制备了 Ag/GO/ITO 器件[156]，其中 GO 薄膜在不同温度下退火，器件的性能与 GO 薄膜的退火温度有很大关系，研究发现在 20°C 的退火温度下，该器件具有典型的存储行为，并且具有较高的开/关电流比；图 4.2(c) 显示了 Ag/GO/ITO 器件在 20°C 退火温度下的 I-V 曲线，插图是该器件的示意图，其中，扫描 1 至扫描 4 为电压扫描过程，可以发现，器件只能打开,不能关闭,这是典型的写一次读多次的行为;图 4.2(d) 显示了高阻态 (HRS) 和低阻态 (LRS) 的保持特性,该器件保持 10^3 s 以上,无明显衰减,开/关电流比为 10^4,可有效避免数据误读。此外,也有研究人员利用基于 GO 的忆阻器成功模拟生物突触,在未来神经形态计算中具有一定的应用潜力[157,158]。

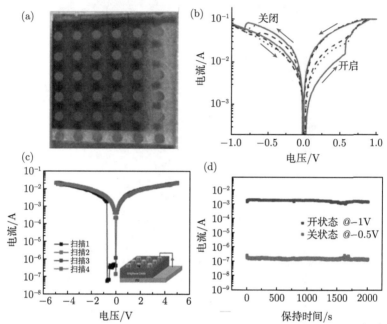

图 4.2 (a) AG/GO/ITO 器件光学照片;(b) Ag/GO/ITO 器件的 I-V 曲线[155];(c) GO 薄膜退火温度为 20℃ 时 Ag/GO/ITO 器件的 I-V 曲线;(d) 器件的保持特性[156](彩图请扫封底二维码)

2. 作为电极工作

石墨烯不仅可以作为电极与介质之间的界面层,还可以作为电极。Lu 等报道了一种基于具有固有选择元件的石墨烯电极的阻变存储器件,如图 4.3(a)~(d) 所示;他们使用多层石墨烯 (MLG) 作为导电电极,以提高电极的导电性[143],发现 MLG/Ta$_2$O$_{5-x}$/TaO$_y$/MLG 和 MLG/TaO$_y$/Ta$_2$O$_{5-x}$/MLG 的沉积顺序导致阻变存储器的两种不同的开关特性;石墨烯材料在器件制备过程中的自发功能化,导致了功能化石墨烯电极的易失性阈值开关和阻变存储器的整体高非线性开关特性。此外,Lu 等在功能化石墨烯电极中发现环氧基团,它作为导电电子的深能级陷阱,导致电导较低的可变范围跳跃。高能电子可以在足够高的电压下进入一个较浅的陷阱,从而引起电导率和电流的迅速增强,对应于从 HRS 到 LRS 的转变。这种开关过程的形成是由于功能化石墨烯底电极 (BE) 中电子捕获/释放的易失性，以及氧化钽层中氧空位传输诱导的双极开关非易失性。

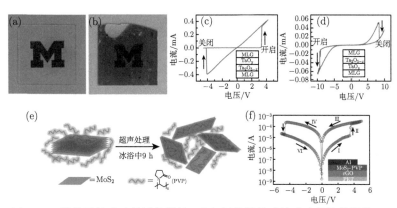

图 4.3 (a) MLG 转移后的玻璃基板的照片；(b) 制备器件后的玻璃芯片的照片；(c) 具有线性开态特性的 MLG/TaO$_y$/Ta$_2$O$_{5-x}$/MLG 器件的双极 RS 特性；(d) 具有非线性开态特性的 MLG/Ta$_2$O$_{5-x}$/TaO$_y$/MLG 器件的双极 RS 特性[143]；(e) 聚合物辅助剥离分散 MoS$_2$ 示意图[144]；(f) I-V 特性曲线[144]

此外，Liu 等提出了一种用聚乙烯吡咯烷酮 (PVP) 剥离分散 MoS$_2$ 的简便方法 (图 4.3(e))[144]；首先，以 MoS$_2$ 为受体，将还原氧化石墨烯 (rGO) 作为 rGO/MoS$_2$-PVP/Al 结构的导电电极；它具有典型的双稳态开关和非易失可写闪存功能；该弹性存储器件的 I-V 特性如图 4.3(f) 所示。这种特殊的电场效应可能与 PVP 模型中 MoS$_2$ 在电场作用下的电荷捕获和释放反应有关。以二维纳米材料作为导电电极和功能材料的闪存具有非易失和可擦除性，在 USB 闪存、柔性存储卡、射频识别固态硬盘、电子标签等方面都有可行的应用。

4.2.2 过渡金属二硫化物

过滤金属二硫化物 (TMD) 表示为 MX$_2$，其中 M 代表过渡金属元素 Mo、W、Sn、Hf、Zr 等，X 代表硫族原子 S、Se、Te，在电学、光学、化学和力学等方面表现出优异的性能[159-161]。作为一种半导体材料，TMD 具有很大的禁带宽度、较高的载流子迁移率、在空气中的稳定性等优点[162,163]。

1. MoS$_2$

MoS$_2$ 作为一种典型的 TMD，由于其独特的力学性质[164]、电学性质[165-168] 和生物学性质[169-172] 而引起了人们的广泛关注。相应地，二维 MoS$_2$ 在催化[173,174]、电子学[175-177]、能源[178,179]、生物医学相关领域[180-183] 等也有广泛的应用前景。二维 MoS$_2$ 在环境方面的应用也是值得期待的。实际上，在自然界中以矿物辉钼矿形式存在的块状 MoS$_2$ 长期以来一直作为环境吸附剂[184] 和催化剂[185]。然而，单层或少层 MoS_2 纳米片的制备一直限制着 MoS_2 在环境中的应用，因为 MoS_2 纳米片具有独特的二维纳米性质。近年来，在获得单层和大规模 MoS_2 纳米片方面的先进技术的发展为 MoS_2 纳米片的制备提供了新的潜力[186,187]。在原有工作的基础上，这里综述了 MoS_2 纳米片的合成方法[188-190]、物理化学调谐[191]、功能化[192]，并展示了基于 MoS_2 光热效应的新的环境应用[193,194]。

然而，与石墨烯及其衍生物/复合材料相比，未加工的 MoS_2 纳米片没有明显的 RS 行为[195]，幸运的是，它可以通过溶液处理或通过与其他种类的材料 (如介电聚合物) 的化学功能化来引入[196]。Cheng 等在 2016 年报道了通过使用 1T 相 MoS_2 纳米片制备出完美的奇对称忆阻器 (图 4.4(a))[197]；他们还证明了 2H 相块状 MoS_2 具有欧姆性质，而 1T 相分离的 MoS_2 表现出特殊的存储响应，导致与电压相关的电阻变化，1T MoS_2 相的电导率是 2H 相的 10^7 倍，电场能引起晶格畸变，增强 1T 相电子的离域性，从而显著增加电导率，导致电阻变化；此外，他们还展示了不对称的 $Ag/MoS_2/Ag$ 记忆器件，在该器件上施加 0 mV→1000 mV→ −1000 mV→0 mV 的偏置电压后，获得了对称的压缩滞后 I-V 回路 (图 4.4(b)、(c))。这些结果也验证了负微分电阻 (NDR) 的存在。

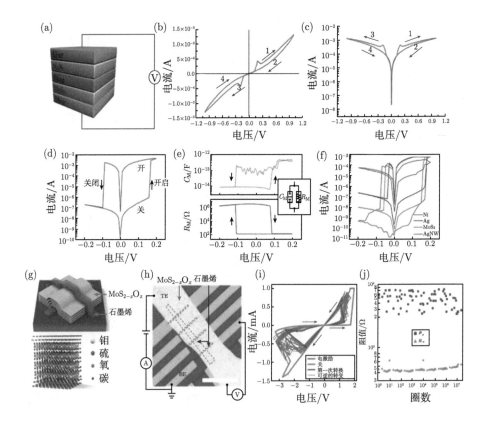

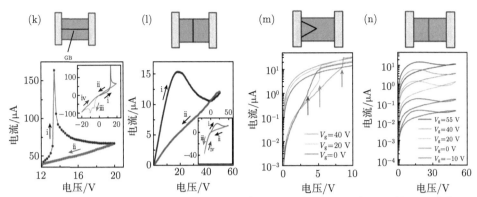

图 4.4 (a) $Ag/MoS_2/Ag/MoS_2/Ag$ 忆阻器示意图;(b) 室温下 (a) 忆阻器的 I-V 曲线;(c) 对数尺度下的 I-V 曲线[197];(d) MoO_x/MoS_2 存储器件的基本 I-V 特性;(e) MoS_2/MoO_x 忆阻器的容性和阻性部分;(f) 基于不同顶电极 (TE) 材料忆阻器的 I-V 曲线[145];(g) 石墨烯/$MoS_{2-x}O_x$/石墨烯 (GMG) 器件示意图 (上) 和横截面图像 (下);(h) 摄影和测量装置;(i) GMG 器件的典型开关特性;(j) 在 1 μs 的脉冲宽度下有 $2×10^7$ 的高耐久性[96];(k)、(l) $V_g = 40$ V 时忆阻器的原理图和局部 I-V 特性;(m) 忆阻器的门电极调谐性;(n) 不同 V_g 下二等分晶界忆阻器的典型 I-V 曲线 ($V > 0$ V)[150] (彩图请扫封底二维码)

随后,一项突破性的实验被报道。由于 MoS_2 基结构易于获得且成本低,Bessonov 等发明了一种用于廉价印刷电子应用的存储器和存储器电容开关[145],他们通过热辅助空气氧化法沉积少层 MoS_2 片来制作该器件。经溶液处理的 WO_x/WS_2 和 MoO_x/MoS_2 异质结具有从 10^2~10^8Ω 的空前宽的可调电阻范围和 0.1~0.2 V 的运行电压 (图 4.4(d))。双极型 RS 由超薄 (小于 3 nm) 氧化物层控制,该氧化物层也具有电容贡献。由于开关动力学的高度非线性,可以通过一系列电脉冲来实现不同的突触柔性机制 (图 4.4(e)、(f))。实验证明,LTP、STP 和二进制开关是一种根据脉冲幅度、脉冲数和输入频率来调节忆阻器阻态的机制。

基于金属氧化物 RRAM 以其可扩展性、温度鲁棒性和机器学习潜力而受到广泛关注[198]。然而,较厚的氧化层导致相对较高的运行电压,而较薄的氧化层导致较小的窗口和较大的漏电流[198]。在此基础上,通过优化氧化层本身,人们提出了一种新的界面工程概念,以降低运行电压,提高开关比。MoS_2-Pd 界面工程的概念被提出,Wang 等制备了透明的 $ITO/HfO_2/MoS_2$-Pd NPs/ITO RRAM[198],采用 MoS_2 和钯纳米粒子的混合结构设计了 HfO_x 基 RRAM 的电极/氧化物界面,通过接口工程,在较厚的 HfO_x 层 (约 15 nm) 处,可以极大地降低置位电压 (从 −3.5 V 到 −0.8 V),增加开关比和良好的均匀性,并将窗口提高 30 倍。此外,即使在较高的温度下,RRAM 通常也可以工作 200 个存储器开关周期,并且 LRS 和 HRS 中的电阻没有明显下降,显示出良好的数据保持性能和稳定性。由于 ITO 的高透过率和 MoS_2 薄膜的原子厚度,该 RRAM 具有较高的可见光透明度、较大的 RS 窗口和较低的置位电压，使其在供电电压相对较低的可穿戴电子器件和透明电路中具有很好的应用前景。

目前的研究大多集中在单一类型的二维材料忆阻器，其存在高温不稳定等缺点[96,153,154]。通过组合不同的二维材料可以实现异质结构，以解决电子器件的缺点，特别是对于需要垂直多层结构的电子器件。Wang 等首次基于全层二维异质结构 (石墨烯/$MoS_{2-x}O_x$/石墨烯) (图 4.4(g)、(h)) 制备了一种鲁棒忆阻器件，显示了可重复的双极型 RS (图 4.4(i))、卓越的开关性能、耐用性高达 10^7 和最高工作温度可达 340℃ (图 4.4(j))[96]；他们发现，转变层 ($MoS_{2-x}O_x$)、电极 (石墨烯) 及其原子锐化界面的原子层结构对器件的鲁棒性和高温热稳定性起着至关重要的作用；高温原位高分辨透射电镜 (HRTEM) 显示，$MoS_{2-x}O_x$ 层具有很高的热稳定性，扫描透射电镜 (STEM) 的研究表明了一种基于氧离子迁移的开关机制和良好的有限导电通道；此外，完全分层的二维材料为柔性电子应用提供了良好的机械灵活性，实验证明了超过 1000 次弯曲循环的优异耐久性。

基于金属–绝缘体–金属 (MIM) 结构的忆阻器，使用 TiO_2 或其他绝缘氧化物，由于缺乏对开关电压的外部控制和对导电细丝形成的控制，其性能受到限制[199-201]。Sangwan 等报道了一种在单层 MoS_2 器件中基于晶界 (GB) 的新型忆阻元件，其三端 MoS_2 忆阻器是门电极可调谐的[150]。特别地，触点产生的晶界电阻易于反复调制，开关比可达约 10^3 (图 4.4(k))。在交叉晶界忆阻器中，MoS_2 忆阻器中的可单独寻址的本地门可以连续调整设定电压 (V_{SET}) 并提供容错结构。通过改变栅极偏置 (0~40 V)，可使 V_{SET} 从 3.5 V 到 8.0 V (图 4.4(m)) 变化，受控 V_{SET} 为设计互补和双极型 RS 提供了额外的灵活性。此外，在零偏置电压下，二等分晶界忆阻器提供超过三个数量级的双稳态的电阻值，而正曲线形状和开关比 (4~6) 与栅压保持相对不变 (图 4.4(l)、(n))。一系列结果表明了基于晶界的忆阻器的优良性能。

2. WS_2

为了实现低功耗运行，科研人员利用先进材料制备出不同于金属离子导电细丝或氧空位两种主要机制的忆阻器[153,154]。WS_2 是一种 TMD，因为它具有可调谐的带隙和独特的半导体特性而不同于石墨烯等其他二维材料[67,202-206]。Yan 等提出了一种基于 2H 相 WS_2 的 Pd/WS_2/Pt 器件结构的低功耗忆阻器。开关时间快至 13 ns (14 ns)，开态下运行电流低至 1 μA，置位 (复位) 能量达到飞焦耳水平 (图 4.5(a)~(d))[41]。TEM 观察表明，WS_2 薄膜中的钨和硫空位比 WS_2 纳米片中的要多。因此，钨和硫空位的形成以及它们之间的电子跳跃是 RS 性能的主要原因，这不同于传统的金属离子导电丝或氧空位。此外，该器件还可以模拟生物体的突触功能，包括负向和正向脉冲刺激训练下的兴奋和抑制，以及 STDP、PPF、STP 向 LTP 过渡时的记忆和学习。

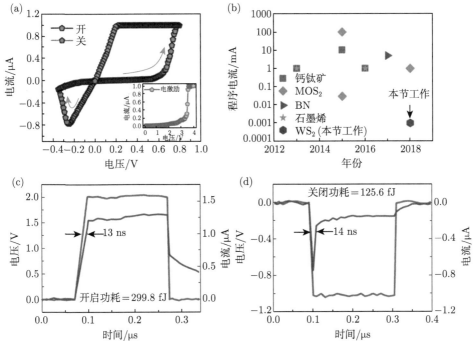

图 4.5 (a) 线性坐标的经典 I-V 特性; (b) 其他文献中报道的杂项运行电流 (钙钛矿[99,151,207]、MoS_2[96,150,197]、BN[91,208] 和石墨烯[209,210]); (c) 以 13 ns 的电压脉冲开启; (d) 以 14 ns 的电压脉冲断开[41] (彩图请扫封底二维码)

此外,Rehman 等在 2017 年报道了一种 $Ag/WS_2/Ag$ 器件,该器件由柔性聚对苯二甲酸乙二醇酯 (PET) 衬底通过全印刷制备技术制备,使用分离的 WS_2 薄片作为可复写双极型存储器件的功能层[147]。他们展示了用全印刷技术的分离 WS_2 薄片的结构、物理、机械和电性能;然后,他们使用一层原子厚的氧化铝 (Al_2O_3) 来封装器件,以防止受潮。总之,$Ag/WS_2/Ag$ 基存储器件可实现低功耗、高开关比 (约 10^3)、非易失性双极型、长保持时间 (约 10^5 s)、低工作电压 (约 2 V)、高耐久性 (1500 次电压扫描)、1500 次循环的机械鲁棒性、低电流顺应性 (50 μA) 以及在环境条件下的独特可重复性,为柔性电子器件集成提供了新的可能性。

3. $MoTe_2$

$MoTe_2$ 的多态性引起了人们对电子应用的极大兴趣[211,212]。Zhang 等展示了基于 $2H$-$MoTe_2$ 和 $Mo_{1-x}W_xTe_2$ 的垂直 RRAM 器件中,在电场和畸变瞬态结构 ($2H_d$) 的作用下,由 2H 半导体到正交 T_d 导电相的结构转变 (图 4.6(a))[213];他们证明了利用电场诱导相变的可能性;$Mo_{1-x}W_xTe_2$ 器件的置位电压单调地依赖于晶片的厚度 (图 4.6(b));通过比较 $2H$-$MoTe_2$ 样品和 $2H$-$Mo_{1-x}W_xTe_2$ 合金的

RRAM 特性，他们发现 2H-$Mo_{1-x}W_xTe_2$ 合金需要较小的 V_{SET} 来触发 RRAM 行为；在 $Mo_{1-x}W_xTe_2$ 合金中，半导体态转变为金属态所需的能量随着 x 的增加而越来越低 (图 4.6(c))；与温度有关的电学测量表明，$2H_d$ 态可以作为半导体金属的行为依赖于瞬态；此外，他们使用了开关电流小于 1 μA 且比值达到 10^6 的 Al_2O_3/$MoTe_2$ 堆栈来实现选择器 RRAM 框架。

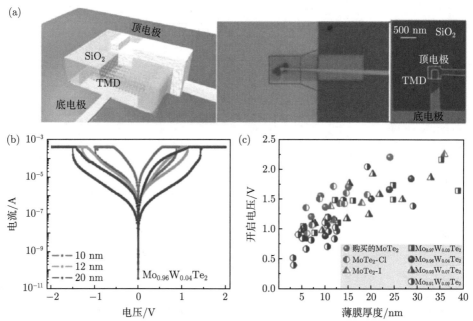

图 4.6 (a) 垂直 $MoTe_2$ 器件的示意图，以及光学和 SEM 图像；(b) 垂直 $Mo_{0.96}W_{0.04}Te_2$ RRAM 器件电激活后的对数刻度 I-V 曲线；(c) V_{SET} 值与 $MoTe_2$、$Mo_{0.97}W_{0.03}Te_2$、$Mo_{0.96}W_{0.04}Te_2$、$Mo_{0.93}W_{0.07}Te_2$ 和 $Mo_{0.91}W_{0.09}Te_2$ 片层厚度的关系 [213] (彩图请扫封底二维码)

4.2.3 其他二维材料

1. BN

BN 具有宽的禁带宽度 (约 5.5 eV)[91,214]，是 RRAM 和其他逻辑器件的理想材料 [215-217]。Pan 等研究了 RS 在多层 BN 反应器中的存在，并开发了一系列以多层 h-BN 为活性 RS 介质的 RRAM 器件 (图 4.7(a))[91]。此外，图 4.7(b) 显示的 V_{SET} 和 V_{RESET} 分别为 0.4 V 和 −0.8 V。通过插入界面石墨烯电极和调节 BN 堆的粒径和厚度来控制器件的性能。基于 BN 的器件首次显示出双极型和阈值 RS 共存，因为它容易产生 B 空位和金属离子的穿透。通过在金属电极和 h-BN 之间插入多层石墨烯来研究 RS 现象。由于石墨烯在平面外的高电阻，石墨烯界面电极已被证明可以缩短

4.2 基于二维材料的忆阻器

RRAM 器件的功耗,保护介质免受环境湿度的影响,通过防止氧扩散来稳定 RS 响应,允许机械弯曲而不损害性能,并提供优良的透明度。

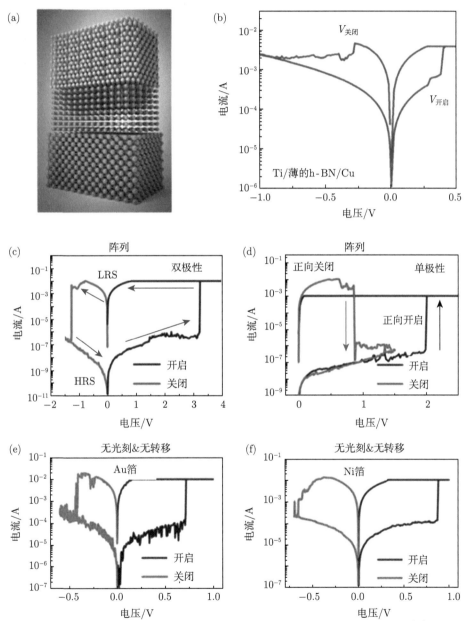

图 4.7 (a) Ti/薄 h-BN/Cu 器件的结构;(b) 忆阻器器件中的经典 I-V 特性[91];单层 h-BN 交叉栅器件的 (c) 双极型和 (d) 单极型 RS 响应的 I-V 曲线;基于 (e) Au 箔和 (f) Ni 箔的单层 h-BN 器件的基本 I-V 特性[90] (彩图请扫封底二维码)

此外，Wu 等报道了垂直夹在金属电极中间用化学气相沉积法生长的单层 BN 的非易失 RS (NVRS) 性能[90]；他们用三种金属电极 (Ni、Au 和 Ag) 研究了两种不同的器件结构；无形状开关在双极和单极工作中都可用，通过脉冲操作实现了快的开关速度 (小于 15 ns) 和高的开关比 (10^7)；单原子 h-BN NVRS 器件的实现将原子电阻扩展到二维绝缘体；单层 h-BN MIM 器件的经典 I-V 特性 (图 4.7(c)~(f)) 分别显示了 Au 箔和 Ni 箔上器件的双极 RS 行为和单极 RS 行为；此外，他们还研究了与温度相关的电子输运，并用第一性原理进行了模拟，揭示了开关过程中通过占据硼空位中的金属离子实现的电导桥接机制，以及界面局域缺陷与金属离子之间的相互作用。这些努力可能为印刷电子、超薄柔性存储器、射频开关和神经形态计算带来新应用。

2. 黑磷

黑磷 (BP) 是 VA 族中第一种也是被研究最多的一种二维材料。2014 年，Li 等首先制作了 BP，并构建了基于 BP 的场效应晶体管[218]。随后的研究还发现了其他一些有潜力的特性，如优越的力学性能[219,220]、高度各向异性的载流子输运[221-225]、负泊松比[226]、应变诱导导带[227]、垂直拓扑特性和优异的光学响应[222,228,229]。此外，BP 的褶皱结构使其具有比平面二维材料更大的面积，并有利于其对药物[230,231]、显像剂[110,232]、金属离子[233]等的负载能力。BP 的这些特性使其在晶体管[234-238]、电池[239-242]、储能[243,244]、光子学[222,245,246]、传感器[247-250]和光热癌症治疗[230,251-255]等方面具有广泛的应用。

处理 BP 的主要困难是其不稳定性，即在环境条件下的降解，这阻碍了使用液相剥离 (LPE) 方法大量生产少层 BP。Hao 等用选定的溶剂从剥离的 BP 纳米片上制备了 BP 膜，并报道了一种环境稳定的 Al/BP 膜/ITO 结构的非易失性 RRAM 器件[107]。柔性 BP 存储器件是由沉积在 BP 膜顶部的 Al 层制成的，并使用降解产生的非晶态顶部降解层 (TDL) 作为 Al 电极绝缘层的底部，以防止对 BP 膜的进一步降解，并确保即使在三个月的环境暴露下仍能保持优异的存储性能。在他们的工作中，BP 存储器表现出双稳态 RS 特性，可以获得高达 3×10^5 的高开关电流比，10^5 s 内的高稳定性和高达 10^4 的高开关比。

许多研究者研究了多种提高 BP 稳定性的技术，包括与碳自由基[256]和芳基重氮[257]的共价键合、碱金属氢化物的插层[258]、氟原子掺杂[259]、磺酸钛配体表面配位[256]、金属离子修饰[233]，以及用六方氮化硼[260,261]、石墨烯壳层[262]、氧化石墨烯[113]、有机单层膜[263]、AlO_x[234,264]等封装。当然，寻找其他稳定的磷烯类光热材料也是一种可能的方法，如硫化锡 (SnS)[104,265,266]。这些策略将为磷烯及其类似材料在环境问题上的光热应用开辟新的潜力。

3. 钙钛矿

有机卤化物钙钛矿材料以其简单廉价的制备工艺、良好的电荷迁移率和优异的光吸收系数，为开发高效、低成本的太阳能电池开辟了新的时代[267-281]。除太阳能电池外，钙钛矿的发展迅速，直接扩展到许多其他领域，包括在光学和电子器件中的应用[272-274]。Yoo 等在 2016 年展示了一种新型的双功能电阻开关存储器件，它在模拟数字电阻开关 FTO (掺杂氟的氧化锡) 衬底上使用简单的 Ag 和 $CH_3NH_3PbI_{3-x}Cl_x$ 钙钛矿双层结构[99]；该器件的 RS 特性是由在较高偏置扫描作用下导电细丝的突然形成和断裂引起的 (图 4.8(a))；双稳态 RS 性能可靠，保持时间为 4×10^4 s，寿命超过 10^3 次，表明 $Ag/CH_3NH_3PbI_{3-x}Cl_x/FTO$ 器件是一种有希望的 RRAM 器件；研究还表明，光照对 RS 性能有一定的影响 (图 4.8(b)、(c))；该器件不仅具有数字开关特性，而且具有模拟开关特性，可用于神经形态学计算器件的应用。

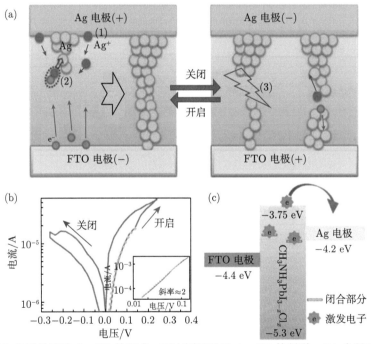

图 4.8 (a) 左图显示了 Ag CF 的形成，而右图显示了 Ag CF 的断裂；(b) 光照下的对数的 I-V 特性；(c) 光照射下 $Ag/CH_3NH_3PbI_{3-x}Cl_x/FTO$ 器件的能级[99]

随后，Liu 等制备了有机和无机杂化钙钛矿 (OIHPS) 忆阻器，验证了 OIHPS 在可复写存储器件中的应用；以甲基丙烯酸甲酯 (PMMA) 为电阻开关层，采用两步旋涂工艺制备钙钛矿型 $CH_3NH_3PbI_3$ 作为电荷捕获层，形成电荷陷阱；通过 I-V 特性研究了 ITO/PMMA/CH$_3$NH$_3$PbI$_3$/PMMA/Ag 器件的导电机理；这种存储器件可以以高达 10^3 的开关比重编程；结果表明，OIHPS 在 RS 存储器中具有良好的电子应用前景。

4. 硒化镓

除 TMD 外，二维层状 Ⅲ-Ⅵ 半导体材料因其优异的光学和电学性能而得到了广泛的研究[275]。其中，GaSe 的直接带隙约为 25 meV，间接带隙约为 2 eV[276]。GaSe 具有优异的电学和光学特性[277,279]，广泛应用于太赫兹产生、非线性光学和光电子学等领域[280-282]。Yang 等采用忆阻器件和场效应晶体管相结合的方法，研究了二维层状 GaSe 纳米片中的三端晶体管[152]；银电极忆阻器具有非易失双极 RS 特性；三端二维 GaSe 晶体管展现出非易失双极 RS 特性，具有开关比大、耐久性好、电阻保持性能好、阈值电场低等优点；暴露于空气一周后，RS 行为显著增强（图 4.9(a)、(c)）；RS 特性显示，V_{SET} 和 V_{RESET} 与初始偏置扫描回路相比没有显著变化（图 4.9(b)）；GaSe 基忆阻器的超低对应置位电场可能是因为 GaSe 中固有的 Ga 空位迁移能量小，这有利于该器件在低功耗非易失性存储器中的实际应用（图 4.9(d)）。

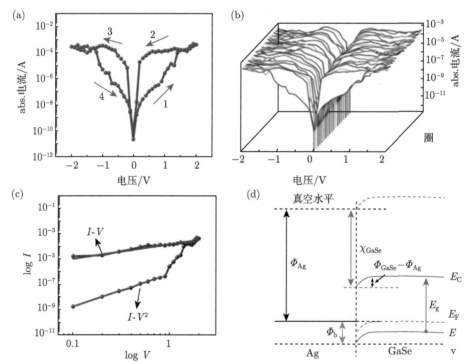

图 4.9　(a) Ag/GaSe/Ag 器件的 I-V 曲线；(b) 器件的 50 个数据周期；(c) 图 (a) 中 I-V 曲线的拟合结果；(d) 金属-半导体结 Ag/GaSe 能带图[152]

4.3 基于二维材料忆阻器的开关机制

4.3.1 细丝

1. 金属细丝

忆阻器具有丰富的阻变机制,如细丝、空位和电子俘获等。其中,细丝是一种被广泛认可的机制,主要分为金属离子细丝和空位细丝。基于金属离子细丝的忆阻器电极一般为活性电极。细丝的形成和断裂是通过氧化/还原反应实现的。Pan等报道了一种具有导电细丝机制的多层六方氮化硼电阻转变[91];由该器件采集的典型 I-V 曲线如图 4.10(a) 所示,显示出无激活双极 RS。当读电压为 0.1 V、循环次数超过 350 次时,周期之间的可变性非常小;图 4.10(b) 和 (c) 显示了在 LRS 下 Ti/薄 h-BN/Cu 忆阻器的非缺陷位置和晶界/细丝位置的电子能量损失谱图,这表明,由于 Ti 和 B 在图 4.10(c) 的晶界/细丝位置的严格运动,信号向深度传播;此外,由于在晶界处的活化能较低,使 B 更容易向 Ti 电极迁移,Ti 离子向阴极迁移形成 Ti 或 TiN 基导电细丝;导电细丝的稳定形成和断开是器件 HRS 和 LRS 之间稳定切换的原因,也是器件优良性能的关键。

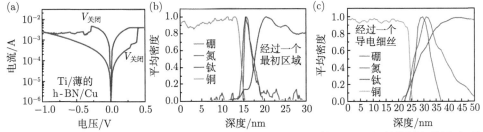

图 4.10 (a) 器件中的 I-V 特性显示出双极 RS;(b)、(c) 器件 (LRS) 初始位置和晶界/细丝位置的电子能量损失谱横截面分析[91] (彩图请扫封底二维码)

此外,Qian 等发现,Cu 箔底电极和 Ag 顶电极形成二维 h-BN 忆阻器的物理机制是银丝[283]。图 4.11 显示了样品横截面的 TEM 图像,这些图像是在置位操作后拍摄的,它清楚地显示了典型的圆锥形的导电细丝,如图 4.11(a)、(d) 所示;细丝将 Cu 箔底电极和 Ag 顶电极连接到 h-BN 层,h-BN 层在靠近 Cu 箔底电极的地方较薄,靠近 Ag 顶电极的地方较宽 (图 4.11(a))。为了研究细丝的化学成分,他们采用扫描透射电镜 (STEM) 模式下的能谱仪 (EDS) 对其进行了分析;图 4.11(c) (图 4.11(b) 中标为 "1" 的圆形区域) 显示 Ag 元素在 h-BN 薄膜中间的浓度较大。此外,他们还用 HRTEM 分析了相同开态的存储单元中的非完整细丝 (图 4.11(e),对应于图 4.11(d));很明显,细丝中的晶格条纹沿着 Ag 顶电极中的晶格条纹生长, 细丝的晶格条纹与 Ag 电极的晶格条纹相同, 如图 4.11(e) 所示。此外, 图 4.11(e) 清楚地显示了在顶电极区测量的间距为 0.23 nm 和 0.20 nm, 分别对应于 (111)Ag 和 (200)Ag 平面。这些结果表明, Ag 是 Ag/h-BN/Cu 箔器件中导电细丝的主要元素。由 Ag 顶电极开始生长的非完整细丝 (在封闭区域内突出) (图 4.11(f)), 细丝的形状也与广泛接受的电化学金属化 (ECM) 理论相矛盾。ECM 理论的核心假设是 M^{z+} 在正极处转化为具有足够电子的 M 原子, 转移层在电子漂移过程中减小而不能被捕获, 这对于具有较高的扩散系数和溶解度的传统固体电解质材料 M^{z+} (Ag^+ 或 Cu^{2+}) 是合理的。ECM 理论得到了 $Ag/H_2O/Pt$ 和 $Ag/SiO_2/Pt$ 结构等直观证据的支持。对于那些阳离子迁移率较低的开关材料, 如 $Cu/ZrO_2/Pt$ 和 Ag/非晶 Si/Pt 存储器件, 会出现从阳极向阴极生长的细丝。此外, h-BN 膜可以为 Ag/hBN/Cu 器件提供更低的 Ag^+ 迁移速率。因此, Ag^+ 可以从底电极中夺走注入的自由电子, 然后在到达阴极之前在 h-BN 膜中转化为 Ag 原子, 如图 4.11(g) 所示。事实上, 即使对于阳离子迁移率高的开关材料, 电子电流仍然是运行电流的主要成分, 而不是离子电流。因此, h-BN 中的阳离子可能被足够的电子还原。

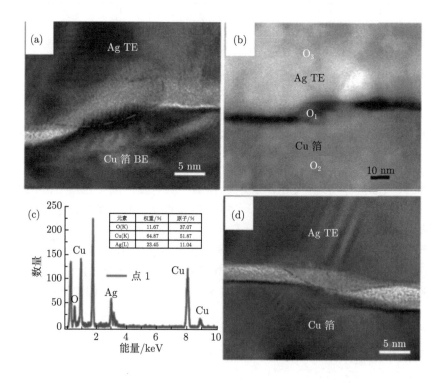

4.3 基于二维材料忆阻器的开关机制

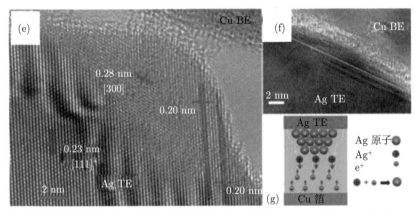

图 4.11 (a) 导电细丝的 TEM 图像；(b) 器件在开态下的 STEM 图像；(c) 从 (b) 中的圆形区域获得的 EDS 图像；(d) 同一器件中细丝的 TEM 图像，但未与 Cu 底电极完全接触；(e) 在 (d) 中 Ag 细丝的 HRTEM 图像；(f) 不完整 Ag 细丝的 HRTEM 图像 (有封闭区域照明)；(g) h-BN 基器件导电细丝生长的原理方案[283]

银细丝机制在金属离子导电细丝机制中最为常见，在二维石墨烯量子点器件中也得到了证实。Yan 等探讨了 $Ag/Zr_{0.5}Hf_{0.5}O_2/$氧化石墨烯量子点$/Zr_{0.5}Hf_{0.5}O_2/Ag$ (简称 AZA) 器件的物理机制是 Ag 细丝[284]。在 HRS 时 AZA 器件的拟合曲线如图 4.12(a) 所示，它将拟合线分为两段。他们采用直接隧穿 (DT) 和福勒–诺德海姆隧穿 (FNT) 模型来描述材料的导电行为。图 4.12(b) 显示了 AZA 的 TEM 分析，揭示了中间电阻态。Ag 细丝向顶电极生长，但细丝没有完全到达顶电极。该器件的相关映射图像如图 4.12(c) 所示，它显示了顶电极与纳米团簇峰之间存在一个间隙，在这个间隙中有一些 Ag 纳米晶。因此，在适当的电场条件下，电子可以很容易地隧穿该间隙，这与 FNT 和 DT 的隧穿机制分析是一致的。因此，通过该研究，进一步发展了金属离子导电细丝机制。

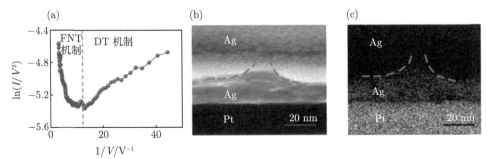

图 4.12 (a) I-V 曲线拟合结果；(b) TEM 图像；(c) AZA 器件的映射图像[284]

2. 空位细丝

空位缺陷在二维材料中广泛存在，并使空位细丝的形成成为可能，从而导致阻态的转变[285,286]。Yang 等报道了一种结构为 Ag/GaSe/Ag 的 GaSe 三端微晶体管[152]。Ga 空位在外电场作用下可以形成导电细丝，并在 LRS 中引起欧姆传导反应。图 4.13(a) 显示了 p 型 GaSe 中一些可移动的 Se 离子。当施加负 V_{ds} 时，Se 离子会向源电极移动。累积的 Se 离子将在 GaSe 和 Ag 电极的界面上形成阳离子空位，如图 4.13(b) 所示。Ga 空位可以作为 p 型细丝核。当负电压增大时，细丝在电场作用下生长，直至形成导电细丝。现在，该器件被切换到 LRS，即置位过程如图 4.13(c) 所示。当使用正 V_{ds} 施加电场时，Se 离子迁移到漏极，导致器件的 Ga 空位细丝断裂，器件处于关态 (图 4.13(d))。因此，空位细丝的形成和断裂使器件稳定开关。

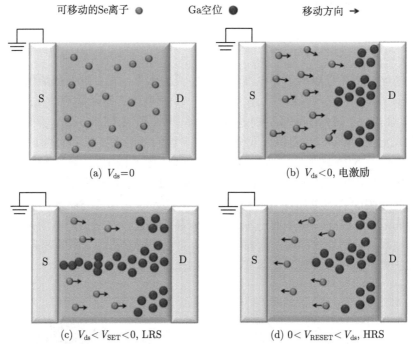

图 4.13　(a) $V_{ds}=0$ 时的初始状态；(b) Ga 空位导电细丝的生长过程；(c) 由细丝连接的源极 (S) 和漏极 (D)；(d) 导电细丝断裂的过程[152]

4.3.2　原子空位

1. 单一型空位

除了细丝机制外，另一种开关机制是基于原子空位。有些空位是二维材料固有的，有些是由外界刺激 (如电场) 产生的。此外，有的材料只形成一种空位，有的材料

4.3 基于二维材料忆阻器的开关机制

具有多个空位,这是稳定开关的关键。Bessonov 等报道了一种 Ag/MoO$_x$/MoS$_2$/Ag 忆阻器件[145],如图 4.14(a) 和 (b) 所示。他们假设 MoS$_2$ 晶体边缘和沿结构位错的缺陷导致了活性非化学计量区,促进了低温氧化,避免了聚合物基底的降解。理想的氧化过程为 $2\text{MoS}_2+7\text{O}_2 \longrightarrow 2\text{MoO}_3+4\text{SO}_2$。然而,根据 X 射线光电子能谱 (XPS) 分析,在 200℃ 退火 3 h 后将产生 MoO$_x$ 层 (小于 3 nm),如图 4.14(c) 所示。由于氧扩散的限制,MoS$_2$ 氧化被限制在 3 nm 左右。推测氧化态较低的 Mo (Mo^{5+}) 和由 Mo^{6+} 组成非化学计量的 MoO$_x$。另外,由于 MoO$_3$ 和 MoS$_2$ 之间的晶格失配,在异质界面上出现了大量的点缺陷和结构畸变。氧化温度越高,则表面 MoO$_3$ 含量越高 (图 4.14(d))。研究表明,该材料具有优异的空位诱导性能。

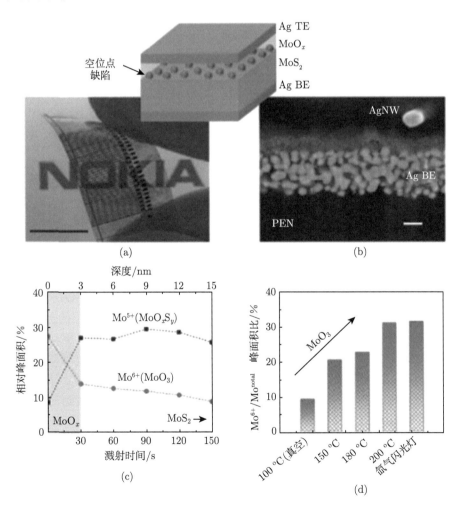

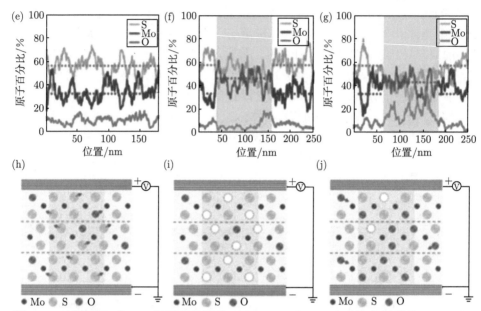

图 4.14 (a) MoO_x/MoS_2 忆阻器示意图; (b) $AgNW/MoO_x/MoS_2$ 忆阻器的 SEM; (c) MoO_x/MoS_2 器件的 XPS 分析; (d) 表面 Mo 的氧化程度[145]。(e)~(g) EDS 行扫描轮廓, (e) 初始 $MoS_{2-x}O_x$ 层原子分布均匀, 其中 Mo:(S+O)≈1:2; (f) S 原子和 Mo:(S+O) 的减少约为 1:1.2; (g) 断电时氧原子数增加, Mo:(S+O) 的比例又回到 1:2; (h)~(j) GMG 器件 RS 机制的原理图; (h) 电激活步骤表明初始状态元素分布均匀并在电激活过程存在硫空位; (i) 处于开态; (j) 处于关态[96]

随后, Wang 等基于氧离子在忆阻器器件横截面上的迁移, 论证了开关机制和导电通道 (图 4.14(e)~(g))[96]。在初始状态下, $MoS_{2-x}O_x$ 层呈现均匀的原子分布, Mo 与 S、O 的平均原子比约为 0.5 (图 4.14(e))。这表明 $MoS_{2-x}O_x$ 中的硫空位在被热氧化后主要被氧占据 (图 4.14(h))。电激活后 Mo:(S+O) 降至 1:1.2, 沟道区 S 原子百分比明显降低 (图 4.14(f))。S 原子的损失 (S 空位的形成) 可能是由于焦耳热对热迁移的影响: 电流的增加导致温度的升高, 进而驱动较轻的离子离开沟道区 (图 4.14(h)、(i))。在热分解作用下, 沟道区氧离子被驱动去填充硫空位, 导致沟道区氧原子增加, 从而使器件切换到关态 (图 4.14(j))。开关机制主要依赖于氧离子的迁移且沟道区的结构变化较小, 这被认为是观察到的优良开关性能的主要原因。

2. 复合型空位

现有的基于金属离子细丝或氧空位机制的忆阻器器件工作电流大, 不能满足低功耗的要求。因此, 为了实现低功率器件, 需要寻找不同于金属离子细丝或氧空位机制的忆阻器器件[287,288]。2019 年, Yan 等提出了新的 RS 传导机制, 即

4.3 基于二维材料忆阻器的开关机制

空位之间的电子跳变以及硫和钨之间空位的形成[41]。原始 WS_2 纳米片和薄膜的 TEM 图像如图 4.15(a)、(b) 所示。薄膜的 TEM 图像 (图 4.15(b)) 显示了比原始 WS_2 纳米片 (图 4.15(a)) 更多的缺陷。放大后的图像显示标准六边形单元和有 W 和 S 空位的单元 (插图,图 4.15(b))。图 4.15(c) 显示了从图 4.15(a) 和 (b) 中所有长方形区域收集的 W 原子的轮廓线,这显示了薄膜中的 W 空位 (V_W) 缺陷。结果表明,该器件中存在 W 和 S 的两个空位,正是由于这两个空位的存在,器件才能实现稳定的开关。

图 4.15 (a) WS_2 纳米片和 (b) 薄膜在 LRS 的 TEM;(c) 长方形区域的 W 原子线分布,上部和下部图像分别对应于 (a) 和 (b)[41]

此外,Yan 等报道了基于 $Al/Ti_3C_2T_x/Pt$ 结构的忆阻器,分析了 Ti 和 O 空位对开关过程的贡献,如图 4.16 所示[45]。图 4.16(a)~(d) 是 HRS 和 LRS 时 $I\text{-}V$ 曲线的拟合结果。为了研究器件的物理机制,他们对器件的微观结构进行了分析。如图 4.16(e) 所示,$Ti_3C_2T_x$ 的微观结构表明,$Ti_3C_2T_x$ 部分氧化为 TiO_2。XPS 光谱也证实了这一点。图 4.16(f) 中的 TEM 图像显示,$Ti_3C_2T_x$ 薄片 (黄色虚线没有 Ti 空位的区域) 中存在 Ti 空位 (白色虚线的黑色区域),转换得到如图 4.16(g) 和图 4.16(h) 所示的间距图。箭头处没有峰表示原子空位的存在。结果表明,原子空位和部分氧化在开关过程中起着重要作用。他们用 XPS 技术测量了氧空位的增强和元素价态的变化。直径 0.1 mm 的 X 射线可以覆盖电极的全部区域,并收集样品垂直方向上所有元素的信息。在低电阻状态下,$Ti_2P_{1/2}$ 的位置向右移动。

此外,他们还分析了 Al 电极下 20 nm、60 nm 和 100 nm 的 HRS 和 LRS 的 XPS 结果。图 4.16(l)~(k)、(m)~(o) 显示了功能层的 XPS 深度剖面。图 4.16(l) 解释了在高电阻和低电阻下 Ti^{3+} 和 Ti^{4+} 的变化趋势,表明 Ti^{4+} 的数量减少而 Ti^{3+} 的数量增加,Ti^{4+} 的减少也表明氧空位的增加。图 4.16(p) 显示了 LRS 和 HRS 中氧元素随片层深度的增加趋势。因此,Ti 和氧空位可以作为载流子跃迁的陷阱位置,较容易的跃迁有利于 LRS 和 HRS 的变化。

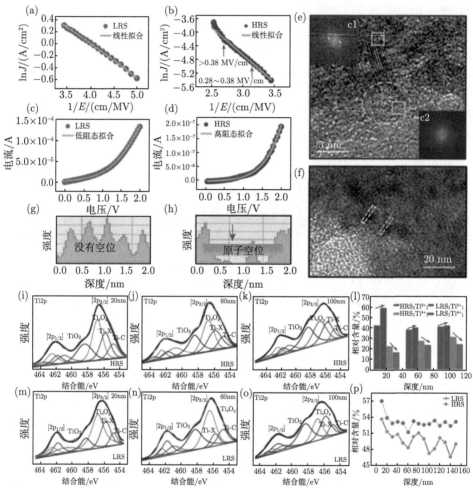

图 4.16 (a) LRS 和 (b) HRS 的拟合曲线；(c) LRS 和 (d) HRS 部分的 I-V 特性；(e) $Ti_3C_2T_x$ 氧化为 TiO_2；(f) 白色虚线区域表示一些原子空位，黄色虚线区域表示没有原子空位；(g) 黄色虚线区域和 (h) 白色虚线区域的线条轮廓。$Ti_3C_2T_x$ 薄片初始和 LRS 的 XPS 深度分析：(i)、(m) 20 nm；(j)、(n) 40 nm；(k)、(o) 60 nm；(l) 不同深度 Ti^{3+}、Ti^{4+} 的相对含量；(p) 薄片不同深度的 HRS 和 LRS 的氧含量 [45] （彩图请扫封底二维码）

4.3.3 电子的捕获和释放

除了细丝和空位机制外，其他研究者发现，器件的开关机制与在界面上电子的捕获和释放有关。He 等报道了 W/MoS_2/p-Si 结构的忆阻器[149]。pn 结的开关反应可以归因于 MoS_2/p-Si 界面上的陷阱位置。在这种随机局域电位涨落 (RLPF) 技术中，低能空穴和电子分布在势阱中，在空间上彼此分离，导致复合时间很长 (图 4.17(a))。由于更多的载流子被激发并重新分布，持续的光电流在高光子时更加明显，并占据势阱位置。以往对 MoS_2 FET 磁滞的研究表明，磁滞主要是由电子的捕获/释放过程引起的。此外，还验证了界面陷阱位置对 RS 的影响。因为器件中存在着本征 SiO_2 层，MoS_2/p-Si 界面上的陷阱位置在存储行为中起着重要作用。当器件正偏置时，电子被捕获在界面陷阱位置，MoS_2 的电导率降低，器件转向 HRS (图 4.17(b))。由于 MoS_2 的自发电子俘获，HRS 可以迅速恢复到中间电阻态 (IRS)。

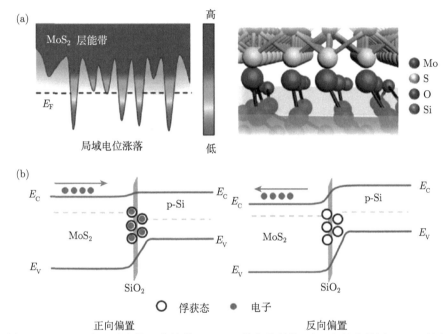

图 4.17 (a) 具有局域电位涨落的单层 MoS_2 的能带结构 (左) 和具有悬浮 Si—O 键的 MoS_2/SiO_2 界面 (右)；(b) MoS_2/p-Si 结的 RS 示意图 [149] (彩图请扫封底二维码)

值得注意的是，He 等观察到 HRS 和 LRS 都是易失性的，它们的电流由于自发的捕获/释放过程而逐渐衰减到 IRS。此外，当器件被反向偏置时，电子脱落并迁移到 MoS_2，器件被设置为 LRS。这一工作证明了一种新的 RS 机制，值得进一步探索。

4.4 现有的挑战

尽管二维材料在器件制备工艺[196]、透明和柔性器件等方面具有巨大优势，但仍存在待解决的问题。第一，针对二维材料的大面积可控制备和转移技术仍存在巨大挑战，由于面积太小而限制了其在大面积电子器件中的应用。第二，二维材料与衬底或环境中其他材料之间的相互作用限制了其应用 (这些相互作用可能改变二维材料的性能或产生不利于研究的新材料)。例如，一些二维材料可能在空气中被氧化而改变其原有性质。第三，二维材料在神经计算中的稳定性问题。神经计算的研究还处于起步阶段，材料也在不断探索中。要实现神经的卓越算力，必须找到一种稳定的材料，这就要求二维材料在稳定性上进一步增强。第四，大多数二维材料的光吸收较低或有限的光谱范围限制了光子和光电器件的能力。由于光学和电子性能依赖于二维结构的横向尺寸，因此横向量子限制也是一个需要解决的问题[289,290]。第五，二维材料的掺杂是一个技术问题。这是由于用传统的三维材料掺杂方法掺杂二维材料时容易形成缺陷，并且二维材料在制备过程中容易形成很多缺陷。

4.5 小　　结

本章介绍了石墨烯、TMD、BN、黑磷、钙钛矿、GaSe 等二维材料在忆阻器领域的应用，强调了其二维体系的优越性，分析了不同二维材料体系的结构和特性。研究发现，各种二维材料体系具有丰富的物理开关机制，包括细丝、原子空位、电子的捕获和释放等；此外，忆阻器将对类脑计算和人工智能产生重大影响。

不可否认，基于二维材料的忆阻器所呈现出的优势使其在未来具有很大的发展前景。在这里，我们列出了一些未来存在的研究方向。

首先，从短期来看，基于二维材料的忆阻器的性能，如开关电流比、循环性能、保持时间、功耗和开关电压等都要进一步提升，使忆阻器能够满足实际存储需求。在这方面，制备二维材料的过程非常关键。二维材料的制备方法很多，一般分为自上而下和自下而上两种制备方法。自上向下的制备方法主要为剥离法，其中最著名的是利用透明胶带从块状层状材料中剥离出薄片的微机械剥离法[290]。此外，还有一种常见的方法是液相剥离法，即将二维材料置于各种溶剂中进行剥离，在此过程中常采用超声波辅助剥离[291-293]。超声波在溶剂中产生气泡，气泡的坍缩产生的力将大块层材料剥离成二维材料。自下而上的制备方法 (包括化学气相沉积、外延生长和小分子化学合成) 是制备高质量单层膜的关键技术，并将成为未来研究的主要方向。

其次，基于二维材料的忆阻器开关机理是一个重要的研究方向。在这方面，利用原位探测技术结合 EELS 的高空间分辨率，可以在不同的开关状态下以原子分辨率观察和识别忆阻器器件中的导电通道。理想情况下，可以在原子尺度上观察到导电通道的形成和湮灭。此外，缺陷的研究对了解材料的基本特性至关重要。研究缺陷的方法有很多，例如，原子力显微镜可以研究材料中的晶界和晶界迁移，这直接影响材料的光学和电子性能；通过像差校正高分辨透射电镜 (AC-HRTEM)

和像差校正扫描透射电镜 (AC-STEM)，也可以在原子尺度上观察材料中由制备工艺、应力和电压等因素而形成的晶界结构。对于原子空位缺陷，需要用精密仪器来表征原子结构，TEM 是一种很好的方法。AC-HRTEM 和 AC-STEM 是观察原子缺失和空位形成的良好手段。除了这些表征技术外，还可以通过第一性原理计算缺陷形成能来确定缺陷形成的难易程度。这些方法对材料缺陷的研究有一定的启示。

最后，基于二维材料的忆阻器器件具有通过光电调制来持久存储和释放信息的能力，这是一个前沿的研究方向。人们设计各种混合系统，增加存储级数量，提高存储性能。在这种情况下，可以添加可见光/紫外线来控制器件，以探索具有不同波长和不同功率的光对器件性能的影响。此外，探索光的影响不应局限于可见光区域，发展工作在红外波段的光存储器件是十分必要的。另一个研究方向是基于二维材料的忆阻器在柔性电子学中的应用。面向柔性可穿戴和未来生物医学应用 (如植入治疗和人机界面) 的电子器件必须具有良好的柔性和力学性能，而具有独特优势的二维材料在这方面有着巨大的应用潜力，有望提高人们的生活水平。此外，基于二维材料的忆阻器近年来在突触模拟和神经形态系统方面的应用取得了长足进展，但要实现类脑计算还有很长的路要走。而目前基于二维材料的神经形态系统只包含一个或几个突触或神经元，系统集成还处于相对较低的水平，未来还有很大的发展空间。最后，在原子级极限薄度下，材料的一些优异性能将会显露出来，材料的特性和功能主要是由表面化学和界面耦合决定的。为了充分发挥二维材料的技术潜力并设计出具有原子精度的新材料体系，二维异质将是未来重要的研究方向。通过结合不同二维材料的宝贵特性，异质可以用来解决单一二维材料无法解决的问题，特别是那些需要垂直多层结构的问题。二维层状材料通过范德瓦耳斯力连接，没有直接化学键。因此，不同的二维材料在组合时不受晶格匹配的限制，容易形成异质结，未来具有发展潜力。在这方面，大面积异质结的制备是一个潜在的发展方向。在实际生产中，需要材料来满足大面积的需要。目前的制备方法可以用于单层二维材料的制备，但连续多层异质结的制备比较困难，因为连续多层异质结的生长往往会破坏底层材料的质量和性能。因此，连续可控地制备多层大面积异质结构是未来发展的一个潜力。总之，二维材料忆阻器的研究才刚刚开始，为了实现这些前景，还需要不断的创新研究。

参 考 文 献

[1] 江之行, 席悦, 唐建石, 等. 忆阻器及其存算一体应用研究进展 [J]. 科技导报, 2024, 42(2): 31-49.

[2] Yan X, Zhou Z, Ding B, et al. Superior resistive switching memory and biological synapse properties based on a simple TiN/ SiO_2/p-Si tunneling junction structure[J]. Journal of Materials Chemistry C, 2017, 5(9): 2259-2267.

[3] Yan X, Zhang L, Yang Y, et al. Highly improved performance in $Zr_{0.5}Hf_{0.5}O_2$ films inserted with graphene oxide quantum dots layer for resistive switching non-volatile memory[J]. Journal of Materials Chemistry C, 2017, 5(42): 11046-11052.

[4] Xiao Z, Zhao J, Lu C, et al. Characteristic investigation of a flexible resistive memory based on a tunneling junction of Pd/BTO/LSMO on mica substrate[J]. Applied Physics Letters, 2018, 113(22): 223501.

[5] Liu Q, Guan W, Long S, et al. Resistive switching memory effect of ZrO_2 films with Zr^+ implanted[J]. Applied Physics Letters, 2008, 92(1): 012117.

[6] Yang Y, Yuan G, Yan Z, et al. Flexible, semitransparent, and inorganic resistive memory based on $BaTi_{0.95}Co_{0.05}O_3$ film[J]. Advanced Materials, 2017, 29(26): 1700425.

[7] Yan X, Li Y, Zhao J, et al. Roles of grain boundary and oxygen vacancies in $Ba_{0.6}Sr_{0.4}TiO_2$ films for resistive switching device application[J]. Applied Physics Letters, 2016, 108(3): 033108.

[8] Yan X, Zhang E, Hao H, et al. Highly transparent bipolar resistive switching memory in $Zr_{0.5}Hf_{0.5}O_2$ films with amorphous semiconducting In-Ga-Zn-O as electrode[J]. IEEE Transactions on Electron Devices, 2015, 62(10): 3244-3249.

[9] Yan X B, Xia Y D, Xu H N, et al. Effects of the electroforming polarity on bipolar resistive switching characteristics of $SrTiO_2$-δ films[J]. Applied Physics Letters, 2010, 97(11): 112101.

[10] Zhu J G. Magnetoresistive random access memory: The path to competitiveness and scalability[J]. Proceedings of the IEEE, 2008, 96(11): 1786-1798.

[11] Apalkov D, Dieny B, Slaughter J M. Magnetoresistive random access memory[J]. Proceedings of the IEEE, 2016, 104(10): 1796-1830.

[12] Engel B N, Akerman J, Butcher B, et al. A 4-Mb toggle MRAM based on a novel bit and switching method[J]. IEEE Transactions on Magnetics, 2005, 41(1): 132-136.

[13] Tehrani S, Slaughter J M, Chen E, et al. Progress and outlook for MRAM technology[J]. IEEE Transactions on Magnetics, 1999, 35(5): 2814-2819.

[14] Hamann H F, O'Boyle M, Martin Y C, et al. Ultra-high-density phase-change storage and memory[J]. Nature Materials, 2006, 5(5): 383-387.

[15] Lacaita A L, Redaelli A. The race of phase change memories to nanoscale storage and applications[J]. Microelectronic Engineering, 2013, 109: 351-356.

[16] Pirovano A, Lacaita A L, Benvenuti A, et al. Electronic switching in phase-change memories[J]. IEEE Transactions on Electron Devices, 2004, 51(3): 452-459.

[17] Wong H S P, Raoux S, Kim S B, et al. Phase change memory[J]. Proceedings of the IEEE, 2010, 98(12): 2201-2227.

[18] Qureshi M K, Srinivasan V, Rivers J A. Scalable high performance main memory system using phase-change memory technology[C]. International Symposium on Computer Architecture, 2009: 24-33.

[19] Govoreanu B, Kar G S, Chen Y Y, et al. 10×10 nm^2 Hf/HfO_x crossbar resistive RAM with excellent performance, reliability and low-energy operation[C]//2011 International

Electron Devices Meeting. IEEE, 2011: 31.6.1-31.6.4.

[20] Yoon J H, Zhang J, Ren X, et al. Truly electroforming-free and low-energy memristors with preconditioned conductive tunneling paths[J]. Advanced Functional Materials, 2017, 27(35): 1702010.

[21] Wu C, Kim T W, Guo T, et al. Mimicking classical conditioning based on a single flexible memristor[J]. Advanced Materials, 2017, 29(10): 1602890.

[22] Zhao H, Dong Z, Tian H, et al. Atomically thin femtojoule memristive device[J]. Advanced Materials, 2017, 29(47): 1703232.

[23] Chang T, Yang Y, Lu W. Building neuromorphic circuits with memristive devices[J]. IEEE Circuits and Systems Magazine, 2013, 13(2): 56-73.

[24] Chua L. Memristor-the missing circuit element[J]. IEEE Transactions on Circuit Theory, 1971, 18(5): 507-519.

[25] Strukov D B, Snider G S, Stewart D R, et al. The missing memristor found[J]. Nature, 2008, 453(7191): 80.

[26] Yang J J, Pickett M D, Li X, et al. Memristive switching mechanism for metal/oxide/metal nanodevices[J]. Nature Nanotechnology, 2008, 3(7): 429.

[27] Jo S H, Chang T, Ebong I, et al. Nanoscale memristor device as synapse in neuromorphic systems[J]. Nano Letters, 2010, 10(4): 1297-1301.

[28] Kim J, Pershin Y V, Yin M, et al. An experimental proof that resistance-switching memory cells are not memristors[J]. Advanced Electronic Materials, 2020, 6(7): 2000010.

[29] Choi B J, Torrezan A C, Strachan J P, et al. High-speed and low-energy nitride memristors[J]. Advanced Functional Materials, 2016, 26(29): 5290-5296.

[30] Torrezan A C, Strachan J P, Medeiros-Ribeiro G, et al. Sub-nanosecond switching of a tantalum oxide memristor[J]. Nanotechnology, 2011, 22(48): 485203.

[31] Zhou J, Cai F, Wang Q, et al. Very low-programming-current RRAM with self-rectifying characteristics[J]. IEEE Electron Device Letters, 2016, 37(4): 404-407.

[32] Kim K H, Hyun Jo S, Gaba S, et al. Nanoscale resistive memory with intrinsic diode characteristics and long endurance[J]. Applied Physics Letters, 2010, 96(5): 053106.

[33] Yu S, Wu Y, Jeyasingh R, et al. An electronic synapse device based on metal oxide resistive switching memory for neuromorphic computation[J]. IEEE Transactions on Electron Devices, 2011, 58(8): 2729-2737.

[34] Yi W, Tsang K K, Lam S K, et al. Biological plausibility and stochasticity in scalable VO_2 active memristor neurons[J]. Nature Communications, 2018, 9(1): 1-10.

[35] Indiveri G, Linares-Barranco B, Legenstein R, et al. Integration of nanoscale memristor synapses in neuromorphic computing architectures[J]. Nanotechnology, 2013, 24(38): 384010.

[36] Kuzum D, Yu S, Wong H S P. Synaptic electronics: materials, devices and applications[J]. Nanotechnology, 2013, 24(38): 382001.

[37] Thomas A. Memristor-based neural networks[J]. Journal of Physics D: Applied Physics, 2013, 46(9): 093001.

[38] Pi S, Li C, Jiang H, et al. Memristor crossbar arrays with 6-nm half-pitch and 2-nm critical dimension[J]. Nature Nanotechnology, 2019, 14(1): 35.

[39] Xia Q, Robinett W, Cumbie M W, et al. Memristor- CMOS hybrid integrated circuits for reconfigurable logic[J]. Nano Letters, 2009, 9(10): 3640-3645.

[40] Khiat A, Ayliffe P, Prodromakis T. High density crossbar arrays with sub-15 nm single cells via liftoff process only[J]. Scientific Reports, 2016, 6(1): 1-8.

[41] Yan X, Zhao Q, Chen A P, et al. Vacancy-induced synaptic behavior in 2D WS_2 nanosheet-based memristor for low-power neuromorphic computing[J]. Small, 2019, 15(24): 1901423.

[42] Yan X, Zhao J, Liu S, et al. Memristors: memristor with Ag-cluster-doped TiO_2 films as artificial synapse for neuroinspired computing[J]. Advanced Functional Materials, 2018, 28(1): 1870002.

[43] Yan X, Zhou Z, Zhao J, et al. Flexible memristors as electronic synapses for neuro-inspired computation based on scotch tape-exfoliated mica substrates[J]. Nano Research, 2018, 11(3): 1183-1192.

[44] Yan X, Pei Y, Chen H, et al. Self-assembled networked PbS distribution quantum dots for resistive switching and artificial synapse performance boost of memristors[J]. Advanced Materials, 2019, 31(7): 1805284.

[45] Yan X, Wang K, Zhao J, et al. A new memristor with 2D $Ti_3C_2T_x$ MXene flakes as an artificial bio-synapse[J]. Small, 2019, 15(25): 1900107.

[46] Li Y, Wang Z, Midya R, et al. Review of memristor devices in neuromorphic computing: materials sciences and device challenges[J]. Journal of Physics D: Applied Physics, 2018, 51(50): 503002.

[47] Xu C, Dong X, Jouppi N P, et al. Design implications of memristor-based RRAM cross-point structures[C]//2011 Design, Automation & Test in Europe. IEEE, 2011: 1-6.

[48] Sharifi M J, Banadaki Y M. General spice models for memristor and application to circuit simulation of memristor-based synapses and memory cells[J]. Journal of Circuits, Systems, and Computers, 2010, 19(02): 407-424.

[49] Yakopcic C, Taha T M, Subramanyam G, et al. A memristor device model[J]. IEEE Electron Device Letters, 2011, 32(10): 1436-1438.

[50] Ho Y, Huang G M, Li P. Nonvolatile memristor memory: Device characteristics and design implications[C]//Proceedings of the 2009 International Conference on Computer-Aided Design, 2009: 485-490.

[51] Kwon D H, Kim K M, Jang J H, et al. Atomic structure of conducting nanofilaments in TiO_2 resistive switching memory[J]. Nature Nanotechnology, 2010, 5(2): 148-153.

[52] Kim S, Choi S H, Lu W. Comprehensive physical model of dynamic resistive switching in an oxide memristor[J]. ACS Nano, 2014, 8(3): 2369-2376.

[53] Miao F, Strachan J P, Yang J J, et al. Anatomy of a nanoscale conduction channel reveals the mechanism of a high-performance memristor[J]. Advanced Materials, 2011, 23(47): 5633-5640.

[54] Yu S, Gao B, Fang Z, et al. A low energy oxide-based electronic synaptic device for neuromorphic visual systems with tolerance to device variation[J]. Advanced Materials, 2013, 25(12): 1774-1779.

[55] Chang T, Jo S H, Lu W. Short-term memory to long-term memory transition in a nanoscale memristor[J]. ACS Nano, 2011, 5(9): 7669-7676.

[56] Novoselov K S, Geim A K, Morozov S V, et al. Electric field effect in atomically thin carbon films[J]. Science, 2004, 306(5696): 666-669.

[57] Bandurin D A, Tyurnina A V, Yu G L, et al. High electron mobility, quantum Hall effect and anomalous optical response in atomically thin InSe[J]. Nature Nanotechnology, 2017, 12(3): 223-227.

[58] Guo Z, Zhang H, Lu S, et al. From black phosphorus to phosphorene: basic solvent exfoliation, evolution of Raman scattering, and applications to ultrafast photonics[J]. Advanced Functional Materials, 2015, 25(45): 6996-7002.

[59] Zhang H, Tang D Y, Zhao L M, et al. Induced solitons formed by cross-polarization coupling in a birefringent cavity fiber laser[J]. Optics Letters, 2008, 33(20): 2317-2319.

[60] Zhang H, Bao Q, Tang D, et al. Large energy soliton erbium-doped fiber laser with a graphene-polymer composite mode locker[J]. Applied Physics Letters, 2009, 95(14): 141103.

[61] Bao Q, Zhang H, Wang B, et al. Broadband graphene polarizer[J]. Nature Photonics, 2011, 5(7): 411-415.

[62] Sun Z, Martinez A, Wang F. Optical modulators with 2D layered materials[J]. Nature Photonics, 2016, 10(4): 227-238.

[63] Schaibley J R, Yu H, Clark G, et al. Valleytronics in 2D materials[J]. Nature Reviews Materials, 2016, 1(11): 1-15.

[64] Yang L, Majumdar K, Liu H, et al. Chloride molecular doping technique on 2D materials: WS_2 and MoS_2[J]. Nano Letters, 2014, 14(11): 6275-6280.

[65] Liu G, Li Y, Li B, et al. High-performance photodetectors based on two-dimensional tin (II) sulfide (SnS) nanoflakes[J]. Journal of Materials Chemistry C, 2018, 6(37): 10036-10041.

[66] Wei X, Yan F G, Shen C, et al. Photodetectors based on junctions of two-dimensional transition metal dichalcogenides[J]. Chinese Physics B, 2017, 26(3): 038504.

[67] Lv Q, Yan F, Wei X, et al. High-performance, self-driven photodetector based on graphene sandwiched GaSe/WS_2 heterojunction[J]. Advanced Optical Materials, 2018, 6(2): 1700490.

[68] Le Lay G. Silicene transistors[J]. Nature Nanotechnology, 2015, 10(3): 202-203.

[69] Iannaccone G, Bonaccorso F, Colombo L, et al. Quantum engineering of transistors based on 2D materials heterostructures[J]. Nature Nanotechnology, 2018, 13(3): 183-191.

[70] Vaziri S, Lupina G, Henkel C, et al. A graphene-based hot electron transistor[J]. Nano Letters, 2013, 13(4): 1435-1439.

[71] Zhou J, Qin J, Zhang X, et al. 2D space-confined synthesis of few-layer MoS$_2$ anchored on carbon nanosheet for lithium-ion battery anode[J]. ACS Nano, 2015, 9(4): 3837-3848.

[72] Sen U K, Mitra S. High-rate and high-energy-density lithium-ion battery anode containing 2D MoS$_2$ nanowall and cellulose binder[J]. ACS Applied Materials & Interfaces, 2013, 5(4): 1240-1247.

[73] Chao D, Zhu C, Song M, et al. A high-rate and stable quasi-solid-state zinc-ion battery with novel 2D layered zinc orthovanadate array[J]. Advanced Materials, 2018, 30(32): 1803181.

[74] Ye L, Zhang S, Zhao W, et al. Highly efficient 2D-conjugated benzodithiophene-based photovoltaic polymer with linear alkylthio side chain[J]. Chemistry of Materials, 2014, 26(12): 3603-3605.

[75] Buscema M, Groenendijk D J, Steele G A, et al. Photovoltaic effect in few-layer black phosphorus PN junctions defined by local electrostatic gating[J]. Nature Communications, 2014, 5: 4651.

[76] Liu Z, Liu Q, Huang Y, et al. Organic photovoltaic devices based on a novel acceptor material: graphene[J]. Advanced Materials, 2008, 20(20): 3924-3930.

[77] Vélez S, Island J, Buscema M, et al. Gate-tunable diode and photovoltaic effect in an organic-2D layered material p-n junction[J]. Nanoscale, 2015, 7(37): 15442-15449.

[78] Liu Z, Lau S P, Yan F. Functionalized graphene and other two-dimensional materials for photovoltaic devices: device design and processing[J]. Chemical Society Reviews, 2015, 44(15): 5638-5679.

[79] Dervin S, Dionysiou D D, Pillai S C. 2D nanostructures for water purification: graphene and beyond[J]. Nanoscale, 2016, 8(33): 15115-15131.

[80] Guan K, Zhao D, Zhang M, et al. 3D nanoporous crystals enabled 2D channels in graphene membrane with enhanced water purification performance[J]. Journal of Membrane Science, 2017, 542: 41-51.

[81] Han Y, Xu Z, Gao C. Ultrathin graphene nanofiltration membrane for water purification[J]. Advanced Functional Materials, 2013, 23(29): 3693-3700.

[82] Novoselov K S, Geim A K, Morozov S V, et al. Electric field effect in atomically thin carbon films[J]. Science, 2004, 306(5696): 666-669.

[83] Geim A K. Graphene: status and prospects[J]. Science, 2009, 324(5934): 1530-1534.

[84] Neto A H C, Guinea F, Peres N M R, et al. The electronic properties of graphene[J]. Reviews of Modern Physics, 2009, 81(1): 109.

[85] Geim A K, Grigorieva I V. Van der Waals heterostructures[J]. Nature, 2013, 499(7459): 419-425.

[86] He C L, Zhuge F, Zhou X F, et al. Nonvolatile resistive switching in graphene oxide thin films[J]. Applied Physics Letters, 2009, 95(23): 232101.

[87] Zhuang X D, Chen Y, Liu G, et al. Conjugated-polymer-functionalized graphene oxide: synthesis and nonvolatile rewritable memory effect[J]. Advanced Materials, 2010, 22(15): 1731-1735.

[88] Sharbati M T, Du Y, Torres J, et al. Low-power, electrochemically tunable graphene synapses for neuromorphic computing[J]. Advanced Materials, 2018, 30(36): 1802353.

[89] Wang R, Li X, Wang Z, et al. Electrochemical analysis graphite/electrolyte interface in lithium-ion batteries: p-toluenesulfonyl isocyanate as electrolyte additive[J]. Nano Energy, 2017, 34: 131-140.

[90] Wu X, Ge R, Chen P A, et al. Thinnest nonvolatile memory based on monolayer h-BN[J]. Advanced Materials, 2019, 31(15): 1806790.

[91] Pan C, Ji Y, Xiao N, et al. Coexistence of grain-boundaries-assisted bipolar and threshold resistive switching in multilayer hexagonal boron nitride[J]. Advanced Functional Materials, 2017, 27(10): 1604811.

[92] Liu M, Zhao N, Liu H, et al. Dual-wavelength harmonically mode-locked fiber laser with topological insulator saturable absorber[J]. IEEE Photonics Technology Letters, 2014, 26(10): 983-986.

[93] Tang Y L, Yin H L, Chen S J, et al. Field test of measurement-device-independent quantum key distribution[J]. IEEE Journal of Selected Topics in Quantum Electronics, 2014, 21(3): 116-122.

[94] Chen Y, Wu M, Tang P, et al. The formation of various multi-soliton patterns and noise-like pulse in a fiber laser passively mode-locked by a topological insulator based saturable absorber[J]. Laser Physics Letters, 2014, 11(5): 055101.

[95] Sangwan V K, Lee H S, Bergeron H, et al. Multi-terminal memtransistors from polycrystalline monolayer molybdenum disulfide[J]. Nature, 2018, 554(7693): 500-504.

[96] Wang M, Cai S, Pan C, et al. Robust memristors based on layered two-dimensional materials[J]. Nature Electronics, 2018, 1(2): 130-136.

[97] Sangwan V K, Lee H S, Hersam M C. Gate-tunable memristors from monolayer MoS_2[C] //2017 IEEE International Electron Devices Meeting (IEDM). IEEE, 2017: 5.1. 1-5.1. 4.

[98] Liu Y, Li F, Chen Z, et al. Resistive switching memory based on organic/inorganic hybrid perovskite materials[J]. Vacuum, 2016, 130: 109-112.

[99] Yoo E, Lyu M, Yun J H, et al. Bifunctional resistive switching behavior in an organolead halide perovskite based $Ag/CH_3NH_3PbI_{3-x}Cl_x/FTO$ structure[J]. Journal of Materials Chemistry C, 2016, 4(33): 7824-7830.

[100] Li Z, Qiao H, Guo Z, et al. High-performance photo-electrochemical photodetector based on liquid-exfoliated few-layered InSe nanosheets with enhanced stability[J]. Advanced Functional Materials, 2018, 28(16): 1705237.

[101] Sang D K, Wang H, Qiu M, et al. Two dimensional β-InSe with layer-dependent properties: band alignment, work function and optical properties[J]. Nanomaterials, 2019, 9(1): 82.

[102] Cao R, Wang H D, Guo Z N, et al. Black phosphorous/indium selenide photoconductive detector for visible and near-infrared light with high sensitivity[J]. Advanced Optical Materials, 2019, 7(12): 1900020.

[103] Wu F, Xia H, Sun H, et al. AsP/InSe van der Waals tunneling heterojunctions with ultrahigh reverse rectification ratio and high photosensitivity[J]. Advanced Functional Materials, 2019, 29(12): 1900314.

[104] Wu L, Xie Z, Lu L, et al. Few-layer tin sulfide: a promising black-phosphorus-analogue 2D material with exceptionally large nonlinear optical response, high stability, and applications in all-optical switching and wavelength conversion[J]. Advanced Optical Materials, 2018, 6(2): 1700985.

[105] Xie Z, Zhang F, Liang Z, et al. Revealing of the ultrafast third-order nonlinear optical response and enabled photonic application in two-dimensional tin sulfide[J]. Photonics Research, 2019, 7(5): 494-502.

[106] Ge Y, Zhu Z, Xu Y, et al. Broadband nonlinear photoresponse of 2D TiS_2 for ultrashort pulse generation and all-optical thresholding devices[J]. Advanced Optical Materials, 2018, 6(4): 1701166.

[107] Hao C, Wen F, Xiang J, et al. Liquid-exfoliated black phosphorous nanosheet thin films for flexible resistive random access memory applications[J]. Advanced Functional Materials, 2016, 26(12): 2016-2024.

[108] Shao J, Xie H, Huang H, et al. Biodegradable black phosphorus-based nanospheres for in vivo photothermal cancer therapy[J]. Nature Communications, 2016, 7(1): 1-13.

[109] Tao W, Zhu X, Yu X, et al. Black phosphorus nanosheets as a robust delivery platform for cancer theranostics[J]. Advanced Materials, 2017, 29(1): 1603276.

[110] Sun Z, Zhao Y, Li Z, et al. TiL_4-Coordinated black phosphorus quantum dots as an efficient contrast agent for in vivo photoacoustic imaging of cancer[J]. Small, 2017, 13(11): 1602896.

[111] Jiang Q, Xu L, Chen N, et al. Facile synthesis of black phosphorus: an efficient electrocatalyst for the oxygen evolving reaction[J]. Angewandte Chemie, 2016, 128(44): 14053-14057.

[112] Tang X, Liang W, Zhao J, et al. Fluorinated phosphorene: electrochemical synthesis, atomistic fluorination, and enhanced stability[J]. Small, 2017, 13(47): 1702739.

[113] Xing C, Jing G, Liang X, et al. Graphene oxide/black phosphorus nanoflake aerogels with robust thermo-stability and significantly enhanced photothermal properties in air[J]. Nanoscale, 2017, 9(24): 8096-8101.

[114] Liang X, Ye X, Wang C, et al. Photothermal cancer immunotherapy by erythrocyte membrane-coated black phosphorus formulation[J]. Journal of Controlled Release, 2019, 296: 150-161.

[115] Yin Y, Cao R, Guo J, et al. High-speed and high-responsivity hybrid silicon/Black-Phosphorus waveguide photodetectors at 2 μm[J]. Laser & Photonics Reviews, 2019, 13(6): 1900032.

[116] Liu S, Li Z, Ge Y, et al. Graphene/phosphorene nano-heterojunction: facile synthesis, nonlinear optics, and ultrafast photonics applications with enhanced performance[J]. Photonics Research, 2017, 5(6): 662-668.

[117] Sang D K, Wang H, Guo Z, et al. Recent developments in stability and passivation techniques of phosphorene toward next-generation device applications[J]. Advanced Functional Materials, 2019, 29(45): 1903419.

[118] Jiang X, Liu S, Liang W, et al. Broadband nonlinear photonics in few-layer MXene $Ti_3C_2T_x$ (T= F, O, or OH)[J]. Laser & Photonics Reviews, 2018, 12(2): 1700229.

[119] Cai Y, Shen J, Ge G, et al. Stretchable $Ti_3C_2T_x$ MXene/carbon nanotube composite based strain sensor with ultrahigh sensitivity and tunable sensing range[J]. ACS Nano, 2018, 12(1): 56-62.

[120] Kim S J, Koh H J, Ren C E, et al. Metallic $Ti_3C_2T_x$ MXene gas sensors with ultrahigh signal-to-noise ratio[J]. ACS Nano, 2018, 12(2): 986-993.

[121] Tao W, Ji X, Zhu X, et al. Two-dimensional antimonene-based photonic nanomedicine for cancer theranostics[J]. Advanced Materials, 2018, 30(38): 1802061.

[122] Tao W, Ji X, Xu X, et al. Antimonene quantum dots: synthesis and application as near-infrared photothermal agents for effective cancer therapy[J]. Angewandte Chemie, 2017, 129(39): 12058-12062.

[123] Song Y, Liang Z, Jiang X, et al. Few-layer antimonene decorated microfiber: ultra-short pulse generation and all-optical thresholding with enhanced long term stability[J]. 2D Materials, 2017, 4(4): 045010.

[124] Lu L, Tang X, Cao R, et al. Broadband nonlinear optical response in few-layer antimonene and antimonene quantum dots: a promising optical kerr media with enhanced stability[J]. Advanced Optical Materials, 2017, 5(17): 1700301.

[125] Xing C, Huang W, Xie Z, et al. Ultrasmall bismuth quantum dots: facile liquid-phase exfoliation, characterization, and application in high-performance UV-Vis photodetector[J]. ACS Photonics, 2018, 5(2): 621-629.

[126] Lu L, Liang Z, Wu L, et al. Few-layer bismuthene: sonochemical exfoliation, nonlinear optics and applications for ultrafast photonics with enhanced stability[J]. Laser & Photonics Reviews, 2018, 12(1): 1700221.

[127] Lu L, Wang W, Wu L, et al. All-optical switching of two continuous waves in few layer bismuthene based on spatial cross-phase modulation[J]. ACS Photonics, 2017, 4(11): 2852-2861.

[128] Chai T, Li X, Feng T, et al. Few-layer bismuthene for ultrashort pulse generation in a dissipative system based on an evanescent field[J]. Nanoscale, 2018, 10(37): 17617-17622.

[129] Huang H, Ren X, Li Z, et al. Two-dimensional bismuth nanosheets as prospective photodetector with tunable optoelectronic performance[J]. Nanotechnology, 2018, 29(23): 235201.

[130] Ji X, Kong N, Wang J, et al. A novel top-down synthesis of ultrathin 2D boron nanosheets for multimodal imaging-guided cancer therapy[J]. Advanced Materials, 2018, 30(36): 1803031.

[131] Xie Z, Xing C, Huang W, et al. Ultrathin 2D nonlayered tellurium nanosheets: facile liquid-phase exfoliation, characterization, and photoresponse with high performance and

enhanced stability[J]. Advanced Functional Materials, 2018, 28(16): 1705833.

[132] Wang Y, Qiu G, Wang R, et al. Field-effect transistors made from solution-grown two-dimensional tellurene[J]. Nature Electronics, 2018, 1(4): 228-236.

[133] Huang W, Zhang Y, You Q, et al. Enhanced photodetection properties of tellurium@selenium roll-to-roll nanotube heterojunctions[J]. Small, 2019, 15(23): 1900902.

[134] Xing C, Huang D, Chen S, et al. Engineering lateral heterojunction of selenium-coated tellurium nanomaterials toward highly efficient solar desalination[J]. Advanced Science, 2019, 6(19): 1900531.

[135] Guo J, Zhao J, Huang D, et al. Two-dimensional tellurium-polymer membrane for ultrafast photonics[J]. Nanoscale, 2019, 11(13): 6235-6242.

[136] Wu L, Huang W, Wang Y, et al. 2D tellurium based high-performance all-optical nonlinear photonic devices[J]. Advanced Functional Materials, 2019, 29(4): 1806346.

[137] Xing C, Xie Z, Liang Z, et al. 2D nonlayered selenium nanosheets: facile synthesis, photoluminescence, and ultrafast photonics[J]. Advanced Optical Materials, 2017, 5(24): 1700884.

[138] Qin J, Qiu G, Jian J, et al. Controlled growth of a large-size 2D selenium nanosheet and its electronic and optoelectronic applications[J]. ACS Nano, 2017, 11(10): 10222-10229.

[139] Huang W, Xie Z, Fan T, et al. Black-phosphorus-analogue tin monosulfide: an emerging optoelectronic two-dimensional material for high-performance photodetection with improved stability under ambient/harsh conditions[J]. Journal of Materials Chemistry C, 2018, 6(36): 9582-9593.

[140] Yan X, Zhang L, Chen H, et al. Graphene oxide quantum dots based memristors with progressive conduction tuning for artificial synaptic learning[J]. Advanced Functional Materials, 2018, 28(40): 1803728.

[141] Romero F J, Toral-Lopez A, Ohata A, et al. Laser-Fabricated reduced graphene oxide memristors[J]. Nanomaterials, 2019, 9(6): 897.

[142] Liu J, Yin Z, Cao X, et al. Fabrication of flexible, all-reduced graphene oxide non-volatile memory devices[J]. Advanced Materials, 2013, 25(2): 233-238.

[143] Yang Y, Lee J, Lee S, et al. Oxide resistive memory with functionalized graphene as built-in selector element[J]. Advanced Materials, 2014, 26(22): 3693-3699.

[144] Liu J, Zeng Z, Cao X, et al. Preparation of MoS_2-polyvinylpyrrolidone nanocomposites for flexible nonvolatile rewritable memory devices with reduced graphene oxide electrodes[J]. Small, 2012, 8(22): 3517-3522.

[145] Bessonov A A, Kirikova M N, Petukhov D I, et al. Layered memristive and memcapacitive switches for printable electronics[J]. Nature Materials, 2014, 14(2): 199-204.

[146] Liu S, Lu N, Zhao X, et al. Eliminating negative-SET behavior by suppressing nanofilament overgrowth in cation-based memory[J]. Advanced Materials, 2016, 28(48): 10623-10629.

[147] Rehman M M, Siddiqui G U, Doh Y H, et al. Highly flexible and electroforming free resistive switching behavior of tungsten disulfide flakes fabricated through advanced

printing technology[J]. Semiconductor Science and Technology, 2017, 32(9): 095001.

[148] Jeong H Y, Kim J Y, Kim J W, et al. Graphene oxide thin films for flexible nonvolatile memory applications[J]. Nano Letters, 2010, 10(11): 4381-4386.

[149] He H K, Yang R, Zhou W, et al. Photonic potentiation and electric habituation in ultrathin memristive synapses based on monolayer MoS_2 [J]. Small, 2018, 14(15): 1800079.

[150] Sangwan V K, Jariwala D, Kim I S, et al. Gate-tunable memristive phenomena mediated by grain boundaries in single-layer MoS_2[J]. Nature Nanotechnology, 2015, 10(5): 403-406.

[151] Yoo E J, Lyu M, Yun J H, et al. Resistive switching behavior in organic-inorganic hybrid $CH_3NH_3PbI_{3-x}Cl_x$ perovskite for resistive random access memory devices[J]. Advanced Materials, 2015, 27(40): 6170-6175.

[152] Yang Y, Du H, Xue Q, et al. Three-terminal memtransistors based on two-dimensional layered gallium selenide nanosheets for potential low-power electronics applications[J]. Nano Energy, 2019, 57: 566-573.

[153] Ki Hong S, Eun Kim J, Kim S O, et al. Analysis on switching mechanism of graphene oxide resistive memory device[J]. Journal of Applied Physics, 2011, 110(4): 044506.

[154] Wang L H, Yang W, Sun Q Q, et al. The mechanism of the asymmetric SET and RESET speed of graphene oxide based flexible resistive switching memories[J]. Applied Physics Letters, 2012, 100(6): 063509.

[155] Porro S, Ricciardi C. Memristive behaviour in inkjet printed graphene oxide thin layers[J]. RSC Advances, 2015, 5(84): 68565-68570.

[156] Yi M, Cao Y, Ling H, et al. Temperature dependence of resistive switching behaviors in resistive random access memory based on graphene oxide film[J]. Nanotechnology, 2014, 25(18): 185202.

[157] Wang L, Wang Z, Zhao W, et al. Controllable multiple depression in a graphene oxide artificial synapse[J]. Advanced Electronic Materials, 2017, 3(1): 1600244.

[158] Pinto S, Krishna R, Dias C, et al. Resistive switching and activity-dependent modifications in Ni-doped graphene oxide thin films[J]. Applied Physics Letters, 2012, 101(6): 063104.

[159] Xie Z, Duo Y, Lin Z, et al. The rise of 2D photothermal materials beyond graphene for clean water production [J]. Advanced Science, 2020, 7(5): 1902236.

[160] Li H, Li Y, Aljarb A, et al. Epitaxial growth of two-dimensional layered transition-metal dichalcogenides: growth mechanism, controllability, and scalability[J]. Chemical Reviews, 2017, 118(13): 6134-6150.

[161] Zhao Q, Xie Z, Peng Y P, et al. Current status and prospects of memristors based on novel 2D materials[J]. Materials Horizons, 2020, 7(6): 1495-1518.

[162] Manzeli S, Ovchinnikov D, Pasquier D, et al. 2D transition metal dichalcogenides[J]. Nature Reviews Materials, 2017, 2(8): 1-15.

[163] Chou S S, Kaehr B, Kim J, et al. Chemically exfoliated MoS_2 as near-infrared photothermal agents[J]. Angewandte Chemie International Edition, 2013, 52(15): 4160-4164.

[164] Yue Q, Kang J, Shao Z, et al. Mechanical and electronic properties of monolayer MoS_2 under elastic strain[J]. Physics Letters A, 2012, 376(12-13): 1166-1170.

[165] Ataca C, Sahin H, Akturk E, et al. Mechanical and electronic properties of MoS_2 nanoribbons and their defects[J]. The Journal of Physical Chemistry C, 2011, 115(10): 3934-3941.

[166] Wu S, Ross J S, Liu G B, et al. Electrical tuning of valley magnetic moment through symmetry control in bilayer MoS_2 [J]. Nature Physics, 2013, 9(3): 149-153.

[167] Xiong F, Wang H, Liu X, et al. Li intercalation in MoS_2: in situ observation of its dynamics and tuning optical and electrical properties[J]. Nano Letters, 2015, 15(10): 6777-6784.

[168] Lee J, Mak K F, Shan J. Electrical control of the valley Hall effect in bilayer MoS_2 transistors[J]. Nature Nanotechnology, 2016, 11(5): 421-425.

[169] Liu T, Wang C, Gu X, et al. Drug delivery with PEGylated MoS_2 nano-sheets for combined photothermal and chemotherapy of cancer[J]. Advanced Materials, 2014, 26(21): 3433-3440.

[170] Yu J, Yin W, Zheng X, et al. Smart MoS_2/Fe_3O_4 nanotheranostic for magnetically targeted photothermal therapy guided by magnetic resonance/photoacoustic imaging[J]. Theranostics, 2015, 5(9): 931.

[171] Yin W, Yu J, Lv F, et al. Functionalized nano-MoS_2 with peroxidase catalytic and near-infrared photothermal activities for safe and synergetic wound antibacterial applications[J]. ACS Nano, 2016, 10(12): 11000-11011.

[172] Liu T, Shi S, Liang C, et al. Iron oxide decorated MoS_2 nanosheets with double PEGylation for chelator-free radiolabeling and multimodal imaging guided photothermal therapy[J]. ACS Nano, 2015, 9(1): 950-960.

[173] Lukowski M A, Daniel A S, Meng F, et al. Enhanced hydrogen evolution catalysis from chemically exfoliated metallic MoS_2 nanosheets[J]. Journal of the American Chemical Society, 2013, 135(28): 10274-10277.

[174] Sun Y, Alimohammadi F, Zhang D, et al. Enabling colloidal synthesis of edge-oriented MoS_2 with expanded interlayer spacing for enhanced HER catalysis[J]. Nano Letters, 2017, 17(3): 1963-1969.

[175] Gao Q, Zhang Z, Xu X, et al. Scalable high performance radio frequency electronics based on large domain bilayer MoS_2[J]. Nature Communications, 2018, 9(1): 1-8.

[176] Chen X, Shinde S M, Dhakal K P, et al. Degradation behaviors and mechanisms of MoS_2 crystals relevant to bioabsorbable electronics[J]. NPG Asia Materials, 2018, 10(8): 810-820.

[177] Chen X, Park Y J, Kang M, et al. CVD-grown monolayer MoS_2 in bioabsorbable electronics and biosensors[J]. Nature Communications, 2018, 9(1): 1-12.

[178] Wang T, Chen S, Pang H, et al. MoS_2-based nanocomposites for electrochemical energy storage[J]. Advanced Science, 2017, 4(2): 1600289.

[179] Wu W, Wang L, Li Y, et al. Piezoelectricity of single-atomic-layer MoS_2 for energy

conversion and piezotronics[J]. Nature, 2014, 514(7523): 470-474.

[180] Wang S, Li K, Chen Y, et al. Biocompatible PEGylated MoS_2 nanosheets: controllable bottom-up synthesis and highly efficient photothermal regression of tumor[J]. Biomaterials, 2015, 39: 206-217.

[181] Yin W, Yan L, Yu J, et al. High-throughput synthesis of single-layer MoS_2 nanosheets as a near-infrared photothermal-triggered drug delivery for effective cancer therapy[J]. ACS Nano, 2014, 8(7): 6922-6933.

[182] Zhang Y, Xiu W, Sun Y, et al. RGD-QD-MoS_2 nanosheets for targeted fluorescent imaging and photothermal therapy of cancer[J]. Nanoscale, 2017, 9(41): 15835-15845.

[183] Xie W, Gao Q, Wang D, et al. Doxorubicin-loaded Fe_3O_4 @ MoS_2-PEG-2DG nanocubes as a theranostic platform for magnetic resonance imaging-guided chemo-photothermal therapy of breast cancer[J]. Nano Research, 2018, 11(5): 2470-2487.

[184] Li H, Zhu L, Wang J, et al. Development of nano-sulfide sorbent for efficient removal of elemental mercury from coal combustion fuel gas[J]. Environmental Science & Technology, 2016, 50(17): 9551-9557.

[185] Raybaud P, Hafner J, Kresse G, et al. Structure, energetics, and electronic properties of the surface of a promoted MoS_2 catalyst: an ab initio local density functional study[J]. Journal of Catalysis, 2000, 190(1): 128-143.

[186] Splendiani A, Sun L, Zhang Y, et al. Emerging photoluminescence in monolayer MoS_2[J]. Nano Letters, 2010, 10(4): 1271-1275.

[187] Smith R J, King P J, Lotya M, et al. Large-scale exfoliation of inorganic layered compounds in aqueous surfactant solutions[J]. Advanced Materials, 2011, 23(34): 3944-3948.

[188] Sun Y, Alimohammadi F, Zhang D, et al. Enabling colloidal synthesis of edge-oriented MoS_2 with expanded interlayer spacing for enhanced HER catalysis[J]. Nano Letters, 2017, 17(3): 1963-1969.

[189] Zheng X, Guo Z, Zhang G, et al. Building a lateral/vertical 1T-2H MoS_2 /Au heterostructure for enhanced photoelectrocatalysis and surface enhanced Raman scattering[J]. Journal of Materials Chemistry A, 2019, 7(34): 19922-19928.

[190] Zhang L, Ji X, Ren X, et al. Electrochemical ammonia synthesis via nitrogen reduction reaction on a MoS_2 catalyst: theoretical and experimental studies[J]. Advanced Materials, 2018, 30(28): 1800191.

[191] Zhao Q, Xie Z, Peng Y P, et al. Current status and prospects of memristors based on novel 2D materials[J]. Materials Horizons, 2020, 7(6): 1495-1518.

[192] Voiry D, Goswami A, Kappera R, et al. Covalent functionalization of monolayered transition metal dichalcogenides by phase engineering[J]. Nature Chemistry, 2015, 7(1): 45-49.

[193] Theerthagiri J, Senthil R A, Senthilkumar B, et al. Recent advances in MoS_2 nanostructured materials for energy and environmental applications–a review[J]. Journal of Solid State Chemistry, 2017, 252: 43-71.

[194] Wang Z, Mi B. Environmental applications of 2D molybdenum disulfide (MoS_2) nanoshe-

ets[J]. Environmental Science & Technology, 2017, 51(15): 8229-8244.
[195] Hui F, Grustan-Gutierrez E, Long S, et al. Graphene and related materials for resistive random access memories[J]. Advanced Electronic Materials, 2017, 3(8): 1600195.
[196] Tan C, Liu Z, Huang W, et al. Non-volatile resistive memory devices based on solution-processed ultrathin two-dimensional nanomaterials[J]. Chemical Society Reviews, 2015, 44(9): 2615-2628.
[197] Cheng P, Sun K, Hu Y H. Memristive behavior and ideal memristor of 1T phase MoS_2 nanosheets[J]. Nano Letters, 2016, 16(1): 572-576.
[198] Wang X F, Tian H, Zhao H M, et al. Interface engineering with MoS_2–Pd nanoparticles hybrid structure for a low voltage resistive switching memory[J]. Small, 2018, 14(2): 1702525.
[199] Waser R, Aono M. Nanoionics-based resistive switching memories[J]. Nature materials, 2007, 6(11): 833-840.
[200] Yang J J, Strukov D B, Stewart D R. Memristive devices for computing[J]. Nature Nanotechnology, 2013, 8(1): 13-24.
[201] Prodromakis T, Peh B P, Papavassiliou C, et al. A versatile memristor model with nonlinear dopant kinetics[J]. IEEE Transactions on Electron Devices, 2011, 58(9): 3099-3105.
[202] Sik Hwang W, Remskar M, Yan R, et al. Transistors with chemically synthesized layered semiconductor WS_2 exhibiting 10^5 room temperature modulation and ambipolar behavior[J]. Applied Physics Letters, 2012, 101(1): 013107.
[203] Liu L, Kumar S B, Ouyang Y, et al. Performance limits of monolayer transition metal dichalcogenide transistors[J]. IEEE Transactions on Electron Devices, 2011, 58(9): 3042-3047.
[204] Kuc A, Zibouche N, Heine T. Influence of quantum confinement on the electronic structure of the transition metal sulfide TS_2[J]. Physical Review B, 2011, 83(24): 245213.
[205] Zeng H, Liu G B, Dai J, et al. Optical signature of symmetry variations and spin-valley coupling in atomically thin tungsten dichalcogenides[J]. Scientific Reports, 2013, 3(1): 1-5.
[206] Iqbal M W, Iqbal M Z, Khan M F, et al. High-mobility and air-stable single-layer WS_2 field-effect transistors sandwiched between chemical vapor deposition-grown hexagonal BN films[J]. Scientific Reports, 2015, 5(1): 1-9.
[207] Gu C, Lee J S. Flexible hybrid organic–inorganic perovskite memory[J]. ACS Nano, 2016, 10(5): 5413-5418.
[208] Pan C, Ji Y, Xiao N, et al. Coexistence of grain-boundaries-assisted bipolar and threshold resistive switching in multilayer hexagonal boron nitride[J]. Advanced Functional Materials, 2017, 27(10): 1604811.
[209] He C, Li J, Wu X, et al. Tunable electroluminescence in planar graphene/SiO_2 memristors[J]. Advanced Materials, 2013, 25(39): 5593-5598.
[210] Lee J, Du C, Sun K, et al. Tuning ionic transport in memristive devices by graphene

with engineered nanopores[J]. ACS Nano, 2016, 10(3): 3571-3579.

[211] Duerloo K A N, Li Y, Reed E J. Structural phase transitions in two-dimensional Mo-and W-dichalcogenide monolayers[J]. Nature Communications, 2014, 5(1): 1-9.

[212] Duerloo K A N, Reed E J. Structural phase transitions by design in monolayer alloys[J]. ACS Nano, 2016, 10(1): 289-297.

[213] Zhang F, Zhang H, Krylyuk S, et al. Electric-field induced structural transition in vertical $MoTe_2$-and $Mo_{1-x}W_xTe_2$-based resistive memories[J]. Nature Materials, 2019, 18(1): 55-61.

[214] Chopra N G, Luyken R J, Cherrey K, et al. Boron nitride nanotubes[J]. Science, 1995, 2695226: 966-967.

[215] Eichler J, Lesniak C. Boron nitride (BN) and BN composites for high-temperature applications[J]. Journal of the European Ceramic Society, 2008, 28(5): 1105-1109.

[216] Li D, Wang X, Zhang Q, et al. Nonvolatile floating-gate memories based on stacked black phosphorus-boron nitride-MoS_2 heterostructures[J]. Advanced Functional Materials, 2015, 25(47): 7360-7365.

[217] Puglisi F M, Larcher L, Pan C, et al. 2D h-BN based RRAM devices[C]//2016 IEEE International Electron Devices Meeting (IEDM). IEEE, 2016: 34.8. 1-34.8. 4.

[218] Li L, Yu Y, Ye G J, et al. Black phosphorus field-effect transistors[J]. Nature Nanotechnology, 2014, 9(5): 372-377.

[219] Tao J, Shen W, Wu S, et al. Mechanical and electrical anisotropy of few-layer black phosphorus[J]. ACS Nano, 2015, 9(11): 11362-11370.

[220] Wei Q, Peng X. Superior mechanical flexibility of phosphorene and few-layer black phosphorus[J]. Applied Physics Letters, 2014, 104(25): 251915.

[221] Miao X, Zhang G, Wang F, et al. Layer-dependent ultrafast carrier and coherent phonon dynamics in black phosphorus[J]. Nano Letters, 2018, 18(5): 3053-3059.

[222] Xia F, Wang H, Jia Y. Rediscovering black phosphorus as an anisotropic layered material for optoelectronics and electronics[J]. Nature Communications, 2014, 5(1): 1-6.

[223] Fei R, Yang L. Strain-engineering the anisotropic electrical conductance of few-layer black phosphorus[J]. Nano Letters, 2014, 14(5): 2884-2889.

[224] Tran V, Soklaski R, Liang Y, et al. Layer-controlled band gap and anisotropic excitons in few-layer black phosphorus[J]. Physical Review B, 2014, 89(23): 235319.

[225] Qiao J, Kong X, Hu Z X, et al. High-mobility transport anisotropy and linear dichroism in few-layer black phosphorus[J]. Nature Communications, 2014, 5(1): 4475.

[226] Jiang J W, Park H S. Negative poisson's ratio in single-layer black phosphorus[J]. Nature Communications, 2014, 5(1): 1-7.

[227] Rodin A S, Carvalho A, Neto A H C. Strain-induced gap modification in black phosphorus[J]. Physical Review Letters, 2014, 112(17): 176801.

[228] Lu S B, Miao L L, Guo Z N, et al. Broadband nonlinear optical response in multi-layer black phosphorus: an emerging infrared and mid-infrared optical material[J]. Optics Express, 2015, 23(9): 11183-11194.

[229] Chen Y, Jiang G, Chen S, et al. Mechanically exfoliated black phosphorus as a new saturable absorber for both Q-switching and mode-locking laser operation[J]. Optics Express, 2015, 23(10): 12823-12833.

[230] Qiu M, Ren W X, Jeong T, et al. Omnipotent phosphorene: a next-generation, two-dimensional nanoplatform for multidisciplinary biomedical applications[J]. Chemical Society Reviews, 2018, 47(15): 5588-5601.

[231] Qiu M, Wang D, Liang W, et al. Novel concept of the smart NIR-light-controlled drug release of black phosphorus nanostructure for cancer therapy[J]. Proceedings of the National Academy of Sciences, 2018, 115(3): 501-506.

[232] Sun C, Wen L, Zeng J, et al. One-pot solventless preparation of PEGylated black phosphorus nanoparticles for photoacoustic imaging and photothermal therapy of cancer[J]. Biomaterials, 2016, 91: 81-89.

[233] Guo Z, Chen S, Wang Z, et al. Metal-ion-modified black phosphorus with enhanced stability and transistor performance[J]. Advanced Materials, 2017, 29(42): 1703811.

[234] Wood J D, Wells S A, Jariwala D, et al. Effective passivation of exfoliated black phosphorus transistors against ambient degradation[J]. Nano Letters, 2014, 14(12): 6964-6970.

[235] Buscema M, Groenendijk D J, Blanter S I, et al. Fast and broadband photoresponse of few-layer black phosphorus field-effect transistors[J]. Nano Letters, 2014, 14(6): 3347-3352.

[236] Wu P, Ameen T, Zhang H, et al. Complementary black phosphorus tunneling field-effect transistors[J]. ACS Nano, 2018, 13(1): 377-385.

[237] He D, Wang Y, Huang Y, et al. High-performance black phosphorus field-effect transistors with long-term air stability[J]. Nano Letters, 2018, 19(1): 331-337.

[238] Xu Y, Yuan J, Zhang K, et al. Field-Induced n-doping of black phosphorus for CMOS compatible 2D logic electronics with high electron mobility[J]. Advanced Functional Materials, 2017, 27(38): 1702211.

[239] Sun L Q, Li M J, Sun K, et al. Electrochemical activity of black phosphorus as an anode material for lithium-ion batteries[J]. The Journal of Physical Chemistry C, 2012, 116(28): 14772-14779.

[240] Xu Z L, Lin S, Onofrio N, et al. Exceptional catalytic effects of black phosphorus quantum dots in shuttling-free lithium sulfur batteries[J]. Nature Communications, 2018, 9(1): 1-11.

[241] Del Rio Castillo A E, Pellegrini V, Sun H, et al. Exfoliation of few-layer black phosphorus in low-boiling-point solvents and its application in Li-ion batteries[J]. Chemistry of Materials, 2018, 30(2): 506-516.

[242] Hembram K, Jung H, Yeo B C, et al. Unraveling the atomistic sodiation mechanism of black phosphorus for sodium ion batteries by first-principles calculations[J]. The Journal of Physical Chemistry C, 2015, 119(27): 15041-15046.

[243] Qiu M, Sun Z T, Sang D K, et al. Current progress in black phosphorus materials and their applications in electrochemical energy storage[J]. Nanoscale, 2017, 9(36): 13384-

13403.

[244] Xie Z, Peng Y P, Yu L, et al. Solar-inspired water purification based on emerging 2D materials: status and challenges[J]. Solar RRL, 2020, 4(3): 1900400.

[245] Lin S, Liu S, Yang Z, et al. Solution-processable ultrathin black phosphorus as an effective electron transport layer in organic photovoltaics[J]. Advanced Functional Materials, 2016, 26(6): 864-871.

[246] Wang Y, Huang G, Mu H, et al. Ultrafast recovery time and broadband saturable absorption properties of black phosphorus suspension[J]. Applied Physics Letters, 2015, 107(9): 091905.

[247] Youngblood N, Chen C, Koester S J, et al. Waveguide-integrated black phosphorus photodetector with high responsivity and low dark current[J]. Nature Photonics, 2015, 9(4): 247-252.

[248] Venuthurumilli P K, Ye P D, Xu X. Plasmonic resonance enhanced polarization-sensitive photodetection by black phosphorus in near infrared[J]. ACS Nano, 2018, 12(5): 4861-4867.

[249] Abbas A N, Liu B, Chen L, et al. Black phosphorus gas sensors[J]. ACS Nano, 2015, 9(5): 5618-5624.

[250] Xu Y, Yuan J, Fei L, et al. Selenium-doped black phosphorus for high-responsivity 2D photodetectors[J]. Small, 2016, 12(36): 5000-5007.

[251] Xie Z, Chen S, Duo Y, et al. Biocompatible two-dimensional titanium nanosheets for multimodal imaging-guided cancer theranostics[J]. ACS Applied Materials & Interfaces, 2019, 11(25): 22129-22140.

[252] Chen J, Fan T, Xie Z, et al. Advances in nanomaterials for photodynamic therapy applications: Status and challenges[J]. Biomaterials, 2020, 237: 119827.

[253] Yang G, Liu Z, Li Y, et al. Facile synthesis of black phosphorus-Au nanocomposites for enhanced photothermal cancer therapy and surface-enhanced Raman scattering analysis[J]. Biomaterials Science, 2017, 5(10): 2048-2055.

[254] Fan T, Xie Z, Huang W, et al. Two-dimensional non-layered selenium nanoflakes: facile fabrications and applications for self-powered photo-detector[J]. Nanotechnology, 2019, 30(11): 114002.

[255] Xie Z, Chen S, Duo Y, et al. Biocompatible two-dimensional titanium nanosheets for multimodal imaging-guided cancer theranostics[J]. ACS Applied Materials & Interfaces, 2019, 11(25): 22129-22140.

[256] Zhao Y, Wang H, Huang H, et al. Surface coordination of black phosphorus for robust air and water stability[J]. Angewandte Chemie, 2016, 128(16): 5087-5091.

[257] Ryder C R, Wood J D, Wells S A, et al. Covalent functionalization and passivation of exfoliated black phosphorus via aryl diazonium chemistry[J]. Nature Chemistry, 2016, 8(6): 597-602.

[258] Tan S J R, Abdelwahab I, Chu L, et al. Quasi-monolayer black phosphorus with high mobility and air stability[J]. Advanced Materials, 2018, 30(6): 1704619.

[259] Lei W, Liu G, Zhang J, et al. Black phosphorus nanostructures: recent advances in hybridization, doping and functionalization[J]. Chemical Society Reviews, 2017, 46(12): 3492-3509.

[260] Doganov R A, O'farrell E C T, Koenig S P, et al. Transport properties of pristine few-layer black phosphorus by van der Waals passivation in an inert atmosphere[J]. Nature Communications, 2015, 6(1): 1-7.

[261] Chen X, Wu Y, Wu Z, et al. High-quality sandwiched black phosphorus heterostructure and its quantum oscillations[J]. Nature Communications, 2015, 6(1): 1-6.

[262] Koenig S P, Doganov R A, Seixas L, et al. Electron doping of ultrathin black phosphorus with Cu adatoms[J]. Nano Letters, 2016, 16(4): 2145-2151.

[263] Zhao Y, Zhou Q, Li Q, et al. Passivation of black phosphorus via self-assembled organic monolayers by van der Waals epitaxy[J]. Advanced Materials, 2017, 29(6): 1603990.

[264] Na J, Lee Y T, Lim J A, et al. Few-layer black phosphorus field-effect transistors with reduced current fluctuation[J]. ACS nano, 2014, 8(11): 11753-11762.

[265] Tian Z, Guo C, Zhao M, et al. Two-dimensional SnS: a phosphorene analogue with strong in-plane electronic anisotropy[J]. Acs Nano, 2017, 11(2): 2219-2226.

[266] Xie Z, Wang D, Fan T, et al. Black phosphorus analogue tin sulfide nanosheets: synthesis and application as near-infrared photothermal agents and drug delivery platforms for cancer therapy[J]. Journal of Materials Chemistry B, 2018, 6(29): 4747-4755.

[267] Kim H S, Im S H, Park N G. Organolead halide perovskite: new horizons in solar cell research[J]. The Journal of Physical Chemistry C, 2014, 118(11): 5615-5625.

[268] Hu X, Zhang X, Liang L, et al. High-performance flexible broadband photodetector based on organolead halide perovskite[J]. Advanced Functional Materials, 2014, 24(46): 7373-7380.

[269] Buin A, Comin R, Xu J, et al. Halide-dependent electronic structure of organolead perovskite materials[J]. Chemistry of Materials, 2015, 27(12): 4405-4412.

[270] Wehrenfennig C, Liu M, Snaith H J, et al. Charge-carrier dynamics in vapour-deposited films of the organolead halide perovskite $CH_3NH_3PbI_{3-x}Cl_x$[J]. Energy & Environmental Science, 2014, 7(7): 2269-2275.

[271] Green M A, Ho-Baillie A, Snaith H J. The emergence of perovskite solar cells[J]. Nature Photonics, 2014, 8(7): 506-514.

[272] Era M, Morimoto S, Tsutsui T, et al. Organic-inorganic heterostructure electroluminescent device using a layered perovskite semiconductor $(C_6H_5C_2H_4NH_3)_2PbI_4$[J]. Applied Physics Letters, 1994, 65(6): 676-678.

[273] Ha S T, Liu X, Zhang Q, et al. Synthesis of organic-inorganic lead halide perovskite nanoplatelets: towards high-performance perovskite solar cells and optoelectronic devices[J]. Advanced Optical Materials, 2014, 2(9): 838-844.

[274] Stoumpos C C, Soe C M M, Tsai H, et al. High members of the 2D Ruddlesden-Popper halide perovskites: synthesis, optical properties, and solar cells of $(CH_3(CH_2)_3NH_3)_2(CH_3NH_3)_4Pb_5I_{16}$[J]. Chem., 2017, 2(3): 427-440.

[275] Yang Z, Jie W, Mak C H, et al. Wafer-scale synthesis of high-quality semiconducting two-dimensional layered InSe with broadband photoresponse[J]. ACS Nano, 2017, 11(4): 4225-4236.

[276] Fan Y, Bauer M, Kador L, et al. Photoluminescence frequency up-conversion in GaSe single crystals as studied by confocal microscopy[J]. Journal of Applied Physics, 2002, 91(3): 1081-1086.

[277] Late D J, Liu B, Luo J, et al. GaS and GaSe ultrathin layer transistors[J]. Advanced Materials, 2012, 24(26): 3549-3554.

[278] Jie W, Chen X, Li D, et al. Layer-dependent nonlinear optical properties and stability of non-centrosymmetric modification in few-layer GaSe sheets[J]. Angewandte Chemie, 2015, 127(4): 1201-1205.

[279] Hu P A, Wen Z, Wang L, et al. Synthesis of few-layer GaSe nanosheets for high performance photodetectors[J]. ACS Nano, 2012, 6(7): 5988-5994.

[280] Mahjouri-Samani M, Gresback R, Tian M, et al. Pulsed laser deposition of photoresponsive two-dimensional GaSe nanosheet networks[J]. Advanced Functional Materials, 2014, 24(40): 6365-6371.

[281] Zhou X, Cheng J, Zhou Y, et al. Strong second-harmonic generation in atomic layered GaSe[J]. Journal of the American Chemical Society, 2015, 137(25): 7994-7997.

[282] Zhou Y, Nie Y, Liu Y, et al. Epitaxy and photoresponse of two-dimensional GaSe crystals on flexible transparent mica sheets[J]. ACS Nano, 2014, 8(2): 1485-1490.

[283] Qian K, Tay R Y, Nguyen V C, et al. Hexagonal boron nitride thin film for flexible resistive memory applications[J]. Advanced Functional Materials, 2016, 26(13): 2176-2184.

[284] Yan X, Zhang L, Chen H, et al. Graphene oxide quantum dots based memristors with progressive conduction tuning for artificial synaptic learning[J]. Advanced Functional Materials, 2018, 28(40): 1803728.

[285] Lin Z, Carvalho B R, Kahn E, et al. Defect engineering of two-dimensional transition metal dichalcogenides[J]. 2D Materials, 2016, 3(2): 022002.

[286] Terrones H, Lv R, Terrones M, et al. The role of defects and doping in 2D graphene sheets and 1D nanoribbons[J]. Reports on Progress in Physics, 2012, 75(6): 062501.

[287] Zhang J J, Sun H J, Li Y, et al. AgInSbTe memristor with gradual resistance tuning[J]. Applied Physics Letters, 2013, 102(18): 183513.

[288] Lenser C, Kuzmin A, Purans J, et al. Probing the oxygen vacancy distribution in resistive switching Fe-SrTiO$_3$ metal-insulator-metal-structures by micro-X ray absorption near-edge structure[J]. Journal of Applied Physics, 2012, 111(7): 076101.

[289] Miró P, Audiffred M, Heine T. An atlas of two-dimensional materials[J]. Chemical Society Reviews, 2014, 43(18): 6537-6554.

[290] Han J H, Lee S, Cheon J. Synthesis and structural transformations of colloidal 2D layered metal chalcogenide nanocrystals[J]. Chemical Society Reviews, 2013, 42(7): 2581-2591.

[291] Novoselov K S, Geim A K, Morozov S V, et al. Electric field effect in atomically thin

carbon films[J]. Science, 2004, 306(5696): 666-669.
[292] Hernandez Y, Nicolosi V, Lotya M, et al. High-yield production of graphene by liquid-phase exfoliation of graphite[J]. Nature Nanotechnology, 2008, 3(9): 563-568.
[293] Coleman J N, Lotya M, O'Neill A, et al. Two-dimensional nanosheets produced by liquid exfoliation of layered materials[J]. Science, 2011, 331(6017): 568-571.

第 5 章　基于莫特绝缘体的忆阻器

5.1　引　言

随着人工智能、物联网、大数据时代的到来，计算机需要存储和处理的数据越来越多，这已经成为应对内存容量不断增加和计算速度不断提高的一项非常重要的任务。同时，由于 Si 基半导体工艺节点的到来和传统冯·诺依曼计算体系的局限性，进一步缩小器件尺寸，实现大规模和高度并行的数据处理是一个巨大的挑战，而传统计算体系架构是将计算单元与存储器分开，称冯·诺依曼瓶颈[1,2]。因此，构建新型器件来进一步提高存储密度、集成内存和计算，用于进行高并行性计算的新体系架构已成为重要的研究方向[3]。研究人员受到人类大脑高存储密度、高并行计算能力与存储和计算一体化的启发，已经构建了一些新型器件[4-7]。其中，最有潜力的候选者之一是忆阻器，它采用简单的金属/绝缘体/金属 (MIM) 夹层结构，具有功耗低[8]、体积小[9]、集成度高[10]、开关速度快[11,12]的优点。特别是在未来的计算体系结构中，忆阻器可以模仿生物神经元[13-16]和突触[9,17-19]，用于边缘计算[20]、深度学习[21]、神经计算芯片[22]等。截至目前，已发现许多材料，包括晶体材料、有机-无机杂化钙钛矿材料和有机材料，可用于忆阻器应用[23-25]。

近年来，一种特殊的莫特绝缘体材料受到关注，它可以根据单电子近似模型中的经典能带理论，通过改变其电子能带结构来实现从绝缘相到金属相的转变。在绝缘体–金属转变 (IMT) 的过程中，莫特绝缘体表现出费米能级在带隙中[26-29]，完全占据价带和空导带之间具有带隙的绝缘体特性。在外界因素刺激下，由于能带的重叠，绝缘体相可能转变为具有半满带的金属相。实际上，该现象是由莫特解释的，他使用了电子和电子之间的相互作用理论，而这类材料通常被称为莫特绝缘体[30]。IMT 可以改变材料的许多物理特性，在电子器件中具有重要的应用前景。在本章中，我们着重于介绍有关莫特绝缘体转变和相关材料的理论知识，莫特绝缘体转变的驱动方法，包括电压驱动、压力驱动、温度驱动，以及莫特绝缘体材料在存储器和突触的非易失性特性、选通管和神经元的易失性特性上的应用、薄膜制备的前景、有争议的机理、器件的性能优化和应用前景。

5.2 莫特绝缘体转变和莫特绝缘体材料

1937 年，研究人员发现存在一类充满 3d 能带的转变金属氧化物 (TMO)，表现出半导体甚至绝缘体的特性[31]。然而，这与单电子近似能带理论给出的结论完全相反。1949 年，莫特提出了金属–绝缘体相变的概念，来描述这类由电子间强关联性引起的绝缘现象[30]。1987 年，安德森提出，莫特绝缘体中电子之间的斥力可能导致高温超导，这使得莫特绝缘体的研究取得重大突破[32]。根据结构分类，莫特绝缘体可分为莫特–哈伯德绝缘体 (Mott-Hubbard insulator) 和电荷转移绝缘体 (charge-transfer insulator)。对于莫特–哈伯德绝缘体，由于三维轨道上的库仑斥力，相互作用下的三维轨道能带分裂为上下哈伯德带 (UHB 和 LHB)，O 的 2p 轨道位于下哈伯德带之下，如图 5.1(a) 所示。如果 O 的 2p 轨道的能级在上下哈伯德带之间，则系统的最低激发能量从 O 的 2p 轨道激发到图 5.1(b) 中的上哈伯德带，称为电荷转移绝缘体。莫特绝缘体–金属转变 (Mott IMT) 具有不同的机制，例如带宽控制 (BC-Mott IMT (图 5.1(c) 和 (d)) 和填充控制 (FC-Mott IMT)。FC-Mott IMT 包括电子掺杂 (ED-Mott IMT，图 5.1(c)~(e)) 和空穴掺杂 (HD-Mott IMT，图 5.1(c)~(f))[33-35]。莫特绝缘体是带宽 W 和库仑排斥能 U 之间排斥的结果。BC-Mott IMT 模型指出，当外部条件变化时，减小原子间距离，然后增加原子间轨道的重叠。FC-Mott IMT 模型指出，通过向系统中添加电子/空穴，降低了某些电子的跳跃能量成本以及有效的库仑排斥能。由于外部条件的变化，莫特绝缘子材料的物理性能发生了巨大变化，引起了科学家的关注。到目前为止，已经发现了几种莫特绝缘体材料。莫特绝缘体材料可以分为几类，例如，二元氧化物 (主要涉及钒氧化物) (如 VO_2[36,37]、Cr 掺杂的 V_2O_3[38-40])、

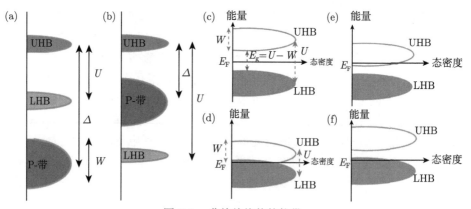

图 5.1 莫特绝缘体的能带

(a) 莫特–哈伯德绝缘体；(b) 电荷转移绝缘体；(c) 绝缘体状态下的能带；(d) BC-Mott IMT；(e) ED-Mott IMT；(f) HD-Mott IMT

铌氧化物 (NbO_2)[41,42]、反铁磁过渡金属氧化物 (如 CoO[43]、MnO[44,45])、钙钛矿 $RNiO_3$ (主要是 $SmNiO_3$[46]、$NdNiO_3$[47] 和 $PrNiO_3$[48,49]) 窄间隙莫特绝缘体 AM_4Q_8 (A = Ga, Ge; M = V, Nb, Ta; Q = Se, S) 结构、$SrCoO_x$[50,51] 和 1T-$TaSe_2$ 材料[52]。在本章中，我们介绍了莫特绝缘体-金属转变的机理，它们之间的区别和一些相关材料。第 6 章将讨论不同莫特绝缘体材料的莫特绝缘体-金属转变机制。

5.3 莫特转变的驱动方法

在 5.2 节中，莫特绝缘体-金属转变 (以下简称莫特转变) 可以分为 BC-Mott IMT、ED-Mott IMT 和 HD-Mott IMT。驱动 IMT 的常用方法是压力、温度和施加电压，这将实现易失性或非易失性存储器的性能。实际上，在 IMT 过程中，莫特绝缘体中的缺陷或离子也能够被很好地驱动[53]，并且浓度 (缺陷或离子) 将被调节，这对应于带绝缘体中的传统价变忆阻效应[54]。不同的记忆性能可用于构建不同的应用，这为研究人员设计不同的存储器提供了良好的理论基础。在本节中，我们将讨论莫特转变的驱动方法。

5.3.1 压力驱动的莫特转变

压力能够驱动莫特跃迁，科学家已经通过拉曼光谱[55]和光学研究[56]在 MnO 上得到证明，它在 90~150 GPa 的压力下会发生相变。其中，在 100 GPa 的压力下，一些特性表明该材料存在一级绝缘体-金属相变[57]，例如磁性[58-60]、结构、光谱[61,62]和反射率[56]。在先前的工作中，研究人员发现 $Fe_{1-x}O$ 在高压下表现出金属化状态。Mitchell 等研究了压力和电导率之间的关系，如图 5.2(a)所示，发现电阻率在 72 GPa 压力下降低，这表明材料内部出现了部分金属化的状态[63]。为深入研究这一结果，研究人员使用了局部密度近似 (LDA) 动态平均场理论 (DMFT) 方法，演示在给定压力下，莫特绝缘体材料中两种状态 (绝缘体和金属状态) 共存[54,58,64]。在高压下是金属化的，这与混合 O 2p-Fe 4s 和 Fe 3d 带的重叠有关。对于 FeO，IMT 具有典型特征，具有明显的哈伯德带和附近强相关金属的窄准粒子峰。t_{2g} 随着压力的增加开始出现明显的尖峰，通过图 5.2(b) 中的占用数和磁矩对压力的影响，曲线在 60 GPa 处扭结[57]，这种现象被认为是由 IMT 和费米能级的重建决定的。因此，FeO 在压力的诱导下会表现出 IMT 现象。除了上述材料之外，一类典型的四面体团簇化合物可以在静水压力下转化为金属，例如 AM_4Q_8 和 ABO_3 (A = Nd、Sm、Pr 等；B = Ni、Mn 等)。在压力驱动 BC-Mott IMT 的机制中，通过施加外部压力来减少原子的距离，然后增加原子之间轨道的重叠[65]。Phuoc 等通过实验和理论证明，当压力近似 6 Pa 时，$GaTa_4Se_8$ 转化为金属，光学间隙闭合，表明基于密度泛函理论 (DFT) 的

Ta 5d 能带带宽随着施加压力的减小而减小[65]。大多数 ABO_3 的电导率主要取决于 Ni-O-Ni 角。例如，$RNiO_3$ (R = Nd、Pr) 的电导率主要取决于 Ni-O-Ni 角。图 5.2 (d) 显示，随着压力的增加，Ni-O-Ni 角变大[66]。O 2p 和 Ni 3d 轨道之间的 Ni-O-Ni 角的扩展增加了重叠，导致带宽增加，电荷转移间隙逐渐减小，最终为零，使其处于半金属状态，实现莫特绝缘体转变[67]。在压力下，晶格常数在压力驱动的 IMT 中也可以发挥重要作用。因此，可以掺杂一些材料来调节晶格常数，以实现压力下的相变。以 $Cr-V_2O_3$ 为例，Lupi 等在整个压力驱动的莫特 IMT 中考虑了掺杂 Cr 的 V_2O_3 中晶格参数的修改，图 5.2(c) 显示，当在 V_2O_3 中掺杂 Cr 时，c/a 在大约 3 kbar 处迅速增加，并且绝缘体状态变为金属状态[68]。掺杂 Cr 的 V_2O_3 的压力驱动 IMT 特性表明了晶格参数的重要性。目前对 $Cr-V_2O_3$ 莫特绝缘体的研究表明，带宽的增加，而不是 V 3d 轨道占据状态的改变，会增加金属相[69,70]。

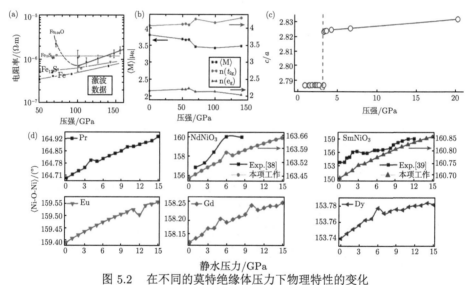

图 5.2 在不同的莫特绝缘体压力下物理特性的变化

(a) $Fe_{0.94}O$ 的电阻率 (经许可转载自文献 [63]。© Elsevier B.V. 1986)；(b) 在 LDA+DMFT(CT-QMC) 计算中得到的占位数和磁矩；(c) 掺铬 V_2O_3 中 c/a 比率的晶格参数[68]；(d) $ANiO_3$ (A = Pr、Nd、Sm、Eu、Gd、Dy) 的平均带角 (⟨Ni-O-Ni⟩) 是静水压力的函数，平均带角 ⟨Ni-O-Ni⟩ 随静水压力增加 (经许可转载自文献 [66]。© Elsevier B.V. 2021)

5.3.2 温度驱动的莫特转变

莫特绝缘体 IMT 可以由温度驱动，并且已经被广泛讨论，例如 $RNiO_3$[81−83]、掺杂和未掺杂的氧化钒[84−86]、Ca_2RuO_4[87]、Sr_2IrO_4[88]、$1T-TaSe_2$[89] 和 NbO_2[90]。表 5.1 总结了不同材料的 IMT 温度。

5.3 莫特转变的驱动方法

表 5.1 相关莫特绝缘体材料的 T_{IMT}

材料	T_{IMT}
Ti_3O_5	448 K[71]
Fe_3O_4	121 K[72]
V_8O_{15}	70 K[73]
VO	126 K[74]
V_5O_9	135 K[75]
V_4O_7	250 K[76]
VO_2	340 K[77]
V_2O_3	165 K[78]
$PrNiO_3$	135 K[67]
$NdNiO_3$	201 K[67]
$SmNiO_3$	403 K[67]
$LaCoO_3$	500 K[79]
NbO_2	1081 K[80]

温度类似于压力,可以通过熵 (entropy) 使 $RNiO_3$ (R = Gd、Nd、Sm 和 Eu) 中的 Ni-O-Ni 角变直 (上升),从而扩大带宽并最终导致 BC-Mott IMT[48,62]。此外,当 R 的原子尺寸增加或平面外晶格常数增加时,$RNiO_3$ 中 BC-Mott IMT 的阈值温度可以降低。这些改变可以增加平面外的 Ni-O-Ni 键角,有利于 $RNiO_3$ 发生 BC-Mott IMT 并降低转变温度。图 5.3(a) 显示,温度的升高可以将金属状态转变为绝缘体状态,并且两者可以在 IMT 过程中共存[91]。如图 5.3(b) 所示,在恒定压力下温度升高会导致 Cr 掺杂的 V_2O_3 发生 IMT[92]。此外,在低温下,纯 V_2O_3 和掺杂的 V_2O_3 是反铁磁绝缘体 (AFI),可以通过升高温度而转变为顺磁金属 (PM) 状态,如图 5.3(c) 所示[93]。Janod 等将这种莫特 IMT 解释为晶体对称破坏和温度控制的绝缘体–金属转变[94]。扫描光发射显微镜可以揭示 PM 相中的不均匀分布[68]。图 5.3(d) 中的 (i) 显示,V_2O_3 在 220 K 的低温下是一种坏金属,同时存在金属 (红色) 畴和绝缘 (蓝色) 域。图 5.3(d) 中的 (ii) 和 (iii) 表明,随着温度升高到 260 K 和 320 K,绝缘子的比例逐渐增加,并且这表明当温度从 220 K 升高到 320 K 时,从低电阻状态 (LRS) 到高电阻状态 (HRS) 的变化。相反,对于从 HRS 到 LRS 的变化,与图 5.3(d) 中的 (i) 相比,(iv) 中的金属和绝缘体相的分布大致相同,表明它们在 Cr 掺杂的 V_2O_3 的特定点周围形成。然而,对于 VO_2,IMT 的潜在物理机制包括派尔斯 (Peierls) 跃迁[95,96] 和莫特跃迁[97,98],两者共存的物理机制。Peierls 跃迁是由结构的变化引起的,而莫特跃迁是由于电子之间的强相关性。Lee 等的工作表明,VO_2 中的温度驱动 IMT 是由空穴载体引起的,这类似于莫特跃迁。然而,Kim 等认为结构的变化是由声子或电子–声子相互作用引起的,其使用密度泛函理论研究声子色散,发现声子软化

不稳定性，这与 Peierls 有关 [98]。最近的一些研究表明，当发生 IMT 时，VO_2 的晶体和电子结构同时发生变化。这种过渡被称为莫特–派尔斯 (Mott-Peierls) 转变。Yao 等使用密度泛函理论计算和原位 X 射线吸收精细结构测量来分析温度驱动 IMT 的机理，其中原子特殊重排，电子结构和电阻之间的剧烈相关性加强了 Mott-Peierls 的 VO_2 的 IMT 机制 [99]。总而言之，在莫特绝缘体中，温度的变化会影响其内部电子结构，这是由材料特有的电子相互作用机制所决定的。从绝缘体到金属，它可以增加带宽并减少带隙。尽管如此，VO_2 在高温金属态和低温绝缘体态之间变化时会发生相变，人们尚未就其内部机制达成共识。

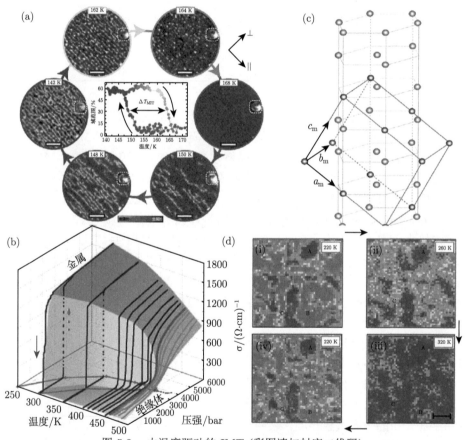

图 5.3　由温度驱动的 IMT (彩图请扫封底二维码)

(a) 绝缘域随着温度的变化而成核和生长 (经许可转载自文献 [91]。©Nature 2016)；(b) 掺铬的 V_2O_3 在温度和压力条件下的导电性变化 (经许可转载自参考文献 [92]。©Science 2003)；(c) 室温 PM 相和低温 AFI 相单元格之间的关系 (经许可转载自参考文献 [93]。©IOP Publishing 2019)；(d) VO_2 带状结构的相空间图的实验结果 (经许可转载自参考文献 [68])

5.3.3 电压驱动的莫特转变

除了温度和压力之外,莫特转变可以由施加的电压驱动。根据机理的不同,它可以分为两种类型:本征材料或掺杂材料。在本征材料中,典型的材料是 VO_2 和 NbO_2[100,101],它们可以通过焦耳热效应,而不是在电压下引起的场击穿效应来诱导莫特绝缘体转变。Mun 等发现莫特绝缘体跃迁可以通过光学显微镜进行实验观察[102]。在图 5.4(a) 中,通过光学显微镜发现,随着电压的增加,绝缘体状态转变为金属状态[103]。在理论上,Del Valle 等已经使用 COMSOL multiphysics 模拟来证明 IMT 过程是由焦耳热导致的局部相变[104],这可以归因于 5.3.1 节中的温度驱动的莫特转变,如图 5.4(b) 所示。另一种类型是外部掺杂剂可以改变钙钛矿的结构并导致莫特转变[105,106],它们是基于 Ti[107-110]、V[111,112]、Cr[35]、Mn[113]、Ni[114] 的转变金属氧化物,Co[115]、Cu[116] 和过渡金属化合物 NiS[117] 的 TMO。另外一种掺杂剂是附加电子,这将导致一个大的单电子轨道的莫特-哈伯德分裂,从而导致莫特跃迁[52],一种典型的材料是 $SmNiO_3$ (SNO)。尽管对 Li 金属施加了足够明显的正偏压,但 Ni^{3+} $t_{2g}^6 e_g^2$ 可以接收额外的电子并将其转化为 Ni^{2+} $t_{2g}^6 e_g^2$。由于 Ni^{2+} $t_{2g}^6 e_g^2$ 单电子轨道之间的强库仑排斥作用,薄膜电阻发生了巨大变化。如图 5.4(c) 和 (d) 所示,电子掺杂剂诱导大的莫特跃迁[118]。另一种掺杂剂是质子,质子迅速对准并与电子耦合以诱导莫特跃迁。Ramadoss 等使用诸如 Pd/Pt 之类的金属电极来催化质子掺杂法[106]。氢在金属/SNO 界面处分为电子和质子 (H === $H^+ + e^-$),电子锚定在镍轨道中,并将价态从 Ni^{3+} 变为 Ni^{2+}。由于半填充时的局部化导致 3 eV 量级的带隙显著增加,质子电子掺杂后 Ni^{2+} 的存在显示出很强的电子相关性,这导致材料电阻大幅增加了几个数量级。其结构示意图如图 5.4(e)[106] 所示。然而,对于其他一些材料,如掺 Cr 的 V_2O_3,Cr 掺杂是一种质子掺杂的方法,而先前的研究表明 V_2O_3 是一种劣质金属,但在 Cr 的掺杂下会转变为绝缘状态,如图 5.4(f) 所示[68],理论表明,与绝缘体状态相比,V_2O_3 中的金属状态是 V 3d 轨道占有率莫特的结构,类似于 FC-莫特跃迁机制。

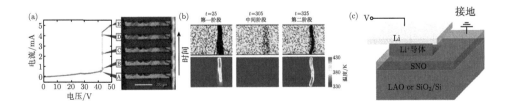

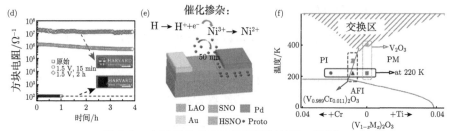

图 5.4 电压下的 IMT (彩图请扫封底二维码)

(a) 电压–电流曲线 (左) 和光学显微镜图像 (右) 作为增加电压的函数 (经许可转载自参考文献 [102]。© Mun, B. S. et al. 2013); (b) 最上面一行, 模拟的金属 (黑色) 和绝缘 (黄色) 域在三个不同时间的二维分布 (经许可转载自参考文献 [104]。© Del Valle, J. et al. 2019); (c) SNO 上 Li 插层的结构示意图; (d) 不同时间内 Li 插层 SNO 的电阻变化和视觉方面的变化 (经许可转载自文献 [118]。© Macmillan Publishers Limited 2014); (e) 利用 Pd/SNO 界面的催化掺杂进行电选择性质子掺杂的莫特记忆器件 (经许可转载自文献 [106]。© IEEE 2018); (f) $(V_{1-x}M_x)_2O_3$ 中 IMT/MIT 的相位图, 作为掺杂浓度和温度的影响 (经许可转载自文献 [68]。© Macmillan Publishers Limited 2010)

5.4 基于莫特绝缘体的忆阻器应用

从 5.2 节和 5.3 节可以发现, 当发生莫特跃迁时, 许多物理性质都会发生变化。因此, 莫特绝缘体在许多领域, 尤其是电子器件领域具有非常重要的应用前景。而不同的机制将导致不同的电阻行为和不同的应用方向。例如, 一些莫特绝缘体可以从稳定的晶体结构过渡到亚稳态晶体结构, 这将导致易失性开关行为 (以 VO_2 和 NbO_2 为例)。VO_2 可以通过电压驱动 IMT, 这是由于在焦耳热驱动下, 从低温单斜相 (M1 相) 转变为高温金红石四方 R 相 [77,119]。在亚稳态 R 相的情况下, VO_2 在去除电压后会回到 M1 状态, 出现易失性的行为 [77]。莫特绝缘体材料 [37,120] 中存在一些缺陷, 并且该缺陷将随着施加电压而产生。这些缺陷将充当掺杂剂并在电压下移动。当存在少量的掺杂剂 (不足以形成导电丝或非常弱的导电丝) 时, 去除电场后自发溶解导电丝, 器件表现出易失性 [121]。然而, 对于莫特绝缘体中的大量缺陷, 以 $SrCoO_x$ 为例, 这些缺陷会在电压下移动并改变功能层材料的组成, 从而产生不可逆的 IMT 并形成稳定的氧空位导电路径, 该路径不随外部电压的抵消而变化, 并且这种状态可以保持 3×10^3 s 以上而没有显著的变化 [122]。由于具有易失性和非易失性这两种特性, 莫特绝缘体在未来的计算系统中具有不同的应用方向。

5.4.1 非易失性特性

1. 非易失性存储器的莫特绝缘体

非易失性忆阻器 (NVM) 具有结构简单、尺寸可扩展、切换速度快、多级数据存储和数据非易失性的特性 [9,123]。目前, 许多材料可以用于忆阻器, 例如 TiO_2[124]、

5.4 基于莫特绝缘体的忆阻器应用

$MnS_2^{[125]}$、$Hf_{0.5}Zr_{0.5}O_2^{[19]}$、$SnSe^{[126]}$、$Ta_2O_5^{[127]}$ 和 $SiC^{[128]}$。大多数电阻开关机制是源于导电空位组成的导电丝的形成和断裂[16]。一些莫特绝缘体 (如 $SrCoO_x$) 具有氧八面体和氧四面体的交替堆叠结构,这是一种独特的晶体结构,通过降低高度有序的氧空位通道 (OVC) 中氧运动的随机性来实现均匀模拟切换[122]。图 5.5(a) 示出了具有可调谐 OVC 拓扑定向相变-随机存取存储器 (TPT-RAM) 的示意图及其相变机制。图 5.5(b) 显示出了器件的电流-电压 (I-V) 曲线,其中器件存在一个电激活过程。图 5.5(c) 显示了 30 个器件测试所得点激活电压的统计分布,通过控制高度有序的 OVC 中的氧离子迁移来提高均匀性器件性能,并且该器件在 85℃ 下具有大于 3000 s 的长保持特性,如图 5.5(d) 所示。Zhao 等报道了类似的工作,通过在 Nb 掺杂的 $SrTiO_3$ (NSTO) 衬底上制备具有窄晶界的高取向结晶 VO_2 薄膜,具有 $Au/VO_2/NSTO$ 结构的忆阻器 (图 5.5(e)) 在负扫描电压下呈现阶梯状的电流-电压曲线,其可以对应于由图 5.5(f) 中的氧空位的单原子接触形成的量化电导状态。在图 5.5(g) 中,电导状态集中在整数和半整数量化电导状态,并且在图 5.5(h) 中,多个量化电导状态可以保持在 2×10^4 s 上,而没有明显的波动[129]。Yang 等制备了 Pt/Cr 掺杂的 $SrZrO_3/SrCuO_3/Si$ 器件[130],如图 5.5(i) 所示。图 5.5(j) 示出了基于图 5.5(i) 得到的电流-电压特性。GaV_4S_8 器件可以达到 65000 多个周期,数据保留时间为 10 年[131]。值得一提的是,许多绝缘体材料可以通过沉积后退火和外部掺杂来调节[130,132-134]。Lee 等使用较小的阳离子 Gd^{3+} 来部分替代 $La_{0.7}Sr_{0.3}MnO_3$ 中的 La^{3+} 阳离子,以控制工作电压,这将导致最大 Sr^{2+} 的 a 位空位失配,而 SrO_y 的面偏析是由 $AMnO_3$ 钙钛矿中的离子而产生的,这种外部掺杂方法将器件的工作电压从 3.5 V 降低到 2.5 V,如图 5.5(k)[135] 所示。基于莫特绝缘体材料的 NVM 可以在低电阻状态和高电阻状态之间切换,如图 5.5(l) 和 (m) 所示。此外,图 5.5(m) 示出了器件可以保持 10^6 个循环,并且低和高电阻状态可以在室温下保持 10 年而不会降解。更重要的是,可以通过施加不同的脉冲宽度电压来调制多级电阻状态特性,如图 5.5(n) 所示,这非常适合于多值存储和增加存储密度。研究人员研究了莫特绝缘体氧化物 $(V_{0.95}Cr_{0.05})_2O_3$ 和 $SrVO_3$,发现该器件具有突出的耐久性能,这证明了 NVM 的巨大潜力,如图 5.5(o) 和 (p) 所示[132-136]。按照上面的描述,可以通过退火和外部掺杂来调节忆阻器的特性。这些可以提供一种提高 NVM 性能的方法。

2. 突触

突触表示前神经元与后神经元之间的连接,具有存储和信息计算的作用[127]。因此,电子器件模仿突触的功能至关重要,这对于开发人工神经网络以实现大脑功能具有重要意义。突触可塑性是指突触对神经元的行为做出反应,可以改变其连接的权重。在过去的几年中,许多研究人员使用常规的互补金属氧化物半导体

(CMOS) 晶体管来模拟人工突触,但是缺乏生物学相似性,电路很复杂。为了简化这种复杂的电路,研究人员提出了使用一种新的原型器件,该器件可以模拟生物神经系统中的 Ca^{2+} 和 Na^+ 迁移。忆阻器和突触晶体管都适用于神经形态计算,以在器件级别模拟突触和神经元的功能。基于莫特绝缘体的两种器件都实现了突触功能,例如尖峰时间依赖性可塑性 (STDP)[137-139]、短期可塑性 (STP)[140-142]、长期可塑性 (LTP)[143] 和双脉冲易化 (PPF)[144]。$Pr_{0.7}Ca_{0.3}MnO_3$(PCMO) 是一种莫特绝缘体材料,场致氧空位驱动表面发生莫特金属-绝缘体转变。作者还发现,Au/PCMO/Pt 的电阻切换现象强烈依赖于金属/PCMO 界面域处 Mn^{4+}/Mn^{3+}

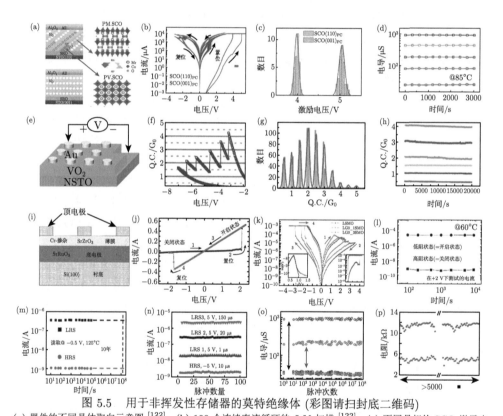

图 5.5　用于非挥发性存储器的莫特绝缘体 (彩图请扫封底二维码)

(a) 器件的不同晶体取向示意图[122];(b) 100 个连续直流循环的 I-V 扫描[122];(c) 不同晶相的 SCO 激励电压的分布情况;(d) 器件在 85℃ 下的保持特性[122];(e) Au/VO$_2$/NSTO 器件的原理图[129];(f) 电导率-电压曲线图[129];(g) 复位点 50 个周期时的量子电导直方图[129];(h) 由几个量子电导率状态保持的多级状态[129];(i) MIM 结构示意图 (经许可转载自参考文献 [130]。© ALP Publishing 2006);(j) 基于图 (e) 结构的 I-V 特性 (经许可转载自参考文献 [130]。© ALP Publishing 2006);(k) La$_{0.7-x}$Gd$_x$Sr$_{0.3}$MnO 薄膜的 I-V 曲线 (经许可转载自参考文献 [135]。© ALP Publishing 2014);(l) 基于图 (f) 的结构的保留特性 (经许可转载自参考文献 [130]。© ALP Publishing 2006);(m) 脉冲宽度对开关特性的依赖性[134];(n) 由几个量子电导率状态的多级状态保留[129];(o) 耐久性测试达 10^8 次[122];(p) 器件中获得的循环耐久性 (经许可转载自文献 [136]。© Elsevier 2018)

5.4 基于莫特绝缘体的忆阻器应用

的混合价态,这可以视为莫特跃迁的证据。Kim 等研究了基于 PCMO 的存储器件 (图 5.6(a)) 以及其突触行为 [146];突触的权重可以通过正 (负) 脉冲来调节,以增加 (减少) 权重,这称为增强 (抑制);然后,可以通过输入脉冲宽度来调节权重变化,其可以用于调节突触可塑性,如图 5.6(b) 所示。此外,已经验证了 STDP,当来自突触前记忆的前尖峰进行到突触后膜的后尖峰时 ($\Delta t > 0$),突触的权重可能是电位;反之,$\Delta t < 0$,突触的权重可能被抑制,如图 5.6(c) 所示。另一个类似突触的器件是一个神经递质。基于 SmNiO$_3$ 的神经递质 (图 5.6(d)) 也可以通过施加不同的电压脉冲来模仿生物突触的增效和抑制,如图 5.6(e) 所示 [147]。STDP 功能是通过改变源极和漏极峰值之间的时间差实现的。当驱动尖峰跟随源尖峰时,电导 (突触权重) 减小。当漏极尖峰跟随源极尖峰时,观察到一个反向的调制,如图 5.6(f) 所示。STP 和 LTP 的突触功能可由 VO$_2$ 神经晶体管中的栅极电压 (gate voltage) 脉冲控制,如图 5.6(g) 和 (h) 所示 [148]。STP-LTP 转变是由质子 (H$^+$) 掺入 VO$_2$ 晶格的电化学反应引起的,它进一步调节了 VO$_2$ 的电导率,而且这种掺入的 H$^+$ 是不挥发的。在 STP 中,PPF 是描述重复刺激时突触后电位增强的一个基本特征。如图 5.6(i) 所示,利用两个重复的栅极脉冲,这一过程已经在一个 VO$_2$ 突触晶体管中被成功模拟。Mou 等的工作实现了使用 TPT-RAM 的自动网络修剪,如图 5.6(j) 所示。图 5.6(k) 和 (l) 显示了在多层感知器和卷积神经网络中模拟结果与用传统非易失性突触制作的基线之间的比较 [122]。

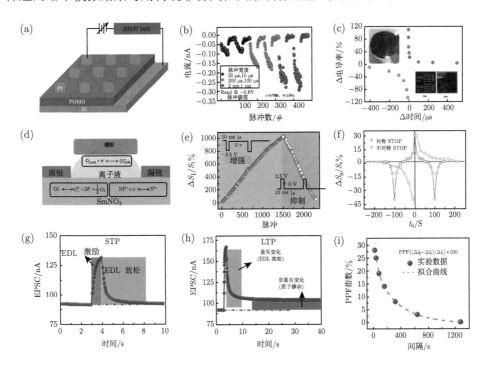

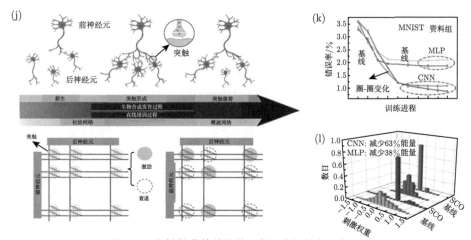

图 5.6　突触的莫特绝缘体 (彩图请扫封底二维码)

(a) 器件结构图；(b) 器件的连续增强和抑制特性 (经许可转载自文献 [146])．©IEEE 2012；(c) 1k 位 RRAM 阵列的 STDP 特性 (经许可转载自文献 [146])．©IEEE 2012；(d) 建议的电阻调制机制 (经许可转载自参考文献 [147])．© Nature 2013；(e) SmNiO$_3$ 的片状电导率调制 (经许可转载自参考文献 [147])．© Nature 2013；(f) 不对称和对称的突触尖峰可塑性突触功能的展示 (经许可转载自参考文献 [147])．© Nature 2013；(g) 基于 VO$_2$ 的突触的 STP 行为 [148]；(h) 基于 VO$_2$ 的突触的 LTP 行为 [148]；(i) PPF 指数 [148]；(j) 基于 SCO 的 TPT-RAM 突触的在线训练对应于人脑中突触发展的选择性稳定；(k) 不同网络模型在训练时的错误率；(l) 不同网络模型的能量减少情况

5.4.2　选择器和神经元的挥发性特性

1. 选通管

在 5.3 节的最后，我们详细分析了莫特忆阻器的不同开关性能。易失性的开关性能可以作为选通管使用，其中，选通性、阈值电压 (V_{th}) 和均匀性是实现高性能选择器的关键。选通管是忆阻器阵列内通路漏电流问题的一个可能的解决方案。在电场刺激下，NbO$_2$ 在约 1081 K 发生 MIT，伴随着从杂乱的金红石结构到整齐金红石结构的一阶相变 [149]。早在 20 世纪 70 年代，Shin 等就报道了 NbO$_2$ 的场开关特性，它可以达到 10^3 的高开关比和小于 0.7 ns 的快速开关速度 [150]，这为制造具有超快开关速度的选通管奠定了良好的实验基础。当选通管被应用于 RRAM 阵列时，在窄读区会出现均匀性差的情况 [151]。当选通管应用于振荡神经元时，低均匀性将导致振荡频率的不确定性 [152]。在 Zhao 等的报告中 [153]，研究比较了 Pt/NbO$_x$/TiN 器件和 Pt/NbO$_x$/Ru，他们发现后者具有更好的选通和均匀性，如图 5.7(a)~(c) 所示，这一结果可能是由于 RuO$_2$ 的形成，它可以抑制电极的氧化作用。此外，电场由厚度决定，V_{th} 取决于开关膜的厚度 [154]。较小的 V_{th} 对于人工神经元的成功至关重要。在 Zhang 等 [155] 的报告中，研究人员制备了 Pt/Nb$_2$O$_5$/NbO$_2$ 器件，并证明随着 NbO$_2$ 薄膜厚度的降低，该器件的 V_{th} 也在降低，如图 5.7(d) 所示。降低 V_{th} 电压也是一个实现高性能选通管有效的手

段[154]。插入调谐层可以降低器件的 V_{th}[156,157],如 TiO_2 和 ZrO_2。Chen 等制备了 $Pt/NbO_x/ZrO_2/TiN$ 器件,发现该器件的 V_{th} 下降,如图 5.7(e) 所示;以及如图 5.7(f) 和 (g) 所示,选择性和均匀性有了明显的改善,其原因是 ZrO_2 层的储氧特性导致 NbO_2 薄膜变薄。总之,对于 NbO_2 来说,适当的开关层厚度可以努力提高性能指标。VO_2 是另一种用于选通管的典型莫特绝缘体。许多结构可以实现如图 5.7(h)~(j) 所示的阈值开关行为[157-160]。值得注意的是,在不同的顶部电极下,器件的 V_{th} 是不同的,如图 5.7(h) 和 (i) 所示。$Pt/VO_2/Pt$ 器件具有快速开关速度和出色的开关均匀性,如图 5.7(j) 和 (k) 所示。图 5.7(l)~(n) 显示,在 $SiO_2/ITO/VO_x/Pt$ 结构下,与原位退火薄膜相比,快速热退火的复合 VO_2 薄

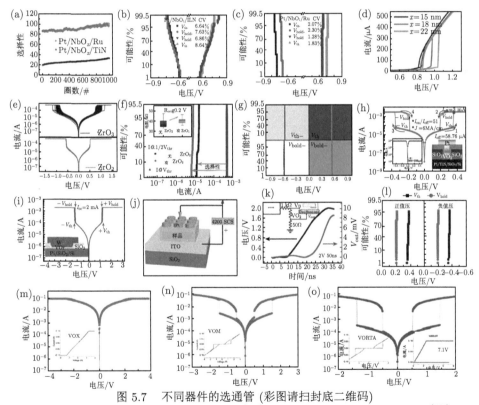

图 5.7 不同器件的选通管 (彩图请扫封底二维码)

(a) $Pt/NbO_x/Ru$ 和 $Pt/NbO_x/TiN$ 的选择性分布[153]; (b) $Pt/NbO_x/TiN$ 的开关电压分布[153]; (c) $Pt/NbO_x/Ru$ 的开关电压分布[153]; (d) 不同厚度的 I-V 曲线 (参考文献 [157]。©AIP Publishing 2021); (e) 有 ZrO_2 和无 ZrO_2 的 I-V 特性; (f) 含 ZrO_2 和不含 ZrO_2 的选择性 (参考文献 [157]。©AIP Publishing 2021); (g) 含 ZrO_2 和不含 ZrO_2 的均匀性; (h) 纳米器件的 I-V 特性[158]; (i) $Pt/VO_2/W$ 的 I-V 特性[159]; (j) 该器件的示意图 (经许可转载自文献 [160]。©RSC Publishing 2019); (k) 器件的开关速度[158]; (l) 纳米级 $Pt/VO_2/Pt$ 器件的 V_{th} 和 V_{hold} 的分布 (经许可转载自文献 [160]。©RSC Publishing 2019); (m) VO_x 的 I-V 曲线; (n) VOM 的 I-V 曲线 (经许可转载自参考文献 [160]©RSC Publishing 2019); (o) VORTA 的 I-V 曲线 (经许可转载自参考文献 [160]。©RSC Publishing 2019)

膜表现出更大的 I_{ON}/I_{OFF} 比,与 VO_x 相比,VORTA 和 VOM 显示出双极阈值开关特性[160]。Xue 等报告了通过电场诱导氧离子迁移在 V_2O_5 薄膜中实现一维 VO_2 纳米通道的金属-绝缘体转变,存在一个电激活过程[161],该器件具有 17 ns 的快速开关速度,8 pJ 的低能量和低变化性。

莫特绝缘体可被用作交叉点非易失性存储器阵列中的选通器件,例如用于 RRAM 或 PCRAM。在交叉点阵列中,由于其固有的结构,一个不可避免的潜行电流路径产生了。这种潜行电流流经附近未被选择的单元,导致读取干扰[162]。为了解决这个问题,基于选择器的莫特绝缘体具有突然的非线性 I-V 特性,以整流半选择单元的电流。图 5.8(a) 显示了 1S1R 的三维堆叠结构[163]。将 HfO_2 电阻随机存取存储器与选通管连接起来,可以组成选通管–存储器 (1S1R) 配置,图 5.8(b) 是 1S1R 的横截面 SEM 图像。图 5.8(c) 显示了 1S1R 结构的 I-V 特性。与没有选择器的对照 RRAM 相比,1S1R 表现出未选择单元 ($\pm 1/2V_{read}$ 区域) 泄漏电流的显著减少。图 5.8(d) 显示了 10^8 个周期的超强耐久性[161]。

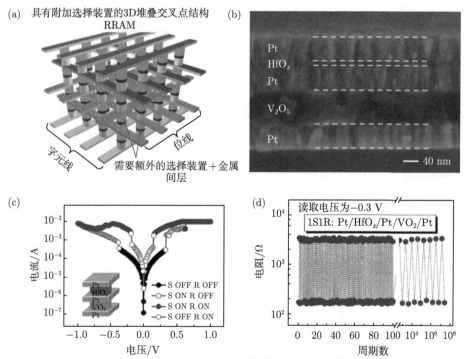

图 5.8　(a) 1S1R 的三维叠加交叉点结构[163];(b) 横截面的 SEM 图像;(c) $Pt/HfO_2/Pt/VO_2/Pt$ 1S1R 结构的 I-V 特性 (经许可转载自参考文献 [161]。©John Wiley and Sons 2017);(d) 1S1R 结构的耐久性能 (经许可转载自参考文献 [161]。©John Wiley and Sons 2017)

2. 神经元

神经元的功能是处理来自突触的输入信号，然后决定是否将该信号发送到邻近的神经元[164]。目前，主要有三种常见的神经元模型：霍奇金-赫胥黎 (HH) 模型、整合与发射 (IF) 模型以及漏整合与发射 (LIF) 模型。这三种模型各有其优点和缺点。例如，HH 模型准确地模拟了膜电压的生物特性，可以很好地与生物神经元的电生理实验结果相一致，然而，其电路结构复杂，往往需要多个器件之间的合作。为了简化电路，有人提出用中频和 LIF 模型来完成神经元的整合和发射功能。然而，这将降低信号的准确性和生物模拟的相似性。与 IF 模型相比，LIF 模型更接近真实的生物神经元，这里 L 表示泄漏。因为细胞膜在膜内和膜外不断地交换离子，当只有一个输入时，电压会自动泄漏并逐渐恢复到静止状态。因此，我们将使用 HH 模型和 LIF 模型进行演示。

HH 模型：Pickett 等提出了一个 HH 模型电路，它有两个相同的 NbO_2 选择器 (图 5.9(b))，每个选择器有一个并联电容，如图 5.9(a) 所示[164]。在这个模型中，两个通道以相反的极性电压通电，以模拟生物神经元中的 Na^+ 和 K^+ 通道，该模型有三种电阻类型，名为输入电阻、输出电阻和负载电阻。为了解释神经元的运行机制，这里通过模拟两个不同的输入脉冲，展示了该模型的全或无动作电位响应。在开始时，由于选择器处于非状态电阻状态，两个通道的电压都低于 V_{th}；当电容的两个通道上的电位高于选择器的 V_{th} 时，电路将被触发并发射一个尖峰；根据实验结果，输出的尖峰信号与生物系统相当，这可以模仿神经元的全或无功能，如图 5.9(c) 所示。在进一步的研究中，Yi 等[14]使用三种类型的电路与两个选通管实现了 23 个生物神经元的尖峰行为，如图 5.9(d) 所示。所有的行为都是由一个单一的强直、相位或混合模式的神经元电路测得的，该电路仅由 2 个 VO_2 忆阻器和 4 个或 5 个电阻、电容元件组成，这为在器件水平上实现类似大脑的神经计算开辟了一条道路。

LIF 神经元模型：该模型主要介绍了神经元的整合和发射过程。神经元系统由两个电阻、一个电容和一个选通管组成，如图 5.10(a) 所示[15]。该电路分为两个电路：充电回路 (CL) 和放电回路 (DL)。当施加一系列的电压脉冲时，电流主要流经带有电容器的支路，电荷在电容器中积累。有一个关键的电路参数：电阻和电容 (RC) 的时间常数由回路的电阻和电容决定。由于器件在开始时处于 HRS，DL 的 RC 时间常数明显大于 CL 的 RC 时间常数，表明通过选择器的电流泄漏是可以忽略不计的。随着电容电荷的逐渐积累，第 2 点的电压逐渐增加并超过了选择器的 V_{th}。然后，选择器的电阻迅速下降，DL 的 RC 时间常数明显小于 CL 的。在这种情况下，电容器中的电荷将放电，输出一个尖峰，如图 5.10(b) 所示[165]。近年来，基于莫特绝缘体的 LIF 神经元已经模拟了一些神经元的行为，

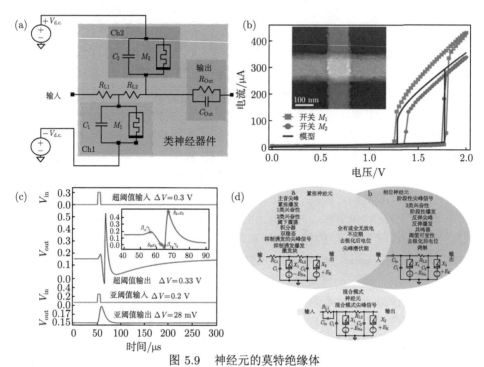

图 5.9 神经元的莫特绝缘体

(a) 神经元电路图 (经许可转载自文献 [164]。©Nature 2013); (b) 两个 NbO_2 器件基于神经元的双稳态 I-V 曲线和 SEM 照片 (经许可转载自参考文献 [164]。©Nature 2013); (c) 全或无反应的神经元和状态变量动态 (转载自参考文献 [164]。©Nature 2013); (d) 神经元功能和应用电路图 (经许可转载自参考文献 [14]。©Nature 2018)

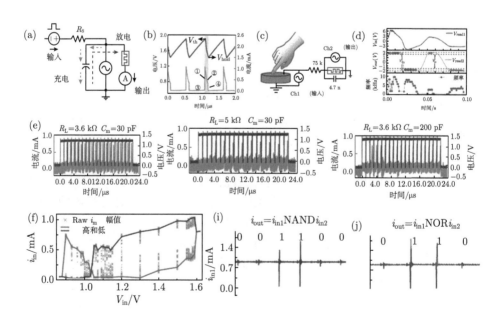

5.4 基于莫特绝缘体的忆阻器应用

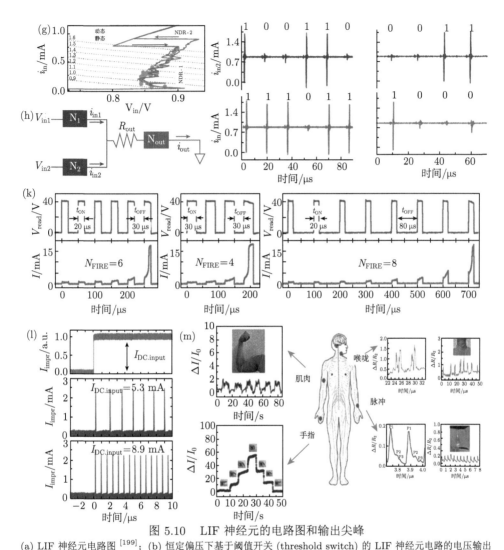

图 5.10 LIF 神经元的电路图和输出尖峰

(a) LIF 神经元电路图 [199]；(b) 恒定偏压下基于阈值开关 (threshold switch) 的 LIF 神经元电路的电压输出波形和电流尖峰 [199]；(c) 人工突发机械感受器系统 (ASMS) 的示意图 (经许可转载自文献 [166])。© Nature 2020；(d) ASMS 的实验结果 (经许可转载自参考文献 [166])。© Nature 2020；(e) 通过改变电路参数而实现不同的输出尖峰 (经许可转载自参考文献 [167])。© Nature 2020；(f) 振荡包络作为 V_{in} 的一个函数 (经许可转载自参考文献 [168])。©Spring Nature 2017；(g) 静态和动态数据的比较 (经许可转载自参考文献 [168])。©Spring Nature 2017；(h) 三个神经形态元素的网络示意图 (经许可转载自参考文献 [169])。©Spring Nature 2020；(i) NAND 操作 (经许可转载自参考文献 [169])。© Spring Nature 2020；(j) NOR 操作 (经许可转载自参考文献 [169])。© Spring Nature 2020；(k) 应用连续脉冲的输出尖峰 (经许可转载自参考文献 [165])。© Wiley-VCH 2017；(l) 用一连串脉冲刺激神经元时的电流与时间的关系 (经许可转载自参考文献 [171])。© Nature 2020；(m) 通过集成在人体不同位置的电子皮肤来监测生理信号和身体的运动 [172]

如全有或全无反应、阈值驱动的尖峰、折返期和强度调制的频率响应 [166]。更重要的是，Zhang 等在 LIF 模型的基础上设计了一个集成了感觉和计算能力的无电源尖峰机械感受器系统。这个系统包括一个压电装置和一个 LIF 神经元,如图 5.10(c) 所示。当对压电装置施加压力时,在顶部电极上产生一个正电压;当压力消除时,则产生一个负电压。输入波形随压力的变化而变化,其反应如图 5.10(d) 所示。当向压电装置施加一个力时,会产生一个类似正弦的信号,该信号被神经元处理并转化为尖峰信号。传入神经在未来开发自我感知的神经机器人方面很有前景。基于图 5.10(a),许多研究表明,可以通过改变突触 RC 等电路参数来控制发射进度,如图 5.10(e) 所示,较小的突触 RC 可以加快整合过程,触发更快的输出尖峰,这就像一个生物神经元[167]。此外,NbO_2 忆阻器可以用于弛豫振荡器。图 5.10(f) 显示了振荡的三重分叉,通过比较不同的 V_{in} 值与器件的静态 i_m-V_m 曲线相交的电压,结果表明准静态和动态测量之间的一致性非常好,如图 5.10(g) 所示[168]。此外,一个由三阶元件和一个电阻耦合器组成的简单网络可以用来创建逻辑门,如图 5.10(h) 所示,利用耦合电阻的差异可以实现与非 (NAND) 和或非 (NOR) 操作,如图 5.10(i) 和 (j) 所示[169]。类似的功能也已经用 AM_4Q_8 和 VO_2 实现[170]。AM4Q8 是窄间隙莫特绝缘体,莫特–哈伯德间隙在 0.1~0.3 eV,它可以通过强相关的窄间隙莫特绝缘体受电脉冲影响的行为而实现 LIF[165]。图 5.10(k) 显示了在不同的 t_{ON} 和 t_{OFF} 下的点火过程,通过调整 t_{ON} 和 t_{OFF},达到点火的电压脉冲数会发生变化[165]。同样地,VO_2 可以利用热传导来模拟神经元的基本功能,图 5.10(l) 显示了速率编码功能,即强刺激产生高频率棘波,而弱刺激产生较慢的模式[171]。重要的是,它可以进一步模拟人形机器人的潜在适用性,如图 5.10(m) 所示[172]。

5.5 小　　结

在本章中,我们回顾了莫特 IMT (BC-IMT 和 FC-IMT) 的机制,莫特 MIT 的驱动方法以及莫特 MIT 在忆阻器、选通器和神经形态计算中的应用。但是,莫特绝缘体材料存在缺点,例如,难以制备薄膜和有争议的机制。下面,我们从几个方面对莫特绝缘体材料进行了展望。

1. 薄膜制备

由于单晶的热力学相位不稳定和多晶结构,莫特绝缘体材料薄膜很难制备,而且单晶薄膜需要后续的退火步骤[160,173,174]。一个合适的制备过程需要进一步优化,因为每种制备方法都有优点和缺点。例如,脉冲激光沉积法可以更好地控制薄膜成分的化学计量比例,具有高沉积率、低基底温度和均匀的薄膜[175,176],然而,它可能需要相对高能量的激光照射,并且形成的薄膜面积小[177];磁控溅射方法对于均匀的薄膜很方便,有很大的可重复性[178],然而,它需要很长的制备

5.5 小结

时间，而且用这种方法时控制薄膜的化学计量比是困难的[179]。应该根据实际应用要求而选择合适的方法。

2. 有争议的机制

关于电子器件中莫特 IMT 的驱动机制仍然没有达成共识，对于电场[180,181]和焦耳加热[182]过程中哪种过程占主导地位，说法相互矛盾。探讨莫特材料在电场中 IMT 的原因至关重要。许多技术已经被用来探索驱动机制。例如，有些工作认为，例如，有些工作通过红外和 X 射线光谱显微镜技术证明驱动机制是焦耳加热过程[183−185]；还有一些工作使用高性能示波器获得非常精细的时间分辨率数据，这可以证明电场是主导过程[186,187]。然而，需要开发新的技术来消除其他因素的影响，确保每次只有一个变量在变化。

3. 电子器件的性能优化

对于莫特绝缘体，由于过渡金属通常有多个价态和不同的化学计量比，可能会有多个晶相，这使得忆阻器具有较小的开关比 (10^2)[155,158,188]。对于数据存储、选择器或神经形态计算来说，改善开关比是很重要的。改善电子器件的开关比的方法之一是提高莫特绝缘体的结晶度并抑制异相的形成。另一种方法是在莫特绝缘体中掺入适当的掺杂剂，这将提供拉伸应力或拉伸应力以增强异相成核。通过在 VO_2 薄膜中掺入 Cr，Zou 等将开关比从 $10^2 \sim 10^3$ 提高到 $10^4 \sim 10^5$ 两个数量级[189]。器件的开关速度直接决定了计算系统的计算速度。在光学激发下，许多莫特绝缘体的开关速度可以达到 $10^2 \sim 10^3$ fs[190]。在不同的工作中，开关速度有很大的变化。有些器件只能达到 0.4 μs 的慢速开关速度[188]，而有些工作可以实现 700 ps 的快速开关速度和 100 fJ 的能量消耗[191]，莫特绝缘体通常具有小的带隙，这导致器件中的泄漏电流更高[157,192]。通过从外部向这种绝缘体掺入其他元素，可以改变其能带结构，例如，利用电负性比镍低的小离子半径实现对镍离子的电子掺杂，可以诱发小带隙，实现较大的电导率调制[47,147]。一些莫特绝缘体的相变温度接近室温，这将面临着由工作环境温度升高而导致器件失效的风险。在不同的应用场景中，研究人员可以通过掺入不同价态的原子来调整相变温度。对于二氧化硅，高价金属离子 (如 W^{6+} 或 Nb^{5+}) 可以用来降低转变温度[105]，而低价金属离子 (如 Cr^{3+}) 可以提高转变温度[93]。对于 $RNiO_3$，IMT 温度随着 R^{3+} 原子的离子半径增加而降低[70]。因为莫特绝缘体的相变会改变材料的体积，这将导致界面上的拉应力或压应力产生空位。这也是导致器件重复性能差的一个重要因素。

4. 应用前景

莫特绝缘体材料在数据存储和神经形态计算方面已显示出良好的前景[54,164,193,194]。然而，对于下一代集成存储和计算系统来说，它仍然处于起步阶段。在神经形态计算中使用莫特绝缘体材料，有一些难点需要被解决。例如，莫特绝缘体与传统 CMOS 工艺的兼容性、集成器件的问题、模拟更多的神经元和突触功能，以及电–突触和电–神经元之间的级联问题[14,195]。近来，莫特绝缘体材料的忆阻器有望识别图像和处理信息[196-198]。莫特绝缘体忆阻器是构建尖峰神经网络的器件之一，由于其在时空领域丰富的神经动力学特性、多样化的编码机制和事件驱动的优势，引起了学者们的关注[166,199]。未来，莫特绝缘体可能成为基于人工智能芯片的尖峰神经网络的最重要材料，以模仿人脑实现信息存储和认知。

参 考 文 献

[1] Wong H S P, Salahuddin S. Memory leads the way to better computing[J]. Nature Nanotechnology, 2015, 10(3): 191-194.

[2] Sugie T, Akamatsu T, Nishitsuji T, et al. High-performance parallel computing for next-generation holographic imaging[J]. Nature Electronics, 2018, 1(4): 254-259.

[3] Zhang X, Huang A, Hu Q, et al. Neuromorphic computing with memristor crossbar[J]. Physica Status Solidi (a), 2018, 215(13): 1700875.

[4] Yuan L, Liu S, Chen W, et al. Organic memory and memristors: from mechanisms, materials to devices[J]. Advanced Electronic Materials, 2021, 7(11): 2100432.

[5] Wang X Y, Zhou P F, Eshraghian J K, et al. High-density memristor-CMOS ternary logic family[J]. IEEE Transactions on Circuits and Systems I: Regular Papers, 2020, 68(1): 264-274.

[6] Hou P, Wang J, Zhong X, et al. A ferroelectric memristor based on the migration of oxygen vacancies[J]. RSC Advances, 2016, 6(59): 54113-54118.

[7] Davies M, Srinivasa N, Lin T H, et al. Loihi: A neuromorphic manycore processor with on-chip learning[J]. IEEE Micro, 2018, 38(1): 82-99.

[8] Ham S, Choi S, Cho H, et al. Photonic organolead halide perovskite artificial synapse capable of accelerated learning at low power inspired by dopamine-facilitated synaptic activity[J]. Advanced Functional Materials, 2019, 29(5): 1806646.

[9] Yan X, Qin C, Lu C, et al. Robust Ag/ZrO$_2$/WS$_2$/Pt memristor for neuromorphic computing[J]. ACS Applied Materials & Interfaces, 2019, 11(51): 48029-48038.

[10] Wang M, Cai S, Pan C, et al. Robust memristors based on layered two-dimensional materials[J]. Nature Electronics, 2018, 1(2): 130-136.

[11] Syu Y E, Chang T C, Lou J H, et al. Atomic-level quantized reaction of HfO$_x$ memristor[J]. Applied Physics Letters, 2013, 102(17): 172903.

[12] Ying J, Wang G, Dong Y, et al. Switching characteristics of a locally-active memristor with binary memories[J]. International Journal of Bifurcation and Chaos, 2019, 29(11): 1930030.

[13] Zhang X, Wang W, Liu Q, et al. An artificial neuron based on a threshold switching memristor[J]. IEEE Electron Device Letters, 2017, 39(2): 308-311.

[14] Yi W, Tsang K K, Lam S K, et al. Biological plausibility and stochasticity in scalable VO$_2$ active memristor neurons[J]. Nature Communications, 2018, 9(1): 1-10.

[15] Zhu J, Wu Z, Zhang X, et al. A flexible LIF neuron based on NbO$_x$ memristors for neural interface applications[C]//2021 5th IEEE Electron Devices Technology & Manufacturing Conference (EDTM). IEEE, 2021: 1-3.

[16] Yan L, Pei Y, Wang J, et al. High-speed Si films based threshold switching device and its artificial neuron application[J]. Applied Physics Letters, 2021, 119(15): 153507.

[17] Yan X, Zhou Z, Zhao J, et al. Flexible memristors as electronic synapses for neuro-inspired computation based on scotch tape-exfoliated mica substrates[J]. Nano Research, 2018, 11(3): 1183-1192.

[18] Yan X, Pei Y, Chen H, et al. Self-assembled networked PbS distribution quantum dots for resistive switching and artificial synapse performance boost of Memristors[J]. Advanced Materials, 2019, 31(7): 1805284.

[19] Yu T, He F, Zhao J, et al. Hf$_{0.5}$Zr$_{0.5}$O$_2$-based ferroelectric memristor with multilevel storage potential and artificial synaptic plasticity[J]. Science China Materials, 2021, 64(3): 727-738.

[20] Zhao H, Liu Z, Tang J, et al. Memristor-based signal processing for edge computing[J]. Tsinghua Science and Technology, 2021, 27(3): 455-471.

[21] Zhang Y, Huang P, Gao B, et al. Oxide-based filamentary RRAM for deep learning[J]. Journal of Physics D: Applied Physics, 2020, 54(8): 083002.

[22] Yan X, He H, Liu G, et al. Robust memristor based in epitaxy vertically aligned nanostructured BaTiO$_3$-CeO$_2$ films on silicon[J]. Advanced Materials, 2022, 34(23): 2110343.

[23] Li Y, Zhang C, Shi Z, et al. Recent advances on crystalline materials-based flexible memristors for data storage and neuromorphic applications[J]. Science China Materials, 2022, 65(8): 2110-2127.

[24] Li Y, Zhang C, Ling S, et al. Toward highly robust nonvolatile multilevel memory by fine tuning of the nanostructural crystalline solid-state order[J]. Small, 2021, 17(19): 2100102.

[25] Zhang C, Li Y, Ma C, et al. Recent progress of organic-inorganic hybrid perovskites in RRAM, artificial synapse, and logic operation[J]. Small Science, 2022, 2(2): 2100086.

[26] Bethe H. Theorie der beugung von elektronen an kristallen[J]. Annalen der Physik, 1928, 392(17): 55-129.

[27] Wilson A H. The theory of electronic semi-conductors[J]. Proceedings of the Royal Society of London. Series A, Containing Papers of a Mathematical and Physical Character, 1931, 133(822): 458-491.

[28] Sommerfeld A. Zur elektronentheorie der metalle auf grund der fermischen statistik[J]. Zeitschrift für Physik, 1928, 47(1): 1-32.

[29] Barlow H M. LIII. A criticism of the electron theory of metals[J]. The London, Edinburgh, and Dublin Philosophical Magazine and Journal of Science, 1929, 7(43): 459-470.

[30] Mott N F. The basis of the electron theory of metals, with special reference to the transition metals[J]. Proceedings of the Physical Society. Section A, 1949, 62(7): 416.

[31] de Boer J H, Verwey E J W. Semi-conductors with partially and with completely filled 3d-lattice bands[J]. Proceedings of the Physical Society (1926-1948), 1937, 49(4S): 59.

[32] Anderson P W. The resonating valence bond state in La_2CuO_4 and superconductivity[J]. Science, 1987, 235(4793): 1196-1198.

[33] Yamauchi T, Hirata Y, Ueda Y, et al. Pressure-induced Mott transition followed by a 24-K superconducting phase in $BaFe_2S_3$[J]. Physical Review Letters, 2015, 115(24): 246402.

[34] Lee H S, Choi S G, Park H H, et al. A new route to the Mott-Hubbard metal-insulator transition: Strong correlations effects in $Pr_{0.7}Ca_{0.3}MnO_3$[J]. Scientific Reports, 2013, 3(1): 1-5.

[35] Zhang K H L, Du Y, Sushko P V, et al. Hole-induced insulator-to-metal transition in $La_{1-x}Sr_xCrO_3$ epitaxial films[J]. Physical Review B, 2015, 91(15): 155129.

[36] Sohn J I, Joo H J, Ahn D, et al. Surface-stress-induced Mott transition and nature of associated spatial phase transition in single crystalline VO_2 nanowires[J]. Nano Letters, 2009, 9(10): 3392-3397.

[37] Shabalin A G, Del Valle J, Hua N, et al. Nanoscale imaging and control of volatile and non-volatile resistive switching in VO_2[J]. Small, 2020, 16(50): 2005439.

[38] Bombardi A, de Bergevin F, Di Matteo S, et al. Precursor symmetry breaking in Cr doped V_2O_3[J]. Physica B: Condensed Matter, 2004, 345(1-4): 40-44.

[39] Metcalf P A, Guha S, Gonzalez L P, et al. Electrical, structural, and optical properties of Cr-doped and non-stoichiometric V_2O_3 thin films[J]. Thin Solid Films, 2007, 515(7-8): 3421-3425.

[40] Kuwamoto H, Honig J M. Electrical properties and structure of Cr-doped nonstoichiometric V_2O_3[J]. Journal of Solid State Chemistry, 1980, 32(3): 335-342.

[41] Cha E, Park J, Woo J, et al. Comprehensive scaling study of NbO_2 insulator-metal-transition selector for cross point array application[J]. Applied Physics Letters, 2016, 108(15): 153502.

[42] Stoever J, Boschker J E, Bin Anooz S, et al. Approaching the high intrinsic electrical resistivity of NbO_2 in epitaxially grown films[J]. Applied Physics Letters, 2020, 116(18): 182103.

[43] Saitoh S, Kinoshita K. Oxide-based selector with trap-filling-controlled threshold switching[J]. Applied Physics Letters, 2020, 116(11): 112101.

[44] Abbas H, Ali A, Jung J, et al. Reversible transition of volatile to non-volatile resistive switching and compliance current-dependent multistate switching in IGZO/MnO RRAM devices[J]. Applied Physics Letters, 2019, 114(9): 093503.

[45] Kuan M C, Yang F W, Cheng C M, et al. Bipolar switching properties of the manganese oxide thin film RRAM devices[J]. Key Engineering Materials, 2014, 602-603: 1056-1059.

[46] Wang Y, Lv Z, Zhou L, et al. Emerging perovskite materials for high density data

storage and artificial synapses[J]. Journal of Materials Chemistry C, 2018, 6(7): 1600-1617.

[47] Oh C, Heo S, Jang H M, et al. Correlated memory resistor in epitaxial NdNiO$_3$ heterostructures with asymmetrical proton concentration[J]. Applied Physics Letters, 2016, 108(12): 122106.

[48] Medarde M, Fernández-Díaz M T, Lacorre P. Long-range charge order in the low-temperature insulating phase of PrNiO$_3$[J]. Physical Review B, 2008, 78(21): 212101.

[49] Zhang K, Si C, Lian C S, et al. Mottness collapse in monolayer 1T-TaSe$_2$ with persisting charge density wave order[J]. Journal of Materials Chemistry C, 2020, 8(28): 9742-9747.

[50] Chowdhury S, Jana A, Mandal A K, et al. Electronic phase switching in the negative charge transfer energy SrCoO$_x$ thin films with the mottronic relevancies[J]. ACS Applied Electronic Materials, 2021, 3(7): 3060-3071.

[51] Wu J, Guo Y, Liu H, et al. Room-temperature ligancy engineering of perovskite electrocatalyst for enhanced electrochemical water oxidation[J]. Nano Research, 2019, 12(9): 2296-2301.

[52] Kuneš J, Anisimov V I. Various scenarios of metal-insulator transition in strongly correlated materials[J]. Annalen der Physik, 2011, 523(8-9): 682-688.

[53] Shao Z, Cao X, Luo H, et al. Recent progress in the phase-transition mechanism and modulation of vanadium dioxide materials[J]. NPG Asia Materials, 2018, 10(7): 581-605.

[54] Wang Y, Kang K M, Kim M, et al. Mott-transition-based RRAM[J]. Materials Today, 2019, 28: 63-80.

[55] Mita Y, Izaki D, Kobayashi M, et al. Pressure-induced metallization of MnO[J]. Physical Review B, 2005, 71(10): 100101.

[56] Patterson J R, Aracne C M, Jackson D D, et al. Pressure-induced metallization of the Mott insulator MnO[J]. Physical Review B, 2004, 69(22): 220101.

[57] Shorikov A O, Pchelkina Z V, Anisimov V I, et al. Orbital-selective pressure-driven metal to insulator transition in FeO from dynamical mean-field theory[J]. Physical Review B, 2010, 82(19): 195101.

[58] Camjayi A, Acha C, Weht R, et al. First-order insulator-to-metal Mott transition in the paramagnetic 3D system GaTa$_4$Se$_8$[J]. Physical Review Letters, 2014, 113(8): 086404.

[59] Frandsen B A, Liu L, Cheung S C, et al. Volume-wise destruction of the antiferromagnetic Mott insulating state through quantum tuning[J]. Nature Communications, 2016, 7(1): 1-8.

[60] Yoo C S, Maddox B, Klepeis J H P, et al. First-order isostructural Mott transition in highly compressed MnO[J]. Physical Review Letters, 2005, 94(11): 115502.

[61] Rueff J P, Mattila A, Badro J, et al. Electronic properties of transition-metal oxides under high pressure revealed by X-ray emission spectroscopy[J]. Journal of Physics: Condensed Matter, 2005, 17(11): S717.

[62] Lee S, Bock J A, Trolier-McKinstry S, et al. Ferroelectric-thermoelectricity and Mott

transition of ferroelectric oxides with high electronic conductivity[J]. Journal of the European Ceramic Society, 2012, 32(16): 3971-3988.

[63] Knittle E, Jeanloz R, Mitchell A C, et al. Metallization of $Fe_{0.94}O$ at elevated pressures and temperatures observed by shock-wave electrical resistivity measurements[J]. Solid State Communications, 1986, 59(7): 513-515.

[64] Craco L, Leoni S. Pressure-induced orbital-selective metal from the Mott insulator $BaFe_2Se_3$[J]. Physical Review B, 2020, 101(24): 245133.

[65] Phuoc V T, Vaju C, Corraze B, et al. Optical conductivity measurements of $GaTa_4Se_8$ under high pressure: Evidence of a bandwidth-controlled insulator-to-metal Mott transition[J]. Physical Review Letters, 2013, 110(3): 037401.

[66] Otsuka A M, Silva Jr R S, dos Santos C, et al. Effect of chemical and hydrostatic pressures on the structural and mechanical properties of orthorhombic rare-earth $RNiO_3$[J]. Computational Materials Science, 2021, 197: 110691.

[67] Torrance J B, Lacorre P, Nazzal A I, et al. Systematic study of insulator-metal transitions in perovskites $RNiO_3$ (R= Pr, Nd, Sm, Eu) due to closing of charge-transfer gap[J]. Physical Review B, 1992, 45(14): 8209.

[68] Lupi S, Baldassarre L, Mansart B, et al. A microscopic view on the Mott transition in chromium-doped V_2O_3[J]. Nature Communications, 2010, 1(1): 1-7.

[69] Corraze B, Janod E, Cario L, et al. Electric field induced avalanche breakdown and non-volatile resistive switching in the Mott Insulators AM_4Q_8[J]. The European Physical Journal Special Topics, 2013, 222(5): 1046-1056.

[70] Catalano S, Gibert M, Fowlie J, et al. Rare-earth nickelates $RNiO_3$: Thin films and heterostructures[J]. Reports on Progress in Physics, 2018, 81(4): 046501.

[71] Rao C N R, Ramdas S, Loehman R E, et al. Semiconductor-metal transition in Ti_3O_5[J]. Journal of Solid State Chemistry, 1971, 3(1): 83-88.

[72] Master R, Choudhary R J, Phase D M. Effect of silver addition on structural, electrical and magnetic properties of Fe_3O_4 thin films prepared by pulsed laser deposition[J]. Journal of Applied Physics, 2012, 111(7): 073907.

[73] Schwingenschlögl U, Eyert V. The vanadium Magnéli phases V_nO_{2n-1}[J]. Annalen der physik, 2004, 13(9): 475-510.

[74] Honig J M, Wahnsiedler W E, Banus M D, et al. Resistivity, magnetoresistance, and hall effect studies in$VO_x(0.82 \leqslant x \leqslant 1.0)$[J]. Journal of Solid State Chemistry, 1970, 2(1): 74-77.

[75] del Valle J, Kalcheim Y, Trastoy J, et al. Electrically induced multiple metal-insulator transitions in oxide nanodevices[J]. Physical Review Applied, 2017, 8(5): 054041.

[76] Marezio M, McWhan D B, Dernier P D, et al. Structural aspects of the metal-insulator transition in V_4O_7[J]. Journal of Solid State Chemistry, 1973, 6(3): 419-429.

[77] Basu R, Sardar M, Dhara S. Origin of phase transition in VO_2[C]//AIP Conference Proceedings. AIP Publishing LLC, 2018, 1942(1): 030003.

[78] Tran R, Li X G, Ong S P, et al. Metal-insulator transition in V_2O_3 with intrinsic

defects[J]. Physical Review B, 2021, 103(7): 075134.

[79] Magnuson M, Butorin S M, Såthe C, et al. Spin transition in LaCoO$_3$ investigated by resonant soft X-ray emission spectroscopy[J]. Europhysics Letters, 2004, 68(2): 289.

[80] Music D, Krause A M, Olsson P A T. Theoretical and experimental aspects of current and future research on NbO$_2$ thin film devices[J]. Crystals, 2021, 11(2): 217.

[81] Scherwitzl R, Zubko P, Lezama I G, et al. Electric-field control of the metal-insulator transition in ultrathin NdNiO$_3$ films[J]. Advanced Materials, 2010, 22(48): 5517-5520.

[82] Yamanaka T, Hattori A N, Pamasi L N, et al. Effects of off-stoichiometry in the epitaxial NdNiO$_3$ film on the suppression of its metal-insulator-transition properties[J]. ACS Applied Electronic Materials, 2019, 1(12): 2678-2683.

[83] Wen H, Guo L, Barnes E, et al. Structural and electronic recovery pathways of a photoexcited ultrathin VO$_2$ film[J]. Physical Review B, 2013, 88(16): 165424.

[84] Wu J M, Liou L B. Room temperature photo-induced phase transitions of VO$_2$ nanodevices[J]. Journal of Materials Chemistry, 2011, 21(14): 5499-5504.

[85] Wei J, Wang Z, Chen W, et al. New aspects of the metal-insulator transition in single-domain vanadium dioxide nanobeams[J]. Nature Nanotechnology, 2009, 4(7): 420-424.

[86] Budai J D, Hong J, Manley M E, et al. Metallization of vanadium dioxide driven by large phonon entropy[J]. Nature, 2014, 515(7528): 535-539.

[87] Gorelov E, Karolak M, Wehling T O, et al. Nature of the Mott transition in Ca$_2$RuO$_4$[J]. Physical Review Letters, 2010, 104(22): 226401.

[88] Moon S J, Jin H, Choi W S, et al. Temperature dependence of the electronic structure of the J eff=1/2 Mott insulator Sr$_2$IrO$_4$ studied by optical spectroscopy[J]. Physical Review B, 2009, 80(19): 195110.

[89] Perfetti L, Georges A, Florens S, et al. Spectroscopic signatures of a bandwidth-controlled Mott transition at the surface of 1T-TaSe$_2$[J]. Physical Review Letters, 2003, 90(16): 166401.

[90] Chen A, Ma G, He Y, et al. Research on temperature effect in insulator–metal transition selector based on NbO$_x$ thin films[J]. IEEE Transactions on Electron Devices, 2018, 65(12): 5448-5452.

[91] Mattoni G, Zubko P, Maccherozzi F, et al. Striped nanoscale phase separation at the metal-insulator transition of heteroepitaxial nickelates[J]. Nature Communications, 2016, 7(1): 1-7.

[92] Limelette P, Georges A, Jérome D, et al. Universality and critical behavior at the Mott transition[J]. Science, 2003, 302(5642): 89-92.

[93] Meneghini C, Di Matteo S, Monesi C, et al. Antiferromagnetic-paramagnetic insulating transition in Cr-doped V$_2$O$_3$ investigated by EXAFS analysis[J]. Journal of Physics: Condensed Matter, 2009, 21(35): 355401.

[94] Janod E, Tranchant J, Corraze B, et al. Resistive switching in Mott insulators and correlated systems[J]. Advanced Functional Materials, 2015, 25(40): 6287-6305.

[95] Goodenough J B. The two components of the crystallographic transition in VO$_2$[J].

Journal of Solid State Chemistry, 1971, 3(4): 490-500.
[96] Qazilbash M M, Brehm M, Chae B G, et al. Mott transition in VO_2 revealed by infrared spectroscopy and nano-imaging[J]. Science, 2007, 318(5857): 1750-1753.
[97] Haverkort M W, Hu Z, Tanaka A, et al. Orbital-assisted metal-insulator transition in VO_2[J]. Physical Review Letters, 2005, 95(19): 196404.
[98] Kim S, Kim K, Kang C J, et al. Correlation-assisted phonon softening and the orbital-selective Peierls transition in VO_2[J]. Physical Review B, 2013, 87(19): 195106.
[99] Yao T, Zhang X, Sun Z, et al. Understanding the nature of the kinetic process in a VO_2 metal-insulator transition[J]. Physical Review Letters, 2010, 105(22): 226405.
[100] Xiao H, Li Y, Fang B, et al. Voltage-induced switching dynamics based on an AZO/VO_2/AZO sandwiched structure[J]. Infrared Physics & Technology, 2017, 86: 212-217.
[101] Yoon J, Kim H, Mun B S, et al. Investigation on onset voltage and conduction channel temperature in voltage-induced metal-insulator transition of vanadium dioxide[J]. Journal of Applied Physics, 2016, 119(12): 124503.
[102] Mun B S, Yoon J, Mo S K, et al. Role of joule heating effect and bulk-surface phases in voltage-driven metal-insulator transition in VO_2 crystal[J]. Applied Physics Letters, 2013, 103(6): 061902.
[103] Mun B S, Chen K, Yoon J, et al. Nonpercolative metal-insulator transition in VO_2 single crystals[J]. Physical Review B, 2011, 84(11): 113109.
[104] Del Valle J, Salev P, Tesler F, et al. Subthreshold firing in Mott nanodevices[J]. Nature, 2019, 569(7756): 388-392.
[105] Ling C, Zhao Z, Hu X, et al. W doping and voltage driven metal-insulator transition in VO_2 nano-films for smart switching devices[J]. ACS Applied Nano Materials, 2019, 2(10): 6738-6746.
[106] Ramadoss K, Zuo F, Sun Y, et al. Proton-doped strongly correlated perovskite nickelate memory devices[J]. IEEE Electron Device Letters, 2018, 39(10): 1500-1503.
[107] Okada Y, Arima T, Tokura Y, et al. Doping-and pressure-induced change of electrical and magnetic properties in the Mott-Hubbard insulator $LaTiO_3$[J]. Physical Review B, 1993, 48(13): 9677.
[108] Katsufuji T, Taguchi Y, Tokura Y. Transport and magnetic properties of a Mott-Hubbard system whose bandwidth and band filling are both controllable: $R_{1-x}Ca_x$ $TiO_{3+y/2}$[J]. Physical Review B, 1997, 56(16): 10145.
[109] Okimoto Y, Katsufuji T, Okada Y, et al. Optical spectra in (La, Y) TiO_3: variation of Mott-Hubbard gap features with change of electron correlation and band filling[J]. Physical Review B, 1995, 51(15): 9581.
[110] Moetakef P, Cain T A. Metal-insulator transitions in epitaxial $Gd_{1-x}Sr_xTiO_3$ thin films grown using hybrid molecular beam epitaxy[J]. Thin Solid Films, 2015, 583: 129-134.
[111] Sage M H, Blake G R, Palstra T T M. Insulator-to-metal transition in(R, Ca)VO_3[J]. Physical Review B, 2008, 77(15): 155121.
[112] Zhang X, Zhang Y, Wang X M, et al. Transport and magnetic properties in the

Gd$_{1-x}$Ca$_x$VO$_3$ system[J]. Japanese Journal of Applied Physics, 2011, 50(10R): 101102.

[113] Ju H L, Sohn H C, Krishnan K M. Evidence for O 2p hole-driven conductivity in La$_{1-x}$Sr$_x$MnO$_3$($0 \leqslant x \leqslant 0.7$) and La$_{0.7}Sr_{0.3}MnO_z$ thin films[J]. Physical Review Letters, 1997, 79(17): 3230.

[114] Zhang Z, Schwanz D, Narayanan B, et al. Perovskite nickelates as electric-field sensors in salt water[J]. Nature, 2018, 553(7686): 68-72.

[115] Mineshige A, Kobune M, Fujii S, et al. Metal–insulator transition and crystal structure of La$_{1-x}$Sr$_x$CoO$_3$ as functions of Sr-content, temperature, and oxygen partial pressure[J]. Journal of Solid State Chemistry, 1999, 142(2): 374-381.

[116] Bringley J F, Scott B A, La Placa S J, et al. Structure and properties of the LaCuO$_3$-δ perovskites[J]. Physical Review B, 1993, 47(22): 15269.

[117] Xu H C, Zhang Y, Xu M, et al. Direct observation of the bandwidth control Mott transition in the NiS$_{2-x}$Se$_x$ multiband system[J]. Physical Review Letters, 2014, 112(8): 087603.

[118] Shi J, Zhou Y, Ramanathan S. Colossal resistance switching and band gap modulation in a perovskite nickelate by electron doping[J]. Nature Communications, 2014, 5(1): 1-9.

[119] Eyert V. The metal-insulator transitions of VO$_2$: A band theoretical approach[J]. Annalen der Physik, 2002, 514(9): 650-704.

[120] Cheng S, Lee M H, Li X, et al. Operando characterization of conductive filaments during resistive switching in Mott VO$_2$[J]. Proceedings of the National Academy of Sciences, 2021, 118(9): e2013676118.

[121] Kalcheim Y, Camjayi A, Del Valle J, et al. Non-thermal resistive switching in Mott insulator nanowires[J]. Nature Communications, 2020, 11(1): 1-9.

[122] Mou X, Tang J, Lyu Y, et al. Analog memristive synapse based on topotactic phase transition for high-performance neuromorphic computing and neural network pruning[J]. Science Advances, 2021, 7(29): eabh0648.

[123] Yan X, Zhang L, Chen H, et al. Graphene oxide quantum dots based memristors with progressive conduction tuning for artificial synaptic learning[J]. Advanced Functional Materials, 2018, 28(40): 1803728.

[124] Yan X, Zhao J, Liu S, et al. Memristor with Ag-cluster-doped TiO$_2$ films as artificial synapse for neuroinspired computing[J]. Advanced Functional Materials, 2018, 28(1): 1705320.

[125] Wang K, Li L, Zhao R, et al. A pure 2H-MoS$_2$ nanosheet-based memristor with low power consumption and linear multilevel storage for artificial synapse emulator[J]. Advanced Electronic Materials, 2020, 6(3): 1901342.

[126] Wang H, Yu T, Zhao J, et al. Low-power memristors based on layered 2D SnSe/graphene materials[J]. Science China Materials, 2021, 64(8): 1989-1996.

[127] Yan X, Cao G, Wang J, et al. Memristors based on multilayer graphene electrodes for implementing a low-power neuromorphic electronic synapse[J]. Journal of Materials

Chemistry C, 2020, 8(14): 4926-4933.

[128] Liu L, Zhao J, Cao G, et al. A memristor-based silicon carbide for artificial nociceptor and neuromorphic computing[J]. Advanced Materials Technologies, 2021, 6(12): 2100373.

[129] Zhao J, Sun Y, Lu W, et al. Realization of long retention properties of quantum conductance through confining the oxygen vacancy diffusion[J]. Applied Physics Reviews, 2022, 9(2): 021419.

[130] Park J W, Park J W, Yang M K, et al. Low-voltage resistive switching of polycrystalline $SrZrO_3$:Cr thin films grown on Si substrates by off-axis rf sputtering[J]. Journal of Vacuum Science & Technology A: Vacuum, Surfaces, and Films, 2006, 24(4): 970-973.

[131] Tranchant J, Janod E, Cario L, et al. Electrical characterizations of resistive random access memory devices based on GaV_4S_8 thin layers[J]. Thin Solid Films, 2013, 533: 61-65.

[132] Lee T J, Kim S K, Seong T Y. Sputtering-deposited amorphous $SrVO_x$-based memristor for use in neuromorphic computing[J]. Scientific Reports, 2020, 10(1): 1-9.

[133] Lim H, Cho H, Kim H, et al. Postdeposition annealing on VO_2 films for resistive random-access memory selection devices[J]. Journal of Vacuum Science & Technology A: Vacuum, Surfaces, and Films, 2018, 36(5): 051501.

[134] Seong D, Park J, Lee N, et al. Effect of oxygen migration and interface engineering on resistance switching behavior of reactive metal/polycrystalline $Pr_{0.7}Ca_{0.3}MnO_3$ device for nonvolatile memory applications[C]//2009 IEEE International Electron Devices Meeting (IEDM). IEEE, 2009: 1-4.

[135] Lee H S, Park C S, Park H H. Effect of La^{3+} substitution with Gd^{3+} on the resistive switching properties of $La_{0.7}Sr_{0.3}MnO_3$ thin films[J]. Applied Physics Letters, 2014, 104(19): 191604.

[136] Querré M, Tranchant J, Corraze B, et al. Non-volatile resistive switching in the Mott insulator $(V_{1-x}Cr_x)_2O_3$[J]. Physica B: Condensed Matter, 2018, 536: 327-330.

[137] Zhao L, Hong Q, Wang X. Novel designs of spiking neuron circuit and STDP learning circuit based on memristor[J]. Neurocomputing, 2018, 314: 207-214.

[138] Williamson A, Schumann L, Hiller L, et al. Synaptic behavior and STDP of asymmetric nanoscale memristors in biohybrid systems[J]. Nanoscale, 2013, 5(16): 7297-7303.

[139] Serrano-Gotarredona T, Masquelier T, Prodromakis T, et al. STDP and STDP variations with memristors for spiking neuromorphic learning systems[J]. Frontiers in Neuroscience, 2013, 7: 2.

[140] Pan Y, Wan T, Du H, et al. Mimicking synaptic plasticity and learning behaviours in solution processed SnO_2 memristor[J]. Journal of Alloys and Compounds, 2018, 757: 496-503.

[141] Zhang X, Liu S, Zhao X, et al. Emulating short-term and long-term plasticity of biosynapse based on Cu/a-Si/Pt memristor[J]. IEEE Electron Device Letters, 2017, 38(9): 1208-1211.

[142] Wang Z Q, Xu H Y, Li X H, et al. Synaptic learning and memory functions achieved using oxygen ion migration/diffusion in an amorphous InGaZnO memristor[J]. Advanced Functional Materials, 2012, 22(13): 2759-2765.

[143] Hu S G, Liu Y, Liu Z, et al. Synaptic long-term potentiation realized in Pavlov's dog model based on a NiO_x-based memristor[J]. Journal of Applied Physics, 2014, 116(21): 214502.

[144] Gong Y, Wang Y, Li R, et al. Tailoring synaptic plasticity in a perovskite QD-based asymmetric memristor[J]. Journal of Materials Chemistry C, 2020, 8(9): 2985-2992.

[145] Kim D S, Kim Y H, Lee C E, et al. Colossal electroresistance mechanism in a Au/$Pr_{0.7}Ca_{0.3}MnO_3$/Pt sandwich structure: Evidence for a Mott transition[J]. Physical Review B, 2006, 74(17): 174430.

[146] Park S, Kim H, Choo M, et al. RRAM-based synapse for neuromorphic system with pattern recognition function[C]//2012 international Electron Devices Meeting. IEEE, 2012: 10.2. 1-10.2. 4.

[147] Shi J, Ha S D, Zhou Y, et al. A correlated nickelate synaptic transistor[J]. Nature Communications, 2013, 4(1): 1-9.

[148] Deng X, Wang S Q, Liu Y X, et al. A flexible mott synaptic transistor for nociceptor simulation and neuromorphic computing[J]. Advanced Functional Materials, 2021, 31(23): 2101099.

[149] Liu X, Sadaf S M, Park S, et al. Complementary resistive switching in niobium oxide-based resistive memory devices[J]. IEEE Electron Device Letters, 2013, 34(2): 235-237.

[150] Shin S H, Halpern T, Raccah P M. High-speed high-current field switching of NbO_2[J]. Journal of Applied Physics, 1977, 48(7): 3150-3153.

[151] Luo Q, Xu X, Lv H, et al. Fully BEOL compatible TaO_x-based selector with high uniformity and robust performance[C]//2016 IEEE International Electron Devices Meeting (IEDM). IEEE, 2016: 11.7. 1-11.7. 4.

[152] Gao L, Chen P Y, Yu S. NbO_x-based oscillation neuron for neuromorphic computing[J]. Applied physics letters, 2017, 111(10): 103503.

[153] Zhao X, Chen A, Ji J, et al. Ultrahigh uniformity and stability in NbO_x-based selector for 3-D memory by using Ru electrode[J]. IEEE Transactions on Electron Devices, 2021, 68(5): 2255-2259.

[154] Lee D, Kwak M, Moon K, et al. Various threshold switching devices for integrate and fire neuron applications[J]. Advanced Electronic Materials, 2019, 5(9): 1800866.

[155] Zhang Z, Chen A, Ma G, et al. Controllable functional layer and temperature-dependent characteristic in niobium oxide insulator-metal transition selector[J]. IEEE Transactions on Electron Devices, 2020, 67(7): 2771-2777.

[156] Jeon D S, Dongale T D, Kim T G. Low power Ti-doped NbO_2-based selector device with high selectivity and low OFF current[J]. Journal of Alloys and Compounds, 2021, 884: 161041.

[157] Chen A, He Y, Ma G, et al. Improved uniformity and threshold voltage in NbO_x-ZrO_2

selectors[J]. Applied Physics Letters, 2021, 119(7): 073503.
[158] Son M, Lee J, Park J, et al. Excellent selector characteristics of nanoscale VO$_2$ for high-density bipolar ReRAM applications[J]. IEEE Electron Device Letters, 2011, 32(11): 1579-1581.
[159] Zhang K, Wang B, Wang F, et al. VO$_2$-based selection device for passive resistive random access memory application[J]. IEEE Electron Device Letters, 2016, 37(8): 978-981.
[160] Zhou X, Gu D, Li Y, et al. A high performance electroformed single-crystallite VO$_2$ threshold switch[J]. Nanoscale, 2019, 11(45): 22070-22078.
[161] Xue W, Liu G, Zhong Z, et al. A 1D vanadium dioxide nanochannel constructed via electric-field-induced ion transport and its superior metal-insulator transition[J]. Advanced Materials, 2017, 29(39): 1702162.
[162] Hua Q, Wu H, Gao B, et al. A threshold switching selector based on highly ordered Ag nanodots for X-point memory applications[J]. Advanced Science, 2019, 6(10): 1900024.
[163] Kim S, Liu X, Park J, et al. Ultrathin (< 10 nm) Nb$_2$O$_5$/NbO$_2$ hybrid memory with both memory and selector characteristics for high density 3D vertically stackable RRAM applications[C]//2012 Symposium on VLSI Technology (VLSIT). IEEE, 2012: 155-156.
[164] Pickett M D, Medeiros-Ribeiro G, Williams R S. A scalable neuristor built with Mott memristors[J]. Nature Materials, 2013, 12(2): 114-117.
[165] Stoliar P, Tranchant J, Corraze B, et al. A leaky-integrate-and-fire neuron analog realized with a Mott insulator[J]. Advanced Functional Materials, 2017, 27(11): 1604740.
[166] Zhang X, Zhuo Y, Luo Q, et al. An artificial spiking afferent nerve based on Mott memristors for neurorobotics[J]. Nature Communications, 2020, 11(1): 1-9.
[167] Duan Q, Jing Z, Zou X, et al. Spiking neurons with spatiotemporal dynamics and gain modulation for monolithically integrated memristive neural networks[J]. Nature Communications, 2020, 11(1): 1-13.
[168] Kumar S, Strachan J P, Williams R S. Chaotic dynamics in nanoscale NbO$_2$ Mott memristors for analogue computing[J]. Nature, 2017, 548(7667): 318-321.
[169] Kumar S, Williams R S, Wang Z. Third-order nanocircuit elements for neuromorphic engineering[J]. Nature, 2020, 585(7826): 518-523.
[170] Xu Z, Bernussi A A, Fan Z. Voltage pulse driven VO$_2$ volatile resistive transition devices as leaky integrate-and-fire artificial neurons[J]. Electronics, 2022, 11(4): 516.
[171] Del Valle J, Salev P, Kalcheim Y, et al. A caloritronics-based Mott neuristor[J]. Scientific Reports, 2020, 10(1): 1-10.
[172] Xia Q, Qin Y, Qiu P, et al. A bio-inspired tactile nociceptor constructed by integrating wearable sensing paper and a VO$_2$ threshold switching memristor[J]. Journal of Materials Chemistry B, 2022, 10(12): 1991-2000.
[173] Wang Y L, Li M C, Zhao L C. The effects of vacuum annealing on the structure of VO$_2$ thin films[J]. Surface and Coatings Technology, 2007, 201(15): 6772-6776.
[174] Zou J, Chen X, Xiao L. Phase transition performance recovery of W-doped VO$_2$ by

annealing treatment[J]. Materials Research Express, 2018, 5(6): 065055.

[175] Ogugua S N, Ntwaeaborwa O M, Swart H C. Latest development on pulsed laser deposited thin films for advanced luminescence applications[J]. Coatings, 2020, 10(11): 1078.

[176] Zhao Y, Chen C, Wang D. The application of pulsed laser deposition in producing bioactive ceramic films[J]. Surface Review and Letters, 2005, 12(03): 401-408.

[177] Bao Q, Chen C, Wang D, et al. Pulsed laser deposition and its current research status in preparing hydroxyapatite thin films[J]. Applied Surface Science, 2005, 252(5): 1538-1544.

[178] Oskirko V O, Zakharov A N, Pavlov A P, et al. Hybrid HIPIMS+ MFMS power supply for dual magnetron sputtering systems[J]. Vacuum, 2020, 181: 109670.

[179] Depla D. Sputter deposition with powder targets: An overview[J]. Vacuum, 2021, 184: 109892.

[180] Yang Z, Hart S, Ko C, et al. Studies on electric triggering of the metal-insulator transition in VO_2 thin films between 77 K and 300 K[J]. Journal of Applied Physics, 2011, 110(3): 033725.

[181] Driscoll T, Quinn J, Di Ventra M, et al. Current oscillations in vanadium dioxide: Evidence for electrically triggered percolation avalanches[J]. Physical Review B, 2012, 86(9): 094203.

[182] Kumar S, Esfandyarpour R, Davis R, et al. Surface charge sensing by altering the phase transition in VO_2[J]. Journal of Applied Physics, 2014, 116(7): 074511.

[183] Okuyama D, Shibuya K, Kumai R, et al. X-ray study of metal-insulator transitions induced by W doping and photoirradiation in VO_2 films[J]. Physical Review B, 2015, 91(6): 064101.

[184] Rodolakis F, Hansmann P, Rueff J P, et al. Inequivalent routes across the Mott transition in V_2O_3 explored by X-ray absorption[J]. Physical Review Letters, 2010, 104(4): 047401.

[185] Singh V R, Jovic V, Valmianski I, et al. Irreversible metal-insulator transition in thin film VO_2 induced by soft X-ray irradiation[J]. Applied Physics Letters, 2017, 111(24): 241605.

[186] Hwang I H, Jin Z, Park C I, et al. Electrical and structural properties of VO_2 in an electric field[J]. Current Applied Physics, 2021, 30: 77-84.

[187] Zhao Z, Li J, Ling C, et al. Electric field driven abnormal increase in conductivity of tungsten-doped VO_2 nanofilms[J]. Thin Solid Films, 2021, 725: 138643.

[188] Yang S, Vaseem M, Shamim A. Fully inkjet-printed VO_2-based radio-frequency switches for flexible reconfigurable components[J]. Advanced Materials Technologies, 2019, 4(1): 1800276.

[189] Zou Z, Zhang Z, Xu J, et al. Thermochromic, threshold switching, and optical properties of Cr-doped VO_2 thin films[J]. Journal of Alloys and Compounds, 2019, 806: 310-315.

[190] Rini M, Tobey R, Dean N, et al. Control of the electronic phase of a manganite by

mode-selective vibrational excitation[J]. Nature, 2007, 449(7158): 72-74.

[191] Pickett M D, Williams R S. Sub-100 fJ and sub-nanosecond thermally driven threshold switching in niobium oxide crosspoint nanodevices[J]. Nanotechnology, 2012, 23(21): 215202.

[192] Mo S K, Kim H D, Denlinger J D, et al. Photoemission study of $(V_{1-x}M_x)_2O_3$ (M = Cr, Ti)[J]. Physical Review B, 2006, 74(16): 165101.

[193] Wang Z, Kang J, Bai G, et al. Self-selective resistive device with hybrid switching mode for passive crossbar memory application[J]. IEEE Electron Device Letters, 2020, 41(7): 1009-1012.

[194] Zhou Y, Ramanathan S. Mott memory and neuromorphic devices[J]. Proceedings of the IEEE, 2015, 103(8): 1289-1310.

[195] Shamsi J, Avedillo M J, Linares-Barranco B, et al. Hardware implementation of differential oscillatory neural networks using VO_2-based oscillators and memristor-bridge circuits[J]. Frontiers in Neuroscience, 2021, 15: 674567.

[196] Li G, Xie D, Zhong H, et al. Photo-induced non-volatile VO_2 phase transition for neuromorphic ultraviolet sensors[J]. Nature Communications, 2022, 13(1): 1-9.

[197] Kang K, Kim D, Kim S H, et al. Doppler velocity characteristics during tropical cyclones observed using ScanSAR raw data[J]. IEEE Transactions on Geoscience and Remote Sensing, 2016, 54(4): 2343-2355.

[198] Velichko A. A method for evaluating chimeric synchronization of coupled oscillators and its application for creating a neural network information converter[J]. Electronics, 2019, 8(7): 756.

[199] Ding Y, Zhang Y, Zhang X, et al. Engineering spiking neurons using threshold switching devices for high-efficient neuromorphic computing[J]. Frontiers in Neuroscience, 2022, 15: 786694.

第 6 章 柔性铁电器件的研究现状和应用

6.1 引 言

近年来，随着人工智能和物联网的快速发展，以及人类日常生活需求的日益多样化，对电子器件提出了更高的要求：更灵活、更小巧、更透明。传统的电子器件基于具有刚性结构的集成电路，这可能会限制它们在某些实际应用领域的兼容性，例如人体皮肤和身体器官[1-3]。为了克服传统电子器件的局限性以及新技术的进步，所做的努力直接导致了柔性电子器件的诞生，以满足人类生活水平的更高需求，并已经成为该领域的研究热潮。柔性电子技术是一种在柔性或可拉伸衬底上制备有机或无机电子器件的新型电子技术。与传统的电子器件相比，柔性电子器件具有高灵活性、适应性、便携性、耐用性、低成本和大面积可加工性。它们能满足器件各种变形的要求，以求更好地实现各种电子产品的功能，使柔性电子器件的形式更加多样化和人性化。因此，柔性电子技术提供了广泛的应用，如柔性电路[4,5]、柔性电池[6,7]、柔性光伏[8]、柔性电子显示器[9,10]、射频识别(RFID)标签[11]、可穿戴电子器件[12,13]、柔性传感器[14,15]、面部识别器[16]等(图6.1)。柔性电子领域的蓬勃发展和创新将直接促进科学和工业的持续发展，并在不久的将来造福全人类。

柔性电子器件需要具有优异性能且适用的柔性材料。目前，许多这类材料已被用于柔性电子器件中。例如，1D 材料包括纳米管[17]、纳米线[18]、纳米带[19]、纳米纤维[20]；2D 材料石墨烯[21,22]、氮化硼(BN)[23]和过渡金属二硫化物(TMD)[24,25]。此外，还使用了一些有机材料，如聚二甲基硅氧烷(PDMS)[26]、聚萘二甲酸乙二醇酯(PEN)[27]、聚对苯二甲酸乙二醇酯(PET)[28]、聚(3,4-乙烯二氧噻吩)：聚(苯乙烯磺酸盐)(PEDOT：PSS)[29]、聚偏二氟乙烯(PVDF)[30,31]和聚乳酸(PLA)[32]等。最近，铁电材料由于其可切换的自发极化、优异的铁电和压电性能、良好的耐久性，以及令人感兴趣的光电性能而引起了人们的极大兴趣。与传统的纳米材料和有机聚合物相比[33]，铁电材料表现出优越的热稳定性和化学稳定性。在实际应用中，柔性电子器件需要实现更快的速度、更高的存储密度和更低的功耗，而铁电材料一直被认为是实现这些目标的最有前途的研究体系。然而，由于铁电材料是刚性材料[34,35]，它们具有较差的延展性和弯曲能力。为了应用于柔性电子，首先需要解决这个关键问题。

图 6.1 典型的柔性电子及其应用

传感器 (经许可转载 [36]。版权 2019, MDPI；经许可转载 [37]。版权 2019, Elsevier Ltd；经许可转载 [38]。版权 2018, American Chemical Society)。RFID(经许可转载 [39]。版权 2014, Elsevier B.V.；经许可转载 [40]。版权 2018, MDPI；经许可转载 [41]。版权 2019, WILEY-VCH Verlag GmbH & Co.KGaA, Weinheim)。显示器 (经许可转载 [42]。版权 2011, WILEY-VCH；经许可转载 [43]。版权 2013, Nature Publishing Group。经许可转载 [10]。版权 2018, WILEY-VCH)。电子皮肤 (经许可转载 [44]。版权 2010, Nature Publishing Group。经许可转载 [45]。版权 2014, Nature Publishing Group)。电池 (经许可转载 [46]。版权 2009, Royal Society of Chemistry。经许可转载 [7]。版权 2012, American Chemical Society。经许可转载 [47]。版权 2016, WILEY-VCH)。电路 (经许可转载 [5]。版权 2012, American Chemical Society。经许可转载 [48]。版权 2014, Society for Information Display)

在本章中，我们将介绍最近开发的柔性铁电薄膜制备技术和新型 2D 柔性铁电材料，以及基于铁电材料的柔性存储器、柔性传感器、柔性光伏器、柔性能量采集器及其在柔性神经形态学中的应用，最后对柔性铁电器件的未来发展进行展望。

6.2 制备柔性铁电器件的途径

6.2.1 通过范德瓦耳斯异质外延制备柔性钙钛矿铁电薄膜

异质外延沉积是指生长的薄膜材料在化学成分或物理结构上不同于衬底材料。异质外延生长需要外延层和衬底之间的晶格匹配、热膨胀系数匹配和化学稳定性匹配。其中，外延层与衬底材料之间的晶格匹配是实现各种异质外延结构生长需要克服的主要问题之一。一些材料的晶格匹配条件很差，例如包括 Si 和 GaAs 在内的四价键合共价半导体材料，因为在它们的衬底表面上有悬空键[49]，如图 6.2(a) 所示。由于它们的共价键的长度和角度难以改变，因此只能附着在具有良好晶格匹配的材料的原子上。但是，当通过范德瓦耳斯 (vdW) 相互作用进行异质外延生长时，晶格匹配条件被极大地放宽。在典型的示例中，一种层状材料在另一种层状材料的裂解表面上生长而没有悬空键，如图 6.2(b) 所示。我们将这种外延生长称为 vdW 外延[50]。有许多材料经常使用 vdW 外延生长方法。例如，早期报道的准 1D 材料 Se 和 Te，它们是 Ⅵ 族元素的半导体。它们具有由原子的螺旋链组成的特殊晶体结构。链中的原子通过强共价键相互结合，而链与链仅通过弱的 vdW 力结合在一起。晶体可以很容易地沿着链被切割，并且不会在其表面产生悬空键。Aoto 等在他们的报告中证明，尽管沿着链轴的晶格失配率高达 20%，但通过 vdW 外延生长仍可以在 Te 的裂解表面上生长具有良好质量的超薄的 Se 薄膜[51]。随后，Te/Se/Te 异质结构也通过相同的方法成功生长。近年来，2D 过渡金属二硫化物 (MX_2, M = 过渡金属, X = 硫族元素) 被报道更适合于 vdW 外延生长。它们是层状晶体，通常具有 X-M-X 三明治结构。单元层中的原子通过强共价键相互结合，而层与层仅通过弱的 vdW 力结合。晶体很容易沿分层被平行切割，并且在切割表面上没有悬空键。因此，可以预期通过 vdW 力在该表面上生长层状材料，从而放宽了晶格匹配的条件。人们还发现，vdW 外延

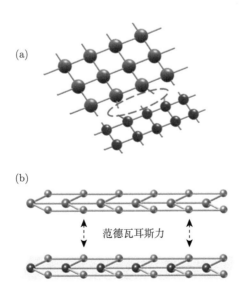

图 6.2　通过 (a) 悬空键 (用虚线圈出) 和 (b) vdW 间隙连接的界面

甚至可以在具有非常不同的晶体结构的层状材料之间实现。例如，Saiki 等展示了 MoS_2 和 $NbSe_2$ 薄膜在云母的裂解表面上的生长 [52]。尽管云母也是层状材料，但其晶体结构比 MX_2 复杂得多，并且它的晶格常数比 MX_2 大 50%。然而，通过 vdW 外延也已经在云母上成功生长了具有良好质量的 MoS_2 和 $NbSe_2$ 层。

云母经常被用作 vdW 外延的衬底，因为它具有三个优点：优异的柔韧性、出色的透明性和高的热稳定性。层状云母的结构单元如图 6.3(a) 所示。铝八面体 (AlO_6) 片将两层硅酸盐四面体 (SiO_4) 片分开。这些单元层通过层间阳离子堆叠在一起。四面体层 (SiO_4) 中的 Si 离子被 Al 离子部分取代，产生净负电荷。四面体层 (SiO_4) 六边形中心下方的顶部氧和羟基组成八面体层。在这种结构中，单元层中存在非常强的共价键，但层间阳离子连接显示出较弱的 vdW 力。云母被 vdW 力分裂成两个大的原子表面，并且具有相等但随机分布的 K^+，保持电中性，这使得云母成为 vdW 异质外延的理想衬底 [53,54]。云母具有非常高的应变力，并且单层云母的厚度可以保持在几微米以下。因其优异的柔韧性它可以完美地应用于柔性电子器件。目前，许多研究证明，高质量的薄膜可以在云母衬底上外延生长并表现出优异的性能 [55−59]。为了促进在云母衬底上进行 vdW 氧化物异质外延以实现柔性透明电子器件的制备，需要开发在云母上生长的功能材料。最近的研究报道了许多关于在柔性云母衬底上通过 vdW 外延生长功能氧化物的结果，如图 6.3(b) 所示。

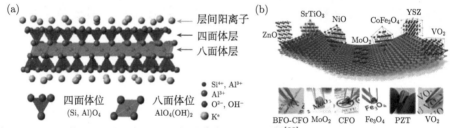

图 6.3　(a) 层状云母 (白云母) 的晶体结构 (经许可转载 [60]。版权 2019，Elsevier)；(b) 在柔性云母上外延生长的功能性氧化物的总结 (经许可转载 [53]。版权 2017，Springer Nature)(彩图请扫封底二维码)

此前，高质量的外延钙钛矿薄膜通常生长在具有相同结构的材料上，如 $SrTiO_3$(STO) 和 $LaAlO_3$ (LAO)。近年来，随着 vdW 外延生长的应用越来越成熟，钙钛矿型氧化物也开始尝试在柔性云母衬底上生长。由于钙钛矿与云母晶格不匹配，需要在云母衬底和功能层之间先生长一层缓冲层才能获得高质量的生长。例如，Jiang 等报道了使用 vdW 异质外延方法可以在云母衬底上获得柔性 $PbZr_xTi_{1−x}O_3$(PZT) 铁电薄膜 [59]。该工作利用脉冲激光沉积 (PLD) 法进行外延生长，首先在云母衬底上生长非常薄的 $CoFe_2O_4$(CFO) 层作为缓冲层，然后生长 $SrRuO_3$(SRO) 层作为用于电学表征的底部电极，最后沉积 PZT 薄膜。

PZT/SRO/CFO/云母的异质结构示意图如图 6.4(a) 所示。他们还对 (001) 云母上的异质结构进行了典型的 X 射线衍射 (XRD) 2θ-θ 面外扫描,结果仅显示 (111) PZT 和 (111) SRO 在 (001) 云母上的衍射峰 (图 6.4(b)),表明 PZT 薄膜在云母衬底上的外延性质。为了从微观上研究 PZT/云母的异质外延结构,他们使用透射电子显微镜 (TEM) 来检测薄膜之间的界面。图 6.4(c) 显示了沿云母的 [010] 晶带轴拍摄的横截面 TEM 图像,显示了薄膜之间的清晰界面。这进一步证实了外延关系与 XRD 结果的一致性,并且界面清晰,没有可见的相互扩散。为了表征其铁电性,他们利用具有导电尖端的压电响应应力显微镜 (PFM) 研究了 PZT 薄膜的局域探测和开关特性。测试结果表明,PZT 层的极化是可切换的,这证实了异质结构的铁电性质和优异的铁电切换性质,如图 6.4(d)、(e) 所示。此外,该器件在 25℃ 时获得了高达约 $75\mu C \cdot cm^{-2}$ 的饱和极化强度和高约 $60\mu C \cdot cm^{-2}$ 的剩余极化强度 (P_r),其矫顽场 (E_C) 约为 100 $kV \cdot cm^{-1}$,表现出优异的铁电性能,如图 6.4(g) 所示。为了拓展这种柔性铁电器件的实际应用,他们进行了一系列循环性能测试。在拉伸和压缩弯曲的情况下,测试了异质结构抗机械弯曲的宏观铁

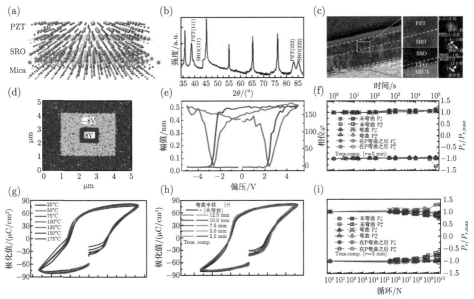

图 6.4 (a) PZT/SRO/云母异质结构示意图;(b) PZT、SRO、CFO 和云母衍射峰的 Φ 扫描;(c) PZT/SRO 和 SRO/CFO/云母界面的横截面 TEM 图像,以及 PZT、SRO 和云母的选定区域的衍射图;(d) 云母上的 PZT 薄膜的 PFM 相位图像;(e) 局部 PFM 幅度和相位滞回线;(f)、(i) 不同应变状态下的保持和疲劳性能测试;(g) 不同温度下的 P-E 滞回线;(h) 各种应变状态下的 P-E 滞回线 (经许可转载[59]。版权 2017,AAAS)

(彩图请扫封底二维码)

电性能，并且从图 6.4(f)、(h)、(i) 中可以明显看出，所制备的 PZT 器件显示出高稳定性和优异的抗机械弯曲的性能，这对于柔性器件的应用是非常理想的。据报道，外延柔性 PZT 器件还可用作信息技术的存储元件、纳米动力学[61]和生物医学应用[62]的能量收集器，并且还具有开发可穿戴和可植入电子领域的可能性。

此外，Yang 等在 2017 年报道了一种基于云母衬底的 SRO/BaTi$_{0.95}$Co$_{0.05}$O$_3$/Au (SRO/BTCO/Au) 的柔性、半透明且经济的阻变存储器[63]。该存储器显示超过 50 的 R_{HRS}/R_{LRS}，当在 1.4 mm 或 2.2 mm 弯曲半径下经过 360000 次写入/擦除循环后仍可工作；此外，在 25~180℃ 的温度范围内或在 500℃ 退火后，信息可以正常写入、擦除和存储。这种柔性存储器具有大尺寸、高温度稳定性、优异的透明性以及制备工艺简单的优点。在此之后，Chu 等通过 vdW 外延在柔性云母衬底上生长了自组装的 BiFeO$_3$ (BFO) - CFO 体异质结，并研究了体异质结构中的磁电耦合[64]；结果表明，这种体异质结构的磁电耦合系数为 74 mV/(cm·Oe)，大于先前报道的柔性衬底上的磁电耦合系数。该方法为性能优化的高灵敏度柔性微电子器件的应用提供了良好的方向。然后，在 2020 年，Sun 等通过适当设计 SRO/BaTiO$_3$(BTO) 双缓冲层，在柔性云母衬底上生长了高质量的 (111) 取向 BFO 铁电薄膜[65]。在所有报道的柔性铁电薄膜中，BFO 铁电薄膜表现出最大的极化 ($P_s \approx 100 \mu C \cdot cm^{-2}$, $P_r \approx 97 \mu C \cdot cm^{-2}$)，并且在 5 mm 半径下的 10^4 个弯曲周期内，铁电极化非常稳定。研究结果还表明，柔性铁电忆阻器可应用于数据存储和处理领域。这些研究还证实，选择云母作为柔性铁电器件的衬底是非常合适的。独立的铁电器件也可以通过直接在一些柔性衬底 (如金属箔或聚合物) 上生长来制备，然而，这种方法具有加工温度低、加工过程中尺寸稳定性差、热膨胀系数大以及薄膜的多晶性质等缺点，这给高质量外延铁电氧化物薄膜的研究和应用带来了挑战和困难[58,66-68]。

通过 vdW 外延制备的柔性铁电薄膜已在许多应用中得到证明。这种方法不需要克服薄膜和衬底之间严格的晶格匹配，从而降低了缺陷密度。此外，衬底与外延膜之间的弱 vdW 力相互作用进一步降低了衬底的夹持效应，有利于器件的更好性能。此外，所制备的柔性铁电器件既保留了外延膜的优异性能，又显示了机械柔韧性、耐久性和热稳定性。当前的研究证明了使用 vdW 异质外延方法制备柔性电子器件的新可能性，这在将来集成基于有机的柔性电子器件方面具有巨大的潜力。

6.2.2 通过蚀刻法制备柔性钙钛矿铁电薄膜

近年来，已经发现转移方法可以有效地获得柔性功能氧化物薄膜，关键问题是薄膜与衬底的分离。一般来说，分离方法可以根据释放机制分为两种类型：干法蚀刻和湿法蚀刻。干法蚀刻，包括离子研磨、反应离子蚀刻、电子回旋共振等离子体蚀刻和电感耦合等离子体蚀刻,由于蚀刻选择性差,不可避免地会对薄膜造成辐射损伤[69,70]。后来,为了减少对薄膜的损伤,通过激光剥离 (LLO) 技术对晶圆或衬底进行激光照射,将功能性氧化薄膜与衬底分离,这被证明是另一种有效的干法蚀刻方法。例如,早在 20 世纪 90 年代末,Tsakalakos 和 Sands 报道了通过 LLO 技术转移高质量外延 $(Pb_{0.90}La_{0.07})(Zr_{0.5}Ti_{0.5})O_3$ (PLZT) 薄膜,并发现在转移后薄膜很好地保留了铁电性能[71]。然而,这种干法蚀刻仍然存在的问题之一是不可避免地会引起结构损坏,尤其是在薄膜的极限范围内。与干法蚀刻不同,湿法蚀刻是一种化学蚀刻技术,选择性地用于去除衬底或牺牲层,通常对功能膜的损伤较小,并保持释放的薄膜的质量。随着各种溶液制备技术的发展,通过湿法蚀刻和转移技术在柔性衬底上制备功能氧化物薄膜变得越来越容易。在早期,Gan 等报道了通过使用特定的化学溶剂蚀刻 STO 衬底获得独立的 SRO 外延薄膜,并证明了应变效应在改变钙钛矿外延薄膜的重要宏观电学和磁性物理性质中起着决定性作用[72]。后来,人们发现了在功能性薄膜和衬底之间蚀刻牺牲层的方法,并且它已经被证明是一种更具成本效益和更快的方法,如图 6.5 所示。需要注意的是,要获得高质量的外延生长的功能薄膜,就需要仔细选择牺牲层,以使其晶体结构和晶格常数与功能氧化薄膜和衬底都能很好地匹配。而且,在功能薄膜的生长过程中,牺牲层应该非常稳定。牺牲层可以通过化学溶液轻松去除,而不会在蚀刻过程中损坏功能氧化薄膜。迄今为止,人们已经通过这种方法成功制备了一些柔性钙钛矿铁电薄膜,在这些研究工作中,最常选择 $La_{0.7}Sr_{0.3}MnO_3$ (LSMO) 和 $Sr_3Al_2O_6$ (SAO) 作为牺牲层。

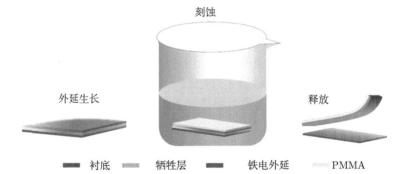

图 6.5 通过蚀刻牺牲层将外延铁电薄膜与衬底分离的示意图 (彩图请扫封底二维码)

1. 使用 LSMO 作为牺牲层的蚀刻方法

在过去的几年中,有关湿法蚀刻 LSMO 牺牲层以获得独立的铁电薄膜的报道引起了研究人员的关注。2016 年,Bakaul 等报道了在室温下通过外延转移将薄的 (薄至一个单元) 单晶复合氧化物薄膜集成到硅衬底上;在他们的工作中,通过

PLD 技术在单晶 STO 衬底上生长 LSMO 牺牲层，然后生长基于 SRO/PZT/SRO 的铁电隧道结 (FTJ)，在释放之前，将聚甲基丙烯酸甲酯 (PMMA) 涂覆在薄膜的表面上作为转移载体，然后用 KI (4 mg) + HCl (5mL) + H_2O (200 mL) 溶液从侧面蚀刻 LSMO，最后，从溶液中释放并收集 FTJ/PMMA 叠层，通过使用微操纵器和柔和的力将 PMMA 层转移到目标衬底上，然后使用丙酮去除 PMMA 层，如图 6.6(a) 所示[73]；同时，SRO/BFO/CFB (CoFeB) /Pt 异质结构和 $CaTiO_3$ (CTO) /STO 超晶格已通过湿法蚀刻 LSMO 牺牲层来成功获得；该研究实现了硅与复合氧化物的直接集成，为铁电薄膜与柔性衬底的集成提供了很大的参考。随后，2017 年，Bakaul 等使用相同的湿法蚀刻和层转移技术在柔性衬底 PET 上集成单晶 PZT 柔性存储器件[66]；转移到柔性衬底上的 PZT 在所有关键参数如铁电性、速度、耐用性和保留时间上都表现出优异的性能；测试过程中，PZT 在平坦和弯曲状态下的剩余极化几乎不变，并显示高达约 $75\mu C \cdot cm^{-2}$ 的极化强度，如图 6.6(b) 所示，它的强度至少是迄今为止在柔性领域上报道的所有其他器件的极化强度的两倍[74-76]；此外，还证明了器件在 4 V 时 57 ns 的开关速度，这比以前在弯曲衬底上的任何演示至少快一个数量级[76]，如图 6.6(c)、(d) 所示；所展示的器件的工作电压小于 1.5 V，这几乎比通常在柔性衬底上使用的电子器件的工作电压小一个数量级 (大于 10 V)[75,77]；实验还证实了超过 10^{10} 个循环的可靠运行次数和大于 10 年的存储保持时间；结果表明，具有最佳存储性能的单晶钙钛矿型铁电薄膜可以有效地集成到柔性衬底上；然而，转移的单晶薄膜的尺寸只有几百微米，相对较小，不便于后续使用，并且 PMMA 层溶解时发生的化学反应可能会产生大量的残留物和分散的碎屑，这可能会导致薄膜破裂。随后，Shen 等提出了一种改进的方法，该方法通过在蚀刻之前将适用的柔性衬底聚酰亚胺 (PI) 条直接粘附到 $LiFe_5O_8$ (LFO) 薄膜上[78]，该方法不仅避免了化学污染，而且大大增加了转移薄膜的面积，将转移薄膜的尺寸增加到了厘米级。此外，弯曲测试的结果表明，转移的 LFO 薄膜具有出色的柔韧性和稳定性。这种简单的转移方法为构建用于制备柔性器件的厘米级单晶薄膜提供了更实用的解决方案。

虽然使用酸性蚀刻溶剂蚀刻 LSMO 牺牲层以获得独立功能层的方法已被广泛使用，但也存在一些缺点，如蚀刻薄膜的酸性蚀刻溶剂选择不当可能导致实验失败，以及使用酸性蚀刻溶剂可能带来安全隐患等。此外，薄膜材料的选择也受到这种方法的限制。因此，需要找到一种不会损坏薄膜的常用牺牲层和蚀刻剂组合，并将其应用于钙钛矿及其异质结构。

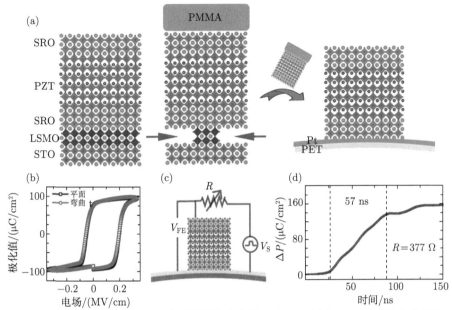

图 6.6 (a) SRO/PZT/SRO 叠层在柔性衬底上的转移过程示意图；(b) 平面和弯曲状态下的 P-E 滞回线；(c) 测量设置的示意图；(d) 弯曲状态下可切换极化的时间演化 (经许可转载 [66]。版权 2017, WILEY-VCH Verlag GmbH & Co. KGaA, Weinheim)

2. 使用 SAO 作为牺牲层的蚀刻方法

除了使用 LSMO 作为牺牲层外，近年来，SAO 作为新兴的牺牲层也受到越来越多的关注。SAO 具有一个晶格常数为 15.844 Å 的立方晶胞，与最具代表性的钙钛矿衬底 STO (a_{STO} = 3.905 Å, 4×a_{STO} = 15.620 Å) 紧密匹配 [79]。尽管结构复杂，但 SAO 晶胞中的 264 个原子与 STO 共享相似的晶格，从投影到 (001) 表面上的两个结构可以清楚地看到，如图 6.7(a) 所示。SAO 具有与钙钛矿相似的晶体结构，这保证了在 STO 上 SAO 和在 SAO 上其他钙钛矿氧化物的高质量外延生长。此外，SAO 易溶于水，不会破坏常见的钙钛矿薄膜，因此通常用作进行层转移技术的理想牺牲层。2016 年，Lu 等开发了一种在去离子水中蚀刻 SAO 牺牲层的方法。在他们的工作中，通过将 SAO/STO 异质结构浸入去离子水中溶解 SAO 层，并且发现 STO 衬底表面完全没有任何残留的 SAO 薄膜 [80]。在此基础上，他们进一步讨论了大型钙钛矿型独立式薄膜的制备。在 SAO 牺牲层上生长了不同的钙钛矿类型的薄膜，包括 STO、LSMO 及其超晶格 (SL)。图 6.7(b) 显示出了释放在 PDMS 支撑层上的毫米尺寸的独立式薄膜的光学显微镜图像，揭示了转移的氧化物薄膜的光滑表面。图 6.7(c)、(d) 显示了 SAO/STO 和 SL/SAO 界面的横截面高角度环形暗场 (HAADF) 扫描透射电子显微镜 (STEM) 图像及其示意图，这与图 6.7(a) 中所示的非常吻合。大尺寸自支撑的 STO 和 LSMO 薄膜的成功制备表明，该方法同样适用于其他钙钛矿氧化物铁电材料，如 BFO、PZT 和 BTO。

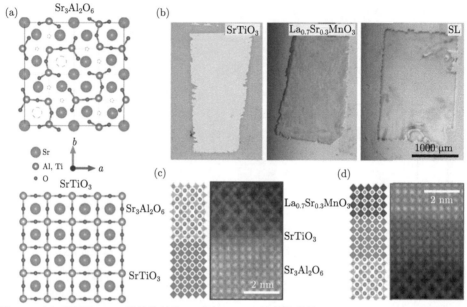

图 6.7　(a) (顶部) SAO 的晶体结构，(底部) STO 的晶体结构；(b) STO、LSMO 及 SL (20 个 STO (5 个单元)/LSMO (5 个单元)) PDMS 上的独立薄膜的光学显微镜图像；(c)、(d) SAO/STO 和 SL/SAO 界面的横截面 HAADF-STEM 图像及其示意图 (经许可转载 [80]。版权 2016, Nature Publishing Group)

尽管该方法在实验中效果很好，但 Baek 等在 2017 年报道了在 SAO 缓冲层上氧化物多层薄膜生长过程中形成的阳离子偏析和沿着晶体缺陷的扩散[81]；此外，他们证明了通过减少生长过程中扩散的热驱动力，可以将缺陷的不利影响降至最低，从而保留了薄膜的固有特性。此外，Hong 等提出了另一个问题，他们指出，如果转移独立的氧化物薄膜后的晶格厚度低于临界厚度，晶格可能会崩塌[82]。然而，该报告随后被 Ji 等证明是错误的[83]，他们通过氧化物分子束外延制备了具有不同厚度的 BFO 薄膜，并通过具有选定区域电子衍射 (SAED) 的 TEM 和具有原子分辨率的平面图和横截面 HAADF 成像对其进行了表征。图 6.8(a) 显示出了观察到的 SAED 衍射点是窄的、尖锐的和圆形的，表明所有的薄膜都是单晶。此外测试图像显示出四、三、二单元厚度的独立 BFO 薄膜的高结晶质量，远低于先前报道的五单元厚度的临界厚度[82]，表明独立 BFO 薄膜没有临界厚度限制。此外，不同厚度的独立式 BFO 薄膜的平面外 PFM 测试都显示出明显的滞回线, 如图 6.8(b) 所示, 这表明即使在两个单元厚度的薄膜中也可以切换极化。有趣的是, 尽管独立式氧化薄膜看起来像块状陶瓷, 但它们可以变得柔软, 可以弯曲甚至折叠到几个单元晶胞的厚度上, 这表明了超薄独立式氧化薄膜在柔性多功能电子应用中的巨大潜力。

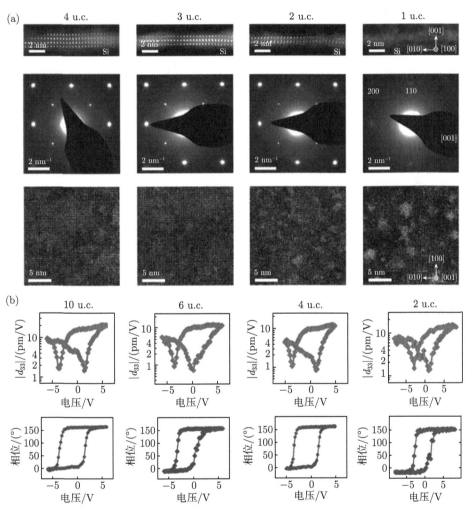

图 6.8 (a) 超薄独立式 BFO 薄膜的横截面 HAADF 图像 (上)、SAED 图案 (中) 和平面图 HAADF 图像 (下); (b) 相应的局部平面外 PFM 滞回线 (经许可转载 [83]。版权 2019, Springer Nature)

此外, 通过采用类似的蚀刻 SAO 牺牲层的方法, Dong 等制备了超弹性的独立式钙钛矿 BTO 铁电薄膜 [84], 图 6.9(a) 显示了这种 BTO 薄膜转移到 PDMS 上的照片; 将独立式 BTO 薄膜转移到 Si 衬底上, 并通过 PFM 测试其铁电性能,结果如图 6.9(b)、(c) 所示。测试过程中，BTO 薄膜可以在原位弯曲测试中折叠大约 180°，并且折叠后的 BTO 薄膜保持完整性和连续性而没有任何裂纹。他们将独立的 BTO 薄膜弯曲成几个多层卷，卷的形成意味着独立的 BTO 膜具有优异的柔韧性，这对于功能氧化物材料来说是很难实现的。此外，他们在研究过程中使用聚焦离子束 (FIB) 制备了 BTO 纳米带，并在 SEM 图像中用纳米管的尖端对其进行了处理，如图 6.9(d) 所示。在此过程中，BTO 纳米带弯曲成不同的曲率。结果表明，BTO 纳米带的最大应变约为 10%，没有任何压裂，显示出该氧化物材料的优异柔韧性，如图 6.9(e) 所示。因此，所报道的 BTO 薄膜的超弹性有望为纳米技术的更广泛应用提供更多的自由度。

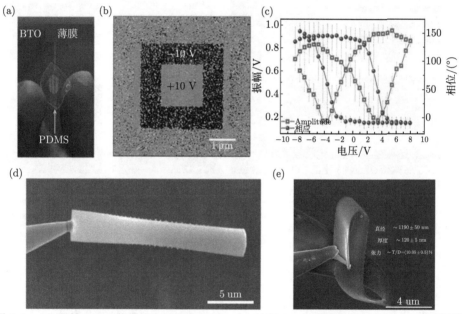

图 6.9　(a) 柔性 BTO 薄膜的光学照片；(b)、(c) 转移 BTO 薄膜的平面外 PFM 相位图像和压电响应；(d) 独立式 BTO 纳米带的 SEM 图像；(e) BTO 薄膜在弯曲过程中应变的计算方法 (经许可转载[84]。版权 2019，AAAS)

后来在 2019 年，基于以 SAO 作为牺牲层的蚀刻工作的成功，Lu 等报道了将 2.8 nm 的单晶 BTO 的铁电隧道结转移到 Si 衬底上[85]。Luo 等报道了基于单晶 BTO/LSMO 铁电隧道结的高质量柔性忆阻器的制备[86]，并发现厚度仅为 3.6 nm 的 BTO 单晶薄膜在剥离后仍能保持可切换的铁电极化，并在室温下表现出高于 500% 的隧穿电阻 (TER)。此外，2020 年，Guo 等通过溶解 SAO 牺牲层将在 STO 上外延生长的 BFO 薄膜转移到柔性衬底上，证明了 BFO 薄膜中光传导的连续可调性[87]。在 2022 年，Zhao 等报道了基于铁电隧道结的人工突触器件, 以 SAO 为牺牲层, BFO 薄膜为功能层[88]。综上所述, 超薄铁电薄膜即使在独立状态下也能保持强铁电性, 这些结果可能会促进基于超薄功能铁电材料的新应用, 这些材料与各种表面兼容, 而不仅限于刚性衬底。

通过选择不同的牺牲层, 可以将独立式薄膜转移到任何衬底上, 以制备具有相应功能的柔性电子器件。然而, 无论选择哪种材料作为牺牲层, 从根本上都需要异质结构的成功外延生长, 这也限制了用于转移的薄膜材料。此外, 蚀刻速率也会影响实验过程。除了在湿法蚀刻过程中出现的随机裂纹外, 在转移过程中薄膜中的裂纹或波纹也应引起更多的注意[89]。

6.2.3 二维柔性铁电材料

2D 材料是指电子仅可在两个维度的纳米尺度 (1~100nm) 上自由运动的材料。自从 Novoselov 等在 2004 年成功分离了石墨烯, 2D 材料得到了广泛的研究[90]。在过去的几十年中, 2D 材料因其独特的物理和化学性质而成为研究的热点。2D 材料分为三类[91]：第一类是层状 vdW 固体, 它们是剥离单层和多层纳米片的主要和最常见的材料；第二类是层状离子固体, 这是另一种重要的层状 2D 材料类型, 并且越来越受欢迎；第三类是表面辅助的非层状固体, 它们是通过化学气相沉积在衬底上合成的。石墨烯的类似物硅烯代表了表面辅助的非层状固体的一个很好的例子。目前, 除了石墨烯之外, 制备具有固定层和大面积的 2D 材料也是必不可少的。可以使用许多用于制备 2D 材料的合成方法, 例如, 使用透明胶带进行微机械剥离。液体剥离是机械剥离的一种替代方法, 是可以大规模生产单层和多层 2D 薄片的更可靠方法。此外, 还有化学气相沉积、在衬底上的 vdW 外延生长以及水热合成。2D 材料也被广泛使用, 例如, MoS_2 具有高开关比和低功耗, 可用于制备低功率的电子器件[91]。据报道, 基于 MoS_2 的三端晶体管表现出电和光学多栅极协同效应, 这归因于 MoS_2 通道中载流子密度的调制[92]。还证明了单层 MoS_2 光电探测器在 561 nm 的波长下具有 880 W^{-1} 的超敏性, 其光的发射范围为 400~680 nm。结果表明, 基于 MoS_2 的积分器在光学传感、光电电路、生物医学成像等领域具有广阔的应用前景[93]。热压形式的 BN 作为耐火材料特别有用。BN 的特性 (例如高抗热震性、高导热性和易于加工) 使其在制备反应器、熔炉及其他高温应用中的价值得以充分体现[94]。硅烯和锗烯是石墨烯的硅和锗基类似物。2014 年发表的一项研究报告称, 硅烯可以在空气中保持 24 h 的稳定性[95]。与石墨烯的六方结构相比, 硅烯具有较低的屈曲结构, 这也有助于基于硅和碳的 2D 材料的稳定性。硅烯和锗烯可能会取代传统的硅衬底技术, 并为电子器件的小型化和功能性做出贡献。同样, 2D 层状氧化物和氢氧化物纳米片的剥离也获得了成功, 这拓宽了纳米片在当前和未来的应用。由于具有极其丰富的结构多样性和电子结构, 这些氧化物和氢氧化物纳米片已被应用于光学、催化、电子学和自旋电子器件 [96,97]。人们已经发现这些 2D 材料具有良好的机械柔韧性，并且相应的柔性电子器件已经被制备并应用于晶体管 [98-100]、传感器 [101]、存储器 [102] 和其他领域。

随着纳米技术的飞速发展，2D 铁电材料因其在高性能、低能耗的微纳智能器件中的应用而被广泛关注。目前，已有几种 2D 铁电材料得到了理论预测和实验验证，如 1T MoS_2[103]、$SnTe$[104]、$GeSe$[105]、$CuInP_2S_6$ [106-109] 和 In_2Se_3[110-112]。首次报道的 2D 铁电材料的实现是在层状 $CuInP_2S_6$ (CIPS) 中 [113]。1997 年，Maisonneuve 等研究了 CIPS，并在 X 射线和中子衍射研究的基础上提出将其作为铁电体 [106]。CIPS 是少数在室温下表现出铁电性的层状化合物之一。CIPS 的原子结构为硫架构，八面体空腔由 Cu、In 和 P-P 三角模式填充。CIPS 块状晶体是由 vdW 弱相互作用垂直堆叠层组成，如图 6.10(a)、(b) 所示。2015 年，Belianinov 等报道了 CIPS 的铁电特性，并说明了尺寸对铁电相稳定性的影响以及目前可以达到的极限 [107]。随后在 2016 年，Liu 等报道了双层 CIPS 的压电响应 (图 6.10(c)~(e)) 和约 4 nm 厚的超薄 CIPS 片的室温铁电性 [108]。由于去极化效应或"死层"效应，超薄铁电薄膜可能会失去铁电性。然而，对于 CIPS，已观察到低于约 4 nm 的铁电开关特性，如图 6.10(f)、(g) 所示。2019 年后期，You 等报道了铁电 CIPS 中巨本征负纵向压电和电拉伸的定量测试 [109]，借助单晶 X 射线晶体学和密度泛函理论计算揭示了该系统中负压电的原子起源。此外，2D vdW 压电材料具有较大的压电响应，这为它们在纳米级柔性机电器件中的应用提供了巨大的潜力，并有望成为环境友好型压电器件的候选者。

此外，人们已经发现了另一种具有平面外极化的 2D 铁电材料硒化铟 (In_2Se_3)。首先对块状 In_2Se_3 晶体的结构进行研究，在 1966 年揭示了其两层晶相，即 α 相和 β 相 [114-116]。在 1990，Abrahams 等在结构分析的基础上提出了 In_2Se_3 族可能的铁电性质 [110]。Ding 等预测层状 α-In_2Se_3 在室温下具有平面外极化铁电，尽管厚度减小到约为 1 nm[111]，他们认为铁电性的起源归因于 Se—In 键的大各向异性，这破坏了系统的对称性。Zhou 等于 2017 年报道了层状 α-In_2Se_3 的面外压电性和铁电性的首次实验 [112]，证实了 α-In_2Se_3 的非对称中心 $R3m$ 的结构，如图 6.11(a) 所示；为了阐明纳米薄片的平面外结构，使用像差校正的扫描透射电子显微镜 (AC-STEM) 直接拍摄气相沉积 (VPD) 生长的多层纳米薄片的横截面，由聚焦离子束分别沿 In_2Se_3 纳米片样品的 [120] 和 [100] 轴切割，如图 6.11(b) 所示，可以从 [120] 横截面 (图 6.11(c)) 获取的环形亮场 (ABF) 图像清楚地识别 Se (1) - In (2) - Se (3) -In (4) - Se (5) 五层之间的 vdW 带隙。有趣的是，图 6.11(d)

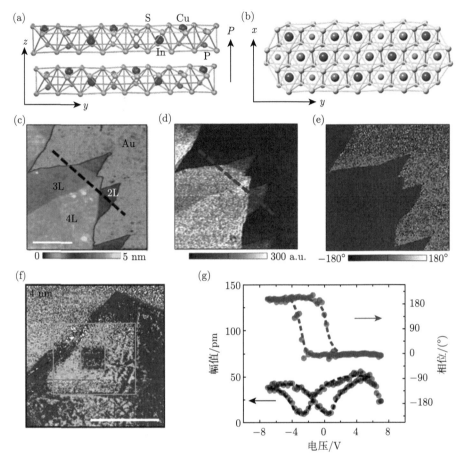

图 6.10　(a) CIPS 晶体结构的侧视图和 (b) 俯视图；(c) AFM 图像；(d) PFM 振幅；(e) 在 Au 涂覆的 SiO_2/Si 衬底上具有不同厚度的 CIPS 的相；(f) 4 nm 厚 CIPS 薄片的 PFM 相位图像；(g) 对于 4 nm 厚的 CIPS 薄片，相应的 PFM 幅度和相位滞回线 (经许可转载[108]。版权 2016，Springer Nature)(彩图请扫封底二维码)

中的 ABF-STEM 强度分布表明，Se (3) 原子偏离中心并向相邻的 In (2) 原子移动，这破坏了五元层中的每一个的反转对称性，并产生了平面外偶极子。这一观察结果与理论计算结果一致。STEM、二次谐波产生 (SHG) 和拉曼测试证实了非中心对称 $R3m$ 晶体结构。他们进行测试以研究 α-In_2Se_3 的铁电性和压电性，如图 6.11(e)~(g) 所示。总之，通过使用具有云母衬底的柔性器件，还证明了电荷传输是通过衬底弯曲来调节的，在压电传感器和纳米机电器件中显示出巨大的应用潜力。随着载流子密度的进一步降低，In_2Se_3 中的铁电性有可能达到单层极限，这对于存储、传感和光伏应用是必须的。

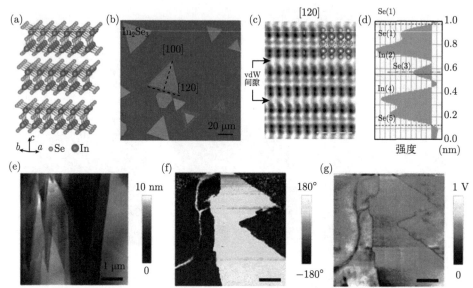

图 6.11 (a) α-In_2Se_3 的晶体结构；(b) In_2Se_3 薄片的 SEM；(c) 沿 [120] 方向切割的 In_2Se_3 薄片的横截面 ABF-STEM 图像；(d) 沿 (c) 中蓝色虚线的强度分布；(e)~(g) α-In_2Se_3 薄片 (大于 100 nm) 的 AFM、PFM 相位和幅度图像 (经许可转载 [112]。版权 2017, American Chemical Society)(彩图请扫封底二维码)

最近的研究报道了几种通过堆叠实现铁电极化的 2D 铁电材料。2018 年，Fei 等报道了 2D 金属 WTe_2 的铁电切换[117]，他们指出，尽管单层中的 WTe_2 是中心对称且非极性的，但堆叠的块状结构是极性的；还发现具有两层或三层的 WTe_2 具有自发的面外极化，可以使用栅极进行切换。2020 年，Xiao 等报道了在几层 WTe_2 中使用电驱动感应的堆叠结构来设计基于贝里曲率的非易失性存储器[118]，正是由于面外电场和静电掺杂之间的相互作用控制了面内层间滑动，从而产生了多极和中心对称的堆叠顺序；此外，基于实验的非线性霍尔输运结果表明，堆叠重排导致了动量空间中层的奇偶选择性贝里曲率的存储，其中贝里曲率及其偶极子的符号反转只发生在奇数层晶体中。另一种通过堆叠表现出铁电性的 2D 材料是六方氮化硼 (h-BN)[119-123]。在自然生长的 h-BN 中，与其他堆叠结构相比，中心对称 vdW 结构具有更低的能量，从而防止极化。2021 年，Yasuda 等[119] 报道了通过 vdW 将 2D 非铁电材料设计成 2D 铁电材料，其堆叠在 h-BN 薄片中，如图 6.12(a) 所示；AA′ 是一种 180° 旋转的自然堆叠结构，它恢复了单层中破坏的反向对称性，并且是非极性的。然而，如果将两个 h-BN 层平行堆叠，则无论是理论上[124,125] 还是实验上[126-129] 都证明 AB 或 BA 的堆叠结构是极性的。在 AB (BA) 堆叠中，上层 B (N) 原子位于下层 N (B) 原子之上，而上层 N (B) 原子位于下六边形中心的空位之上。N 和 B 的 $2p_z$ 轨道的垂直排列扭转了 N 的轨

6.2 制备柔性铁电器件的途径 · 167 ·

道并产生电偶极矩。因此，AB 和 BA 叠层呈现出相反方向的面外极化[130]。同时，Stern 等[120] 报道了在两个自然生长的 h-BN 薄片的界面处出现了稳定的铁电性规则，他们为六种不同的双层 h-BN 提出了高对称结构，如图 6.12(b) 所示。堆叠结构分为两组，称为"平行"和"反平行"扭转方向[124]，在每组中，原子间距离的相对横向移动以循环方式切换堆叠结构。其中，反平行结构 (AA′、AB1′、AB2′) 在空间反演下都是对称的。在平行结构中，AA 堆叠是不稳定的，因为迫使成对的 N 原子堆叠在一起会导致空间排斥力的增加[125]，亚稳态 AB 或 BA 堆叠结构经历层间侧移，只有一半的原子重叠，而另一半与相邻层中六边形的空心对齐[126,128]，这与 Yasuda 等报道的一致。为了实现 AB (BA) 堆叠的 BN 双层薄

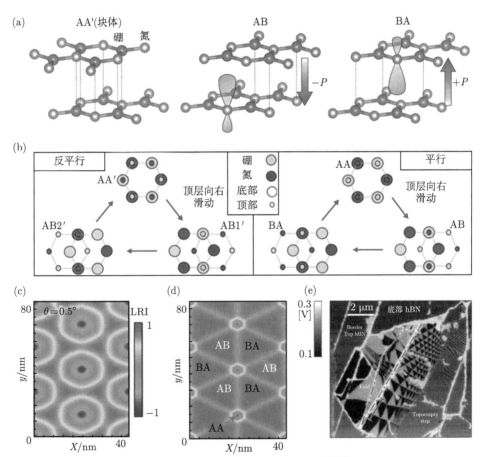

图 6.12 (a) AA′、AB 和 BA 堆叠的原子排列图 (经许可转载[119]。版权 2021, AAAS)；(b) 双层 h-BN 的堆叠结构；(c) 计算的本地注册索引 (LRI) 映射；(d) 几何松弛后计算的 LRI 图；(e) 各种堆叠结构上方的电位图 (经许可转载[120]。版权 2021, AAAS)(彩图请扫封底二维码)

膜,剥离的 BN 单层或将几层 h-BN 片分成两片,然后以精确控制的扭转角将它们压印在一起。零扭转曲角度支持 AB (BA) 堆叠,而小的有限扭转促进层间平移以形成由 AB、BA 和 AA 的三个近比例堆叠结构组成的莫尔图案(图 6.12(c))。结构弛豫将该莫尔图案重建为一个大的三角形区域,该区域由对称的 AB 和 BA 堆叠组成,由不对称的畴壁和拓扑 AA 缺陷隔开,以适应全局失真(图 6.12(d))。扭转 BN 的交错极化通过 PFM 和开尔文探针力显微镜 (KPFM) 得到证实,如图 6.12(e) 所示。Yasuda 等通过测试沉积在单个 AB(BA) 域上的石墨烯薄膜的电阻来检测极化切换。同时,Stern 等通过扫描单个 AB (BA) 域上方的偏置尖端来探测偏振切换。这些发现将合成铁电体确定为一个新兴的研究领域,也为基于 2D 铁电材料的电子器件提供了更多的选择性。

除了上述介绍的本征铁电 2D 材料和人工堆叠的 2D 铁电材料外,非本征 2D 铁电材料的稳定铁电极化也可以通过多种外界因素被有效地诱导,如电子/空穴掺杂[131]、表面功能化[132]、缺陷工程[133]和应变工程[134]。此外,2D 铁电材料的多种材料特性,从众所周知的压电效应、热电效应和体光电效应 (BPVE) 到新探索的谷极化和自旋极化[135],有助于研究人员研究具有各种功能的智能器件,比如隧道结[136]、能量采集器[137]、场效应晶体管[138]和基于 2D 铁电性的光电探测器[139]。近年来,基于 2D 铁电材料的柔性器件也有更多的报道,引起了很大的研究兴趣。2018 年,Xue 等成功开发了一种使用 2D 多层 α-In$_2$Se$_3$ 的高柔性压电纳米发电机[137],在 0.76% 应变下,其输出电压为 35.7 mV,输出电流为 47.3 pA,并证明了其在电子皮肤和双机械能量收集中的潜在应用。2019 年,Dai 等提出了一种基于 2D 多层 γ-InSe 的柔性自供电光电探测器[139];该器件在 400 nm 光照下显示出 824 mA·W^{-1} 的超高光子响应性;另外,当施加 0.62% 的单轴拉伸应变时,该光电检测器的响应性和响应速度分别提高了 696% 和 1010%;此外,该器件在 6 个月测试期间表现出很高的稳定性和可靠性。同年,Dai 等提出,2D 层状 α-In$_2$Se$_3$ 材料的压电系数随层数增加而提高,同时,该材料具有出色的机械耐久性。因此,他们成功制备了用于实时健康监测的柔性自供电压电传感器[140],并且验证了其可靠性。2021 年,Ding 等报道了使用扩散促进的空间有限元法生长大面积、高质量和厚度可控的 2D 铁电 EA$_4$Pb$_3$Br$_{10}$ 单晶薄膜[141];基于 EA$_4$Pb$_3$Br$_{10}$ 单晶薄膜的柔性光电探测器通过简单的模板–剥离工艺实现,该器件具有优异的光检测性能和灵活性;更重要的是,具有面内铁电极化和有效压电系数的光电 EA$_4$Pb$_3$Br$_{10}$ 单晶膜在观察压电光伏效应方面显示出巨大的前景;柔性光电探测器的响应性可以通过压电光伏效应进行调制,在施加的应变下有效增强高达 284%;这项研究还极大地丰富了 2D 钙钛矿铁电体的功能,并探索了新型铁电/压电器件的新可能性。作为一种重要的功能材料,2D 铁电材料有望在即将到来的纳米时代占据主导地位。上述发现将铁电性的研究扩展到了 2D 材料家族，并为柔性电子学的应用开辟了新途径。

6.3 柔性铁电器件的应用

6.3.1 柔性铁电存储器

近年来，柔性电子器件因其柔性和可穿戴性而备受关注，并向智能、多方面和交互系统发展。人们已经生产了不同类型的柔性电子器件，例如柔性电池、电子显示器和集成电路。尽管研究显示了这些柔性电子器件的可行性，但每个电子器件都需要组装成一个集成的电子器件，以便在柔性系统中执行其自身的功能。为了实现集成的柔性电子系统，则开发柔性存储器对于信息存储、数据处理和通信至关重要。最近，柔性存储器因其高机械柔性、强可扩展性、低成本和存储器开关特性而被广泛报道[63,75,142−145]。到目前为止，制备柔性存储器最常用的材料是有机材料和二元氧化物[144,146−153]。然而，基于有机材料的柔性存储器在实际应用中存在一些缺点，例如高温稳定性差、光照下易氧化[154,155]等。同样，基于二元氧化物的柔性存储器也有一些缺点，例如固有的不均匀性、细丝形成的随机性以及器件间和周期间的可变性，这限制了它们的实际应用。为了克服制备柔性存储器所面临的挑战，有必要提高现有存储器材料的可靠性和可加工性，或者开发更合适、更优秀的新型存储器材料。

在过去的几十年中，铁电存储器因其开关速度快、循环耐久性好、物理机制简单而得到了广泛的研究[156−160]。重要的是，铁电存储器的存储机制是通过电可切换的铁电极化产生的，其中，通过正负剩余极化产生对应于存储 "0" 和 "1" 的两个铁电极化态。它是非易失性的[161]，读/写过程可以达到纳秒。与许多其他类型的存储器相比，铁电存储器的优点之一是它可以在恶劣的环境中工作。研究人员广泛研究了它的高温稳定性、耐腐蚀性和耐压性[162,163]。铁电场效应晶体管 (FeFET) 作为快速、低功耗和非易失性存储器技术的长期竞争者，几十年来一直引起人们的极大兴趣[164]。在 FeFET 器件中，信息被存储为栅极绝缘体的永久极化状态，并且可以作为阈值电压的偏置被非破坏性地读取[165]，具有铁电存储器的重要功能。2004 年，Sakai 和 Ilangovan 首次报道了 Pt/SrBi$_2$Ta$_2$O$_9$ (SBT)/HfAlO/Si 基 FeFET，并表明该器件具有出色的数据保留 (大于 12 天) 和耐久性能 (约 10^{12})[166]。尽管这是一项重大成就，但是器件面临着可扩展性差的困难。基于 HfO$_2$ 的铁电材料由于其更好的 CMOS 兼容性和更高的矫顽力场而被认为是解决该问题的理想选择[167]。在 2011，Böscke 等报道了基于 HfSiO 薄膜的 CMOS 兼容的 FeFET，其栅极绝缘层小于 10 nm[165]。器件的横向尺寸和介电厚度至少比以前报道的基于传统铁电材料的器件小一个数量级[166,168]。在 2012 年，Müller

等报道了基于铁电 HfO_2 材料的具有纳秒极化开关 (20 ns) 和长期数据保持 (大于 17 天) 的 FeFET 器件[169]。2018 年,Florent 等证明了基于 HfO_2 的 FeFET 可以集成到 3D-NAND 架构中,为高速、高密度、非易失性存储器的发展铺平道路[170]。2019 年,Xiao 等报道了 ZrO_2 层的添加可以有效提高基于 HZO 薄膜的 FeFET 的性能,这对 FeFET 的应用具有更广泛的研究意义[171]。在 2020,Chen 等首次实现了基于 HZO 和 2D WS_2 的超薄铁电三端突触晶体管[172],该器件通过铁电栅极堆叠磁滞工程实现了大存储窗口,电开关比高达 10^6;同时,铁电突触晶体管器件能够很好地模拟生物突触的基本功能,具有良好的数据保留能力,在神经形态计算领域显示出潜在的应用前景。

近年来,由于铁电薄膜制备技术的进步,超薄高质量铁电薄膜也被成功制备,这使得 FTJ 受到更多关注[173−177]。FTJ 中铁电薄膜的极化转换导致结电阻的大变化,这为电阻的电切换提供了一种新方法,可用于非电荷型存储器和逻辑器件。此外,FTJ 使用非破坏性读出方法,即通过施加远小于铁电薄膜矫顽力电压的电压来读取结电阻,这解决了传统铁电随机存取存储器的主要问题,成为下一代非易失性存取存储器的新候选者。2013 年,Wang 等报道了一个基于 Co/BTO/LSMO 结构的 FTJ,测试证明厚度仅为 2 nm 的铁电薄膜的铁电性仍然很好[178]。FTJ 器件的 TER 在室温下高达 100。此外,FTJ 器件功耗低,写入速度高达纳秒级,满足高性能计算系统的要求。2019 年,Lu 等报道了一种基于 BTO 薄膜的 FTJ 铁电存储器,该器件表现出优异的性能,包括隧道电流密度、开关电压、开关比和稳定性[85]。此外,据报道,与铁电材料仅表现出二进制开关行为的普遍观点不同,忆阻行为已在铁电存储器中实现,并且在这种类似神经元的铁电忆阻器中,它们被发现能够模拟生物突触行为[179−185]。2012 年,Chanthbouala 等报道了一种基于 BTO/LSMO FTJ 的忆阻器[179];他们使用了一个简单的非均匀介质中的畴成核和生长模型来模拟电阻开关行为;此外,他们还得出了控制忆阻反应的解析表达式,这体现了在设计忆阻系统时采用公认的物理现象 (如铁电性) 的优势;他们的工作为下一代神经形态计算架构中的铁电体开辟了不可预见的前景。2017 年,Boyn 等[181] 报道了一种基于 Co/BFO/ (Ca,Ce) MnO_3 (CCMO) FTJ 的忆阻器,并证明了可以通过非均匀极化开关实现尖峰时间依赖性可塑性 (STDP) 突触行为;他们还通过以成核为主的磁畴翻转模型证明了对电导变化的描述;基于该物理模型,铁电纳米突触阵列被证明能够以可预测的方式自主学习识别模式,从而为神经网络的无监督学习开辟了道路。2020 年,Ma 等报道了一种基于 $Ag/BTO/Nb:SrTiO_3$ (NSTO) FTJ 的高性能忆阻器[185],该器件具有最快的运行速度 (600 ps) 和最多的电阻状态 (32 个状态);此外,超快操作还获得了 STDP 作为固体突触器件的功能。总之,基于 FTJ 的忆阻器作为一种新型忆阻器具有诸多优势。第一,它的开关速度非常快。第二,由于铁电畴的尺寸非常小，基于 FTJ 的忆阻器可以微调电阻，从而显示出多级存储的能力[181,186,187]。第三，基于 FTJ 的忆阻器不需要电激活的过程。由于这些特性，基于 FTJ 的忆阻器已成为应用柔性神经形态学的理想选择。

随着独立薄膜制备技术的发展，柔性铁电薄膜也已成功制备，例如通过 vdW 外延方法，蚀刻 LSMO/SAO 牺牲层，以及利用某些 2D 柔性铁电材料，如 6.2 节所述。近年来，柔性铁电存储器的研究被广泛报道[188–192]。2015 年，Ghoneim 等报道了具有硅基 PZT 的柔性非易失性铁电电容器[76]，器件表现出与其他柔性铁电存储器相当的优异的柔性、电容和极化值、优异的耐久性 (1×10^9 的写擦除周期) 和保持特性，可以在 5 mm 半径下弯曲 1300 次。2017 年，Gao 等报道了在 10 μm 厚的云母衬底上生长 SRO/PZT 薄膜，以制备大规模和高质量的柔性铁电电容器[192]，在该柔性电容器中，可以获得约 1200 pm·V^{-1} 的放大压电 d_{33}；此外，在 2.2 mm 半径处反复弯曲之后，电容器的极化、介电可调性和压电响应不会显著劣化，这证明了可穿戴器件和微机电系统在未来发展中的潜在应用。

2017 年，Bakaul 等报道了将基于 PZT 的单晶存储器件通过层转移技术集成到柔性衬底上 (图 6.13(a))[66]；柔性存储器在 4 V 下表现出约 75μC·cm^{-2} 的剩余极化值和 57 ns 的开关速度 (图 6.6(b)、(d))；为了检测这种柔性铁电存储器的可靠性，他们进行疲劳、保持和弯曲循环测试，如图 6.13(b) 所示，在具有 10mm 半径的器件的弯曲状态下，在 10^{10} 开关周期之后，极化保持在 50% 以上，图 6.13(c) 展示出了在相同的弯曲条件下，该器件在 10^5 s 之后没有表现出任何极化损失，表明有可能实现 10 年数据保留的行业标准；此外，图 6.13(d) 示出了在器件弯曲 100 次之后极化性能保持在误差内；他们的工作表明，所制备的柔性铁电存储器具有最佳的存储性能。随后，Yang 等报道了一种基于 BTCO 的柔性且大规模的存储器[63]，如图 6.13(e) 所示。器件的 R_{HRS}/R_{LRS} 超过 50，并且在弯曲和 360000 个写/擦除循环下具有 10^4 s 的保留时间 (图 6.13(f))。即使在 150 mW·cm^{-2} 光下，R_{HRS}/R_{LRS} 在 10^4 s 和 5000 写入/擦除周期之后也可以保持稳定 (图 6.13(g))。重要的是，该器件可以在 25~180°C 的温度范围内安全工作 (图 6.13(h))，这比所研究的大多数有机存储器的性能要好得多。2019 年，Gao 等报道了一种基于云母/Ag-ITO/Bi$_{3.25}$La$_{0.75}$Ti$_3$O$_{12}$ (BLT)/ ITO 结构的柔性、透明、耐疲劳、光读写的铁电非易失性存储器[190]；经过超过 10^8 个写入/擦除循环后，制备的存储器显示出大约 80% 的可见光透射率和无疲劳性能，这些特性在反复弯曲到 3mm 曲率半径后保持稳定；此外，存储器的 "1/0" 状态可以通过读出光伏电流而不是通过破坏性极化开关来访问，这是许多应用中的新兴特征；而且这种透明的抗疲劳存储器可以与隐形眼镜一起使用。类似地，Yang 等报道了在云母衬底上的 (Mn、Ti) 共掺杂的 BFO 存储器[191]，该器件可以弯曲到较小的曲率半径，最高可达 2mm，或者在 4mm 半径处反复经受 10^3 个循环的压缩/拉伸机械弯曲而没有极化劣化,并且在 10^5 s 的保持时间和 10^9 的开关循环之后,抗疲劳性没有任何明显的劣化。

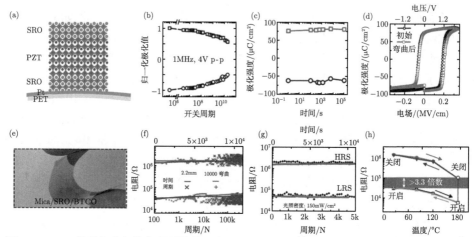

图 6.13　(a) 单晶复合氧化物金属结构示意图;(b) 3×10^{10} 开关循环的疲劳试验;(c) 10^5 s 的保留测试;(d) 弯曲 100 次前后的极化 ((a)~(d) 经许可转载 [66]。版权 2017,WILEY-VCH);(e) 弯曲云母/SRO/BTCO 的光学图像;(f) R_{HRS} 和 R_{LRS} 的保持力和疲劳试验;(g) R_{HRS} 和 R_{LRS} 在光照下的保持力和疲劳试验;(h) R_{HRS} 和 R_{LRS} 的温度依赖性测试 ((e)~(h) 经许可转载 [63]。版权 2017,WILEY-VCH)(彩图请扫封底二维码)

如上所述,基于 FTJ 的存储器具有许多优势,并且性能优于传统存储器。因此,FTJ 对于柔性存储器的研究具有重要意义。近年来,基于 FTJ 的柔性存储器也被广泛报道。2019 年,Luo 等 [86] 报道了一种基于 BTO/LSMO FTJ 的高质量柔性铁电忆阻器,如图 6.14(a) 所示,在报告中,他们介绍了该器件的忆阻特性。如图 6.14(b) 所示,高电阻和低电阻状态之间的电阻滞后循环很明显,这意味着结电阻是施加电压脉冲的历史,即记忆行为 [192];然后,Luo 等表明柔性铁电忆阻器的电阻可以通过电压脉冲的数量、幅度和持续时间来调节。图 6.14(c) 显示了在连续施加不同幅度的脉冲后,结电阻从低电阻态演变为高电阻态。这些结果表明,柔性 FTJ 对于柔性高密度存储器具有多级数据存储能力。此外,他们还对器件的可靠性进行了弯曲和保持测试,这些器件在 5 mm 的曲率半径下循环 100 次,性能没有明显下降,如图 6.14(d) 所示。此外,器件的电阻状态在 OFF、中间态和 ON 状态下保持数小时,这表明它作为非易失性存储器具有优异的保持特性。重要的是,这项研究还促进了基于超薄铁电材料的柔性存储器的应用。2022 年,Sun 等报道了通过设计 SRO/BTO 缓冲层在云母衬底上成功生长柔性 FTJ 忆阻器 [194],它由 BFO 超薄铁电势垒、ZnO 半导体层和 SRO 电极组成 (图 6.14(e));在这种 FTJ 器件中,当极化指向 ZnO 的半导体层时,ZnO 中

6.3 柔性铁电器件的应用

的负屏蔽电荷导致肖特基势垒降低，电阻降低；相反，当极化指向 SRO 时，肖特基势垒将增强，FTJ 将切换到高电阻状态；制备的柔性 FTJ 忆阻器具有稳定的电压调谐多电阻状态，电阻切换在不同弯曲半径下经过 10^3 次弯曲循环后表现出良好的稳定性，如图 6.14(f)、(h) 所示。

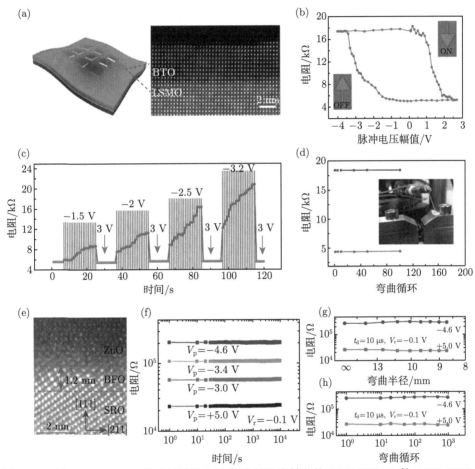

图 6.14 (a) BTO/LSMO 柔性忆阻器的示意图 (左) 和转移的 BTO/LSMO 的 ADF-STEM 图像 (右)；(b) 器件电阻随脉冲电压幅值的变化曲线；(c) 器件电阻状态的演变；(d) 器件分别在高电阻和低电阻状态下的弯曲测试 ((a)~(d) 经许可转载[86]。版权 2019，American Chemical Society)；(e) ZnO/BFO/SRO 界面的 HAADF-STEM 图像；(f) ON 和 OFF 状态之间电阻状态的保持特性；在不同 (g) 弯曲半径和 (h) 弯曲循环下，ON 和 OFF 状态之间的电阻切换 ((e)~(h) 经许可转载[194]。版权 2022，Chinese Ceramic Society)

此外，众所周知，人脑由神经元 (约 10^{11} 个) 和突触 (约 10^{15} 个) 的密集网络组成[195]。神经元负责产生和传递神经网络中的动作电位，突触是一个神经元与另一个神经元或与其他细胞接触和交流的特殊连接。在生物神经网络中，信息的处理和存储同时发生在突触处。因此，突触掌握了大脑的结构，并负责其大规模的并行性，结构可塑性和鲁棒性[196]。最近，神经形态计算已被认为是人工智能应用的一种有前途的技术，这是一种受大脑启发的高度并行的神经元网络模型[197]。为了在硬件上实现神经形态计算，重要的是实现能够真实地模拟生物突触的人工突触和神经元器件[198]。基于此，Zhao 等报告了一种基于 BFO FTJ 的柔性人工突触器件[88]（图 6.15(a)）。这种人工突触器件的工作机制是，器件中的向下极化状态降低了 BFO 势垒的势能差，这有利于电子在 FTJ 中的传输，并导致低电阻状态；相反，在向上极化状态下，BFO 势垒的势能差增大，阻碍了电子传输，产生了高电阻状态。Zhao 等的器件表现出不同的电阻状态，并显示出了超过 1000 的 R_{HRS}/R_{LRS}，电阻状态均保持在 10^4 s 以上，并且没有发现明显的衰减，表明其作为人工突触的潜在应用。此外，较大的开关比也为模拟突触信息传递的刺激和抑制过程提供了有力的支持。Zhao 等还探索了忆阻器多态存储的特点，为实现高密度存储提供了一定的基础。如图 6.15(b) 所示，器件的电流在每次正电压扫描后都会增加，它们做出与生物突触处的突触增强类似的响应[199]。当对器件施加相反极性的连续负电压扫描时，每次扫描后电流减小。结果表明，电导的逐渐变化模拟了连续可变的突触权重，即可变的连接强度。此外，我们利用了脉冲调节器件的行为，发现器件特性和突触现象是一致的。突触权重在早期学习阶段最为明显，而在后期学习阶段则逐渐饱和。重要的是，通过测试在平滑和弯

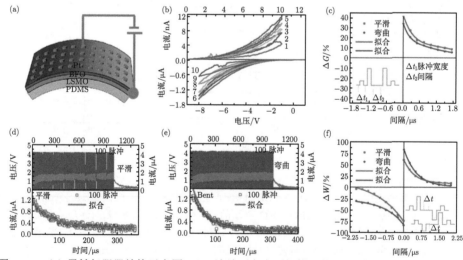

图 6.15 (a) 柔性忆阻器结构示意图；(b) 连续正负电压扫描下的 I-V 曲线；(c) 两种状态下的成对脉冲易化；在 (d) 平滑和 (e) 弯曲状态下，正脉冲的 EPSC 响应；(f) 器件的 STDP 行为测试 (经许可转载[88]。版权 2021，Springer Nature)(彩图请扫封底二维码)

曲状态下的忆阻器，我们已经成功地模拟了生物突触的学习和记忆功能，如兴奋性突触后电流 (EPSC)、成对脉冲易化 (PPF) 和 STDP (图 6.15(c)~(f))。我们小组制备的柔性铁电忆阻器表现出良好的可控性和稳定性，这有利于其在未来柔性神经形态学中的潜在应用，也为柔性脑启发神经形态计算提供了重要参考。

除上述工作外，近期，基于 HfO_2 的柔性铁电存储器的研究也有报道，并受到广泛关注。众所周知，基于 HfO_2 的铁电薄膜表现出优异的 CMOS 兼容性[200,201]，并且铪基材料对环境友好[202]。例如，2019 年，Xiao 等报道了通过原子层沉积 (ALD) 技术在云母衬底上制备 $Hf_{0.5}Zr_{0.5}O_2$ (HZO) 铁电薄膜，用于柔性非易失性存储器[203]；他们将器件置于温度、弯曲应力和辐射等各种条件下，发现器件保持稳定的电学性能；他们的工作表明，基于 HfO_2 的柔性铁电存储器可以在恶劣的条件下稳定工作。Wang 等报道了一种通过低温 ALD 技术制备的基于 HZO 的柔性忆阻器[204]，在该忆阻器中展示了电导逐渐切换的过程；基于梯度电阻切换的特点，模拟了典型的突触可塑性，包括长时程增强 (LTP)、长时程抑制 (LTD)、PPF、双脉冲抑制 (PPD) 和遗忘曲线。

总之，柔性铁电存储器不仅保持了铁电性能，而且在弯曲状态下也能保持稳定。此外，柔性铁电存储器的可穿戴设计使其在大多数情况下便于携带且质量轻。因此，可以预期柔性铁电存储器在未来也可以应用于可穿戴计算机和类脑神经计算等领域，为柔性电子系统的未来发展奠定了坚实的基础。

6.3.2 柔性铁电传感器

传感器的目标是在各种环境条件下，尤其是在纳米级、苛刻和高度柔性的要求下，检测、转换和连接测量的物理量与易于处理的能量形式的信号，柔性传感器自提出以来就引起了极大的关注。小型化趋势下的工业应用和可穿戴设备的需求推动了柔性传感器的发展。铁电材料具有很高的压电系数，可以将微小的振动转化为电信号，因此，基于铁电材料的多功能柔性铁电传感器已成功制备，并用于各种多功能和特殊用途的操作环境[62,205-208]。在 2013 年，Tseng 等报道了基于柔性 PZT 薄膜的触觉传感器，该柔性 PZT 薄膜通过溶胶–凝胶沉积技术在柔性不锈钢衬底上制备，用于生物医学监测[62] (图 6.16(a))；该传感器具有 $0.798\ mV\cdot g^{-1}$ 的灵敏度，并且可以测量不同拓扑区域中的人类脉搏，包括颈动脉、顶端区域、肱动脉、桡动脉、手指和脚踝 (图 6.16(b))；图 6.16(c) 展示出了使用基于 PZT 的柔性触觉传感器，可以成功地检测到人体不同部位的相应信号，测得的脉冲波形可用于诊断心力衰竭和高血压等疾病；作者估计了人类脉搏波从心脏顶点到颈动脉、肱动脉、手指和脚踝动脉的速度，结果是一致的；最后，对封装的 PZT 柔性触觉传感器进行了安全性测试，该器件在人体皮肤上上下移动了 100 次以上，没有发现任何损伤或划痕。此外，2014 年，Dagdeviren 等报道了一种基于 PZT 的柔性皮肤压力传感器，该传感器可用于日常生活[205] (图 6.16(d)~(f))；该传感器使用硅 n 沟道金属氧化物半导体场效应晶体管 (MOSFET) 的纳米膜来放大 PZT 的压电电压响应，并通过电容耦合将其转换为电流输出；该器件提供高水平的压力灵敏度 (约 0.005 Pa) 和快速响应时间 (约 0.1 ms)；他们的工作表明，柔性传感器在健康监测和临床医学中具有重要的应用潜力。

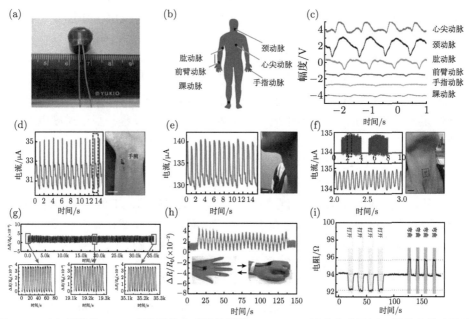

图 6.16 (a) 封装的 PZT 薄膜触觉传感器的照片；(b) PZT 触觉传感器用于测量人体动脉脉搏面积；(c) 人体多个部位的传感器测量值 ((a)~(c) 经许可转载[62]。版权 2013, MDPI)；(d)~(f) IDS-时间图和分别安装在手腕、颈部和喉咙上的传感器的照片 (经许可转载[205]。版权 2014, Nature Publishing Group)；(g) 基于 BNTO/云母的柔性应变传感器的长期弯曲耐久性试验；(h) 基于 BNTO/云母的柔性应变传感器在 1×10^{-4} A 的感应电流下对手指弯曲/释放运动的 $\Delta R/R_0$ 响应；(i) 基于 BNTO/云母的柔性应变传感器的电阻响应在四个光开/关周期和四个弯曲/释放周期下 ((g)~(i) 经许可转载[208]。版权 2019, WILEY-VCH)

此外，在 2019 年，Yang 等提出了一种基于 PLD 技术制备的 $BaNb_{0.5}Ti_{0.5}O_3$ (BNTO) /云母异质结构的柔性应变传感器[208]。BNTO 薄膜的电阻 (R) 在各种弯曲状态下测量为温度 (T) 的函数。在弯曲和释放状态下测量的电阻随着温度的降低而单调增加，表明沉积在云母上的 BNTO 薄膜在整个测量温度范围 (20~350 K) 内具有半导体传输行为。该器件在 20~773.15 K 的宽温度范围内表现出优异的热稳定性，表明应变传感器在极端温度工业领域具有潜在的应用前景。此外，研究表明，柔性 BNTO/云母传感器的应变灵敏度远高于其他常用的大多数柔性材料，如石墨烯、纳米线和纳米管。为了检查该传感器的弯曲稳定性，Yang 等在 5 mm 的弯曲半径下进行了 5000 次弯曲/释放的重复循环，持续时间为 35000 s。如图 6.16(g) 所示，随着弯曲/释放周期的增加，R_0 和 $\Delta R/R_0$ 没有显著变化。这些结果表明，柔性 BNTO/云母传感器具有良好的稳定性和可恢复性。由于柔性传感器的优良特性，它被用于关节运动监测。一个这样的 BNTO/云母传感器用聚酰亚胺胶带连接到手指关节，并监测关节运动产生的信号 (图 6.16(h))。此外，Yang 等证明了柔性的 BNTO/云母异质结构对光和应变都有响应，如图 6.16(i) 所示。结果表明，该器件在 3D 打印等多功能传感产品中也具有广阔的应用前景。

6.3.3 柔性铁电光伏器件

自 19 世纪首次发现光伏效应以来，光伏效应在太阳能电池中的应用受到广泛关注。然而，传统的 pn 结太阳能电池产生的光电压受到半导体带隙宽度和热力学 Shockley-Queisser (S-Q) 的限制。2018 年，Yang 等报道了 STO、TiO_2 和单晶硅的挠曲电诱导的体光伏效应[209]，克服了传统 pn 结太阳能电池的缺点。挠曲电是一种机电特性，指的是应变梯度和电极化之间的耦合[210]。铁电材料的极化开关用于切换光电压/光电流。随着柔性铁电薄膜制备技术的成熟，可以利用其可弯曲性逐渐改变薄膜的弯曲半径，实现应变梯度的连续调谐。最近，柔性铁电薄膜在光伏效应中的应用也有报道。2020 年，Guo 等[87] 发表了一篇关于使用宏观挠曲电效应实现自支撑 BFO 薄膜的连续和可控光电导的报告，这项工作展示了转移到 PDMS 柔性衬底上的独立单晶 BFO 薄膜的可调光伏效应，见图 6.17(a) 和 (b)。他们对器件的铁电光电压特性进行了实验测试，在明暗条件下测试了电流密度电压 (J-V) 曲线，清楚地显示了铁电光伏效应，见图 6.17(c)。在实验过程中，通过不断弯曲器件，由于晶格畸变不均匀，会产生相当大的应变梯度。向下极化的自支撑 BFO 中的应变状态如图 6.17(d) 所示。应变梯度的可控性通过挠曲电效应产生的附加电场与 BFO 极化相关的内部电场之间的耦合进行调制。因此，确定了 BFO 的光电压/光电流。应变梯度可以作为一个自由度来调整光伏响应以实现多级电导。具有向下极化的 BFO 的能带图如图 6.17(e) 所示。由于挠曲电效应，柔性器件向上弯曲会产生有效的向下挠曲电极化。因此，额外的内置电场 $\Delta E_{z,\text{flexo}}$ 与单畴 BFO 薄膜的偏振相关内场 E_{in} 耦合以确定其光伏效应。当 BFO 向上极化时，由挠曲电场产生的合成 $\Delta E_{z,\text{flexo}}$ 将与向下的 E_{in} 抵消，导致光伏效应减弱；而当 BFO 向下极化时，产生的 $\Delta E_{z,\text{flexo}}$ 将加强向上的 E_{in}，从而增强光伏效应。因此，独立铁电薄膜中的多电导态可以通过机械写入来实现，而读取过程可以通过对光伏响应的简单测量来完成。更重要的是，通过控制应变梯度 (SG) 来连续调整光电导是可重现的。如图 6.17(f) 所示，柔性器件在弯曲 300 次后仍能获得多级光电压/光电流，证明了柔性电可调光伏效应在独立式 BFO 薄膜中的稳定性和可重复性。这项研究为柔性电子中的挠曲电效应开辟了研究视野，并可能激发更多利用应变梯度。

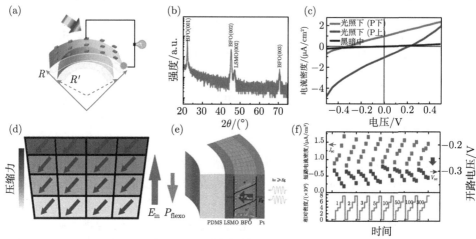

图 6.17　(a) 通过弯曲自支撑 BFO 薄膜获得的应变梯度的示意图；(b) Si 衬底上的自支撑 BFO/LSMO 薄膜的 XRD 结果；(c) 转移的独立式 Pt/BFO/LSMO 光伏电池在光照和黑暗条件下的 J-V 曲线；(d) 具有向下极化的独立 BFO 薄膜中的应变状态的示意图；(e) BFO 薄膜在向下极化至 LSMO 的弯曲状态下的能带弯曲图；(f) 不同弯曲循环后，V_{oc} 和 J_{sc} 随引入的面内应变梯度的变化 (经许可转载 [87]。版权 2020，Springer Nature)(彩图请扫封底二维码)

6.3.4　柔性铁电能量采集器

可再生能源有利于可持续发展，例如太阳能、水能、风能、生物质能、波浪能、潮汐能、海洋热能、地热能等。关于微机电系统，如可穿戴电子器件，对具有自给自足、可持续性和低成本供电的能量发生器有很大的需求 [211,212]。此外，能源管理也成为能源研究领域在提高能源的有效利用率和避免能源收集浪费方面的前沿课题。铁电材料由于其强大的压电特性而被认为是最有希望进行压电振动能量收集的材料系统，它可以在施加电场的情况下产生有效的机电耦合和高机械应变 [213,214]。近年来，基于铁电薄膜的柔性能量采集器被广泛研究。它们已经成为电池的替代品，尤其是应用于体内植入。2014 年，Dagdeviren 等建立了基于 PZT 的集成模块系统，从不同的器官运动中获取和存储能量 (图 6.18(a))[216]，该系统可以在几种动物模型中有效地将肺、心脏和隔膜的自然收缩和舒张运动的机械能转换为电能 (图 6.18(b)、(c))，这是进一步开发基于柔性铁电材料的体内生物医学能量采集器的可能方向之一。2018 年，Wang 等报道了一种在云母衬底上通过溶胶–凝胶法制备的基于 PZT 的柔性能量采集器 [217]；在实验中，对该器件的机械–电能转换性能进行了测试 (图 6.18(d)、(e))，该器件表现出较高的机电能量转换性能，如其所证明的开路输出电压约为 120 V、电流密度约为 150 μA·cm^{-2} 和功率密度 42.7 mW·cm^{-3}，这些结果比之前在压电纳米纤维和有机薄膜基的能量采集器中报道的结果大了 1 ～ 4 个数量级 [215,217,218]，并且也与基于"生长–转移"多步技术的柔性能量采集器相当 [61,219]；并且该柔性能量采集器具有优异的灵活性，可以在高应变下连续弯曲和不弯曲 40000 次 (图 6.18(f))，而不会明显降低输出性能，表现出优异的机械和电耐久性。2019 年，Won 等 [221] 研究了基于应变调谐 PZT 薄膜的柔性振动压电能量采集器；通过阻抗和振动相位角的测量显示了基于柔性 PZT 薄膜的能量收集器件的谐振频率，它的最大功率为 5.6 μW；相应的峰值电压为 690 mV；为了实现柔韧性，合理选择奥氏体金属箔作为衬底，旋转涂层沉积的 PZT 极化为 50 μC·cm^{-2}，与相同条件下刚性 Pt/Si 衬底上的薄膜相比有所改善；结果表明，该器件可广泛应用于小规模能量采集应用。

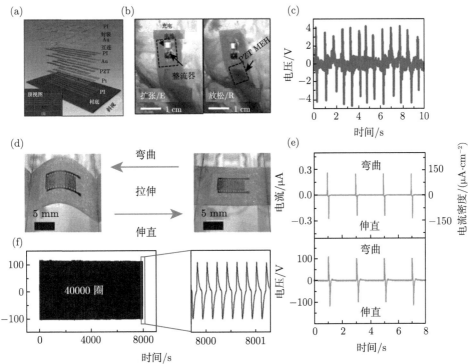

图 6.18　(a) 柔性 PZT 机械能采集器 (MEH) 的示意图；(b) 与整流器和可充电电池共集成的 PZT MEH，安装在牛心脏的右心室上；(c) 安装在右心室上的 PZT MEH 的开路电压与时间的关系 ((a)~(c) 经许可转载 [215]。版权 2014, National Academy of Sciences)；(d) 弯曲和平坦结构的柔性能量采集器的照片；(e) 柔性能量采集器在拉伸模式下产生的机械能–电能转换性能结果；(f) PZT 柔性能量收集器的疲劳测试 ((d)~(f) 经许可转载 [216]。版权 2017, Elsevier Ltd)

6.4 柔性铁电器件应用的潜在研究领域

神经形态学计算具有加速高性能并行和低功耗内存计算,人工智能和自适应学习的潜力。近年来,已有许多基于铁电材料的柔性电子突触器件的报道。这些人工电子突触器件具有许多优异的突触特性,包括多级状态、快速切换速度、高耐久性、高保持率、良好的 CMOS 兼容性,以及可靠的突触学习和记忆功能[221-225]。最近,关于柔性铁电器件的神经形态学计算的研究越来越多。2020 年,Sun 等报道了一种用于神经形态模式识别的基于 BFO 的柔性铁电忆阻器[65];在他们的工作中,该器件显示出最大极化 ($P_s \approx 100\mu C \cdot cm^{-2}$,$P_r \approx 97\mu C \cdot cm^{-2}$)。在弯曲半径为 5mm 的 10^4 次的弯曲试验中,铁电性能保持稳定;通过控制铁电畴翻转,该器件显示出连续可调的电阻,并且考虑到存储特性,对 STDP 函数进行了仿真;重要的是,除了在单个忆阻器中实现 STDP 的学习规则外,他们还提出了基于柔性 BFO 异质结构的深度神经网络 (DNN)。这在数字图像识别,信号处理等方面具有很大的应用潜力;他们模拟了一个双层感知器,如图 6.19(a) 所示;在 1500 个

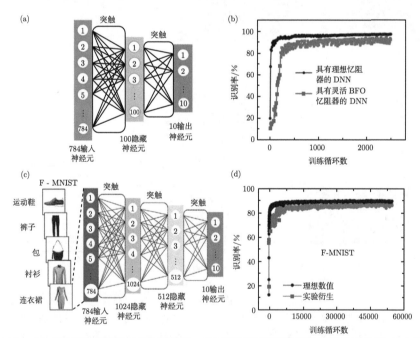

图 6.19 (a) 两层感知器神经网络示意图;(b) 基于理想和真实器件的仿真结果 ((a) 和 (b) 经许可转载[65]。版权 2020, American Chemical Society);(c) 三层感知器神经网络示意图;(d) 基于理想器件和实验器件的仿真结果 ((c) 和 (d) 经许可转载[194]。版权 2022, Chinese Ceramic Society)

6.4 柔性铁电器件应用的潜在研究领域

训练周期之后,基于柔性 BFO 忆阻器的 DNN 对于修改的国家标准技术研究院 (MNIST) 手写数字显示约 90% 的模式识别准确度 (图 6.19(b));这项研究证明了基于 BFO 的柔性铁电忆阻器在神经形态学计算中的潜在应用。然后,2022 年,Sun 等提出了一种用于神经形态学计算的柔性的 BFO FTJ 忆阻器[194];他们利用两层感知器神经网络来识别 MNIST 手写数字;测试结果表明,根据实际的器件结果,经过 3000 个训练周期后,识别精度可以达到 92.8%,这相当于使用理想突触器件进行人工神经网络 (ANN) 仿真获得的 97% 的识别率;除了用于简单手写数字识别的两层人工神经网络之外,他们还设计了三层感知器神经网络,用于 Fashion-MNIST (F-MNIST) 图像数据集的识别任务,如图 6.19(c) 所示;在 55000 个训练周期之后,图像识别的准确度可以达到 86.2% (图 6.19(d));这接近于基于理想突触器件的三层感知器的 89% 的识别精度,甚至更接近于 Berdan 等报道的四层感知器神经网络的 87.6% 的识别精度[222]。

此外,2020 年,Zhong 等开发了一种基于云母/SRO/PZT/铟镓锌氧 (IGZO) 异质结的高稳定性、高性能、全无机的柔性铁电突触晶体管[223]。它由 SRO 栅电极、PZT 铁电层以及顶部带有 Au 源极和漏极的 IGZO 通道组成。栅极电压 V_{GS} 通过部分极化开关来调节半导体沟道的载流子浓度,从而控制电流 I_{DS} 并使人工突触的多个导通状态的访问和编程成为可能。Zhong 等还发现该器件在突触功能上表现出良好的电压脉冲调制电导转换的特性。为了展示 FeFET 突触器件的潜力,他们设计了一个优化的人工神经网络电路,如图 6.20(a) 所示。在模拟学习过

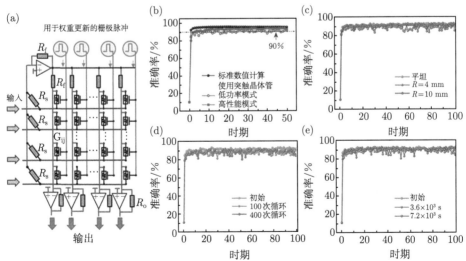

图 6.20 (a) 优化的电路设计;(b) 与标准数值计算方法相比,使用该 FeFET 设计的两层网络的 MNIST 数字识别精度;(c)~(e) 不同弯曲半径、不同弯曲周期、不同弯曲持续时间的 MNIST 数字识别精度 (经许可转载[222]。版权 2020,AIP Publishing)(彩图请扫封底二维码)

程中，人工神经网络在每个周期的 60000 个训练数据中随机选择 10000 张图像进行训练。随后，在测试数据中的另一组 10000 张图像上测试识别精度。测试结果表明，其对 MNIST 手写数字的识别准确率可以达到 94.4%(图 6.20(b))，证实了 FeFET 突触器件的优越性能和所构建的神经网络的更高效率。此外，该器件的主要优点是其对外部干扰的鲁棒性。当弯曲半径减小到 4 mm，弯曲周期为 400 次，弯曲持续时间为 7.2×10^3 s 时，器件的突触行为几乎没有变化，对神经形态网络的识别精度没有显著影响，如图 6.20(c)~(e) 所示。这项工作为实现柔性、稳定、高效、低功耗的大脑模拟神经形态计算提供了一种有前景的方法。

6.5 小　　结

在柔性电子蓬勃发展的大趋势下，迄今为止实现的高质量铁电薄膜的研究成果，使人们坚信柔性铁电器件具有更有价值的研究意义和更广阔的应用前景。在本章中，我们首先总结了近年来制备柔性铁电器件的几种主流方法；在柔性云母衬底上进行 vdW 异质外延，可以获得高质量的柔性铁电薄膜；并且通过结合适当的蚀刻和层转移技术，可以实现更加多样化和功能化的柔性铁电器件的制备；此外，新型 2D 铁电材料已经成为柔性电子器件的理想候选材料。然后，详细讨论了基于柔性铁电器件的应用，包括高速、高密度和可存储的存储器；高灵敏度、抗干扰、适应性强的传感器；连续可控和高度可重复的光伏器件；以及可持续和低成本的能量采集器。最后，我们探索了柔性铁电器件在神经形态计算领域的潜在应用，为类脑计算的未来研究和发展提供了新的思路。

目前，柔性铁电器件的研究越来越成熟，随着时间的推移，在该领域遇到的一些挑战需要得到合理的解决。第一，大多数报道的柔性铁电器件处于分立的状态，将它们与其他电子器件集成是一个亟待解决的问题。第二，柔性铁电器件与 CMOS 的兼容性。对于层状钙钛矿材料 (如 SBT) 或复杂钙钛矿材料 (如 PZT)，集成到 CMOS 工艺中限制了它们在柔性电子器件中的实际应用。第三，大多数报道的柔性铁电器件仍处于实验室阶段，技术的发展和对柔性铁电器件的需求有望加速从实验室到工业化的进程。在柔性铁电器件的蓬勃发展过程中，有效能量，累积能量使用量和应用服务质量被认为是高性能器件的必要指标。第四，铁电材料的界面质量、高漏电流、高介电损耗等问题的存在也会导致器件的低性能，而铁电复合纳米材料的发展可能是解决这些问题的一种有希望的方法。性能更好的铁电复合纳米材料可以弥补单一材料在其自身方面的缺点，并且当应用于柔性电子器件时可能会表现出增强的性能。先进材料的合理设计和合成的最新进展使柔性电子领域的许多应用成为现实。随着未来人工智能、大数据分析和物联网发展的热潮，实现基于柔性铁电器件的脑启发计算将很快成为一种新的研究趋势。柔性电子器件在此类研究中表现出独特的优势，因为它们具有与内源性神经组织相当的软机械性能，类似于神经元和神经突触的小特征尺寸以及类似于神经网络的 3D 互联大孔结构。就铁电器件而言，基于铁电材料的内存计算在解决数据密集型问题的方面具有不可忽视的潜力，还将提高计算效率、解决通信瓶颈、降低漏电流等。最近还报道了基于忆阻器的边缘计算的研究，这些研究为数据计算和信号处理的爆炸性增长提供了有希望的解决方案。目前基于柔性铁电器件的这一领域的研究还处于起步阶段，人们需要为此付出更多的努力。此外，柔性电子器件的功能与神经组织之间的紧密联系有望为神经科学研究和神经病学的临床应用提供新的方向。随着研究和开发的不断努力，柔性铁电器件的创新应用将取得更丰硕的成果。

参 考 文 献

[1] Liu X, Wei Y, Qiu Y. Advanced flexible skin-like pressure and strain sensors for human health monitoring[J]. Micromachines, 2021, 12(6): 695.

[2] Rogers J A, Someya T, Huang Y. Materials and mechanics for stretchable electronics[J]. Science, 2010, 327(5973): 1603-1607.

[3] Song K, Kim J, Cho S, et al. Flexible-device injector with a microflap array for subcutaneously implanting flexible medical electronics[J]. Advanced Healthcare Materials, 2018, 7(15): 1800419.

[4] DiStefano T, Fjelstad J. Novel uses of flexible circuit technology in high performance electronic applications[J]. Microelectronics International, 1996, 13(1): 11-15.

[5] Wang C, Chien J C, Takei K, et al. Extremely bendable, high-performance integrated circuits using semiconducting carbon nanotube networks for digital, analog, and radio-frequency applications[J]. Nano Letters, 2012, 12(3): 1527-1533.

[6] Shi J, Guo X, Chen R, et al. Recent progress in flexible battery[J]. Progress in Chemistry, 2016, 28(4): 577.

[7] Koo M, Park K I, Lee S H, et al. Bendable inorganic thin-film battery for fully flexible electronic systems[J]. Nano Letters, 2012, 12(9): 4810-4816.

[8] Lin Q, Huang H, Jing Y, et al. Flexible photovoltaic technologies[J]. Journal of Materials Chemistry C, 2014, 2(7): 1233-1247.

[9] Koo J H, Kim D C, Shim H J, et al. Flexible and stretchable smart display: materials, fabrication, device design, and system integration[J]. Advanced Functional Materials, 2018, 28(35): 1801834.

[10] Fukagawa H, Sasaki T, Tsuzuki T, et al. Long-lived flexible displays employing efficient and stable inverted organic light-emitting diodes[J]. Advanced Materials, 2018, 30(28): 1706768.

[11] Lemey S, Agneessens S, van Torre P, et al. Wearable flexible lightweight modular RFID

tag with integrated energy harvester[J]. IEEE Transactions on Microwave Theory and Techniques, 2016, 64(7): 2304-2314.

[12] Lam Po Tang S. Recent developments in flexible wearable electronics for monitoring applications[J]. Transactions of the Institute of Measurement and Control, 2007, 29(3-4): 283-300.

[13] Xu K, Lu Y, Takei K. Multifunctional skin-inspired flexible sensor systems for wearable electronics[J]. Advanced Materials Technologies, 2019, 4(3): 1800628.

[14] Nag A, Mukhopadhyay S C, Kosel J. Wearable flexible sensors: A review[J]. IEEE Sensors Journal, 2017, 17(13): 3949-3960.

[15] Gao W, Ota H, Kiriya D, et al. Flexible electronics toward wearable sensing[J]. Accounts of Chemical Research, 2019, 52(3): 523-533.

[16] Wang T Y, He Z Y, Liu H, et al. Flexible electronic synapses for face recognition application with multimodulated conductance states[J]. ACS Applied Materials & Interfaces, 2018, 10(43): 37345-37352.

[17] Park S, Vosguerichian M, Bao Z. A review of fabrication and applications of carbon nanotube film-based flexible electronics[J]. Nanoscale, 2013, 5(5): 1727-1752.

[18] Lee P, Lee J, Lee H, et al. Highly stretchable and highly conductive metal electrode by very long metal nanowire percolation network[J]. Advanced Materials, 2012, 24(25): 3326-3332.

[19] Zhu Z T, Menard E, Hurley K, et al. Spin on dopants for high-performance single-crystal silicon transistors on flexible plastic substrates[J]. Applied Physics Letters, 2005, 86(13): 133507.

[20] Chinnappan A, Baskar C, Baskar S, et al. An overview of electrospun nanofibers and their application in energy storage, sensors and wearable/flexible electronics[J]. Journal of Materials Chemistry C, 2017, 5(48): 12657-12673.

[21] Yu A, Roes I, Davies A, et al. Ultrathin, transparent, and flexible graphene films for supercapacitor application[J]. Applied Physics Letters, 2010, 96(25): 253105.

[22] Kim S J, Choi K, Lee B, et al. Materials for flexible, stretchable electronics: graphene and 2D materials[J]. Annual Review of Materials Research, 2015, 45: 63-84.

[23] Fang H, Bai S L, Wong C P. Thermal, mechanical and dielectric properties of flexible BN foam and BN nanosheets reinforced polymer composites for electronic packaging application[J]. Composites Part A: Applied Science and Manufacturing, 2017, 100: 71-80.

[24] Zheng L, Wang X, Jiang H, et al. Recent progress of flexible electronics by 2D transition metal dichalcogenides[J]. Nano Research, 2022, 15(3): 2413-2432.

[25] Velusamy D B, Kim R H, Cha S, et al. Flexible transition metal dichalcogenide nanosheets for band-selective photodetection[J]. Nat. Commun., 2015, 6: 8063.

[26] Sekitani T, Noguchi Y, Hata K, et al. A rubberlike stretchable active matrix using elastic conductors[J]. Science, 2008, 321(5895): 1468-1472.

[27] Aliane A, Fischer V, Galliari M, et al. Enhanced printed temperature sensors on

flexible substrate[J]. Microelectronics Journal, 2014, 45(12): 1621-1626.

[28] Jilani S F, Alomainy A. Planar millimeter-wave antenna on low-cost flexible PET substrate for 5G applications[C]//2016 10th European Conference on Antennas and Propagation (EuCAP). IEEE, 2016: 1-3.

[29] Fan X, Nie W, Tsai H, et al. PEDOT:PSS for flexible and stretchable electronics: modifications, strategies, and applications[J]. Advanced Science, 2019, 6(19): 1900813.

[30] Stadlober B, Zirkl M, Irimia-Vladu M. Route towards sustainable smart sensors: ferroelectric polyvinylidene fluoride-based materials and their integration in flexible electronics[J]. Chemical Society Reviews, 2019, 48(6): 1787-1825.

[31] Lopez-Encarnacion J M, Burton J D, Tsymbal E Y, et al. Organic multiferroic tunnel junctions with ferroelectric poly (vinylidene fluoride) barriers[J]. Nano Letters, 2011, 11(2): 599-603.

[32] Rodríguez-Roldán G, Suaste-Gómez E, Reyes-Cruz H. Fabrication and characterization of a PVDF/PLA membrane made by electrospinning as a flexible temperature sensor[C]//2018 15th International Conference on Electrical Engineering, Computing Science and Automatic Control (CCE). IEEE, 2018: 1-4.

[33] Lu Y, Wang Q, Chen R, et al. Spin-dependent charge transport in 1D chiral hybrid lead-bromide perovskite with high stability[J]. Advanced Functional Materials, 2021, 31(43): 2104605.

[34] Chen X, Zhou X, Cheng R, et al. Electric field control of Néel spin-orbit torque in an antiferromagnet[J]. Nature Materials, 2019, 18(9): 931-935.

[35] Chen X, Shi S, Shi G, et al. Observation of the antiferromagnetic spin Hall effect[J]. Nature Materials, 2021, 20(6): 800-804.

[36] Han T, Nag A, Simorangkir R B V B, et al. Multifunctional flexible sensor based on laser-induced graphene[J]. Sensors, 2019, 19(16): 3477.

[37] Yun J, Song C, Lee H, et al. Stretchable array of high-performance micro-supercapacitors charged with solar cells for wireless powering of an integrated strain sensor[J]. Nano Energy, 2018, 49: 644-654.

[38] Zhao G, Zhang X, Cui X, et al. Piezoelectric polyacrylonitrile nanofiber film-based dual-function self-powered flexible sensor[J]. ACS Applied Materials & Interfaces, 2018, 10(18): 15855-15863.

[39] Li W, Li W, Wei J, et al. Preparation of conductive Cu patterns by directly writing using nano-Cu ink[J]. Materials Chemistry and Physics, 2014, 146(1-2): 82-87.

[40] Yang L, Rida A, Tentzeris M M. Design and development of radio frequency identification (RFID) and RFID-enabled sensors on flexible low cost substrates[J]. Synthesis Lectures on RF/Microwaves, 2009, 1(1): 1-89.

[41] Wang Y, Yan C, Cheng S Y, et al. Flexible RFID tag metal antenna on paper-based substrate by inkjet printing technology[J]. Advanced Functional Materials, 2019, 29: 1902579.

[42] Kim S, Kwon H J, Lee S, et al. Low-power flexible organic light-emitting diode display

device[J]. Advanced Materials, 2011, 23(31): 3511-3516.

[43] Wang C, Hwang D, Yu Z, et al. User-interactive electronic skin for instantaneous pressure visualization[J]. Nature Materials, 2013, 12(10): 899-904.

[44] Takei K, Takahashi T, Ho J C, et al. Nanowire active-matrix circuitry for low-voltage macroscale artificial skin[J]. Nature materials, 2010, 9(10): 821-826.

[45] Kim J, Lee M, Shim H J, et al. Stretchable silicon nanoribbon electronics for skin prosthesis[J]. Nature Communications, 2014, 5(1): 1-11.

[46] Tominaka S, Nishizeko H, Mizuno J, et al. Bendable fuel cells: on-chip fuel cell on a flexible polymer substrate[J]. Energy & Environmental Science, 2009, 2(10): 1074-1077.

[47] Wongrujipairoj K, Poolnapol L, Arpornwichanop A, et al. Suppression of zinc anode corrosion for printed flexible zinc-air battery[J]. Physica Status Solidi (b), 2017, 254(2): 1600442.

[48] Tajima R, Miwa T, Oguni T, et al. Truly wearable display comprised of a flexible battery, flexible display panel, and flexible printed circuit[J]. Journal of the Society for Information Display, 2014, 22(5): 237-244.

[49] Koma A. Van der Waals epitaxy for highly lattice-mismatched systems[J]. Journal of Crystal Growth, 1999, 201: 236-241.

[50] Koma A, Sunouchi K, Miyajima T. Summary abstract: Fabrication of ultrathin heterostructures with van der Waals epitaxy [J]. Journal of Vacuum Science & Technology B: Microelectronics Processing and Phenomena, 1985, 3(2): 724.

[51] Aoto N, Sunouchi K, Koma A. Heteroepitaxy onto surfaces with no dangling bonds-heteroepitaxy of selenium on cleaved faces of tellurium[C]//Extended Abstracts of the Conference on Solid State Devices and Materials. Publication Office, Business Center for Academic Society Japan, 1983, 15: 309.

[52] Saiki K, Ueno K, Shimada T, et al. Application of van der Waals epitaxy to highly heterogeneous systems[J]. Journal of Crystal Growth, 1989, 95(1-4): 603-606.

[53] Chu Y H. van der Waals oxide heteroepitaxy[J]. NPJ Quantum Materials, 2017, 2(1): 1-5.

[54] Chang L, You L, Wang J. The path to flexible ferroelectrics: Approaches and progress [J]. Japanese Journal of Applied Physics, 2018, 57(9): 0902A3.

[55] Zhou Y, Nie Y, Liu Y, et al. Epitaxy and photoresponse of two-dimensional GaSe crystals on flexible transparent mica sheets[J]. ACS Nano, 2014, 8(2): 1485-1490.

[56] Wu P C, Chen P F, Do T H, et al. Heteroepitaxy of Fe_3O_4/muscovite: A new perspective for flexible spintronics[J]. ACS Applied Materials & Interfaces, 2016, 8(49): 33794-33801.

[57] Liu Y, Tang M, Meng M, et al. Epitaxial growth of ternary topological insulator Bi_2Te_2Se 2D crystals on mica[J]. Small, 2017, 13(18): 1603572.

[58] Ren C, Tan C, Gong L, et al. Highly transparent, all-oxide, heteroepitaxy ferroelectric thin film for flexible electronic devices[J]. Applied Surface Science, 2018, 458: 540-545.

[59] Jiang J, Bitla Y, Huang C W, et al. Flexible ferroelectric element based on van der Waals heteroepitaxy[J]. Science Advances, 2017, 3(6): e1700121.

[60] Yen M, Bitla Y, Chu Y H. van der Waals heteroepitaxy on muscovite[J]. Materials Chemistry and Physics, 2019, 234: 185-195.

[61] Park K I, Son J H, Hwang G T, et al. Highly-efficient, flexible piezoelectric PZT thin film nanogenerator on plastic substrates[J]. Advanced Materials, 2014, 26(16): 2514-2520.

[62] Tseng H J, Tian W C, Wu W J. Flexible PZT thin film tactile sensor for biomedical monitoring[J]. Sensors, 2013, 13(5): 5478-5492.

[63] Yang Y, Yuan G, Yan Z, et al. Flexible, semitransparent, and inorganic resistive memory based on $BaTi_{0.95}Co_{0.05}O_3$ film[J]. Advanced Materials, 2017, 29(26): 1700425.

[64] Chu Y H, Amrillah T, Bitla Y, et al. Flexible multiferroic bulk heterojunction with giant magnetoelectric coupling via van der Waals epitaxy[J]. ACS Nano, 2017, 11(6): 6122-6130.

[65] Sun H, Luo Z, Zhao L, et al. $BiFeO_3$-based flexible ferroelectric memristors for neuromorphic pattern recognition[J]. ACS Applied Electronic Materials, 2020, 2(4): 1081-1089.

[66] Bakaul S R, Serrao C R, Lee O, et al. High speed epitaxial perovskite memory on flexible substrates[J]. Advanced Materials, 2017, 29(11): 1605699.

[67] Zuo Z, Chen B, Zhan Q, et al. Preparation and ferroelectric properties of freestanding Pb (Zr, Ti) O_3 thin membranes[J]. Journal of Physics D: Applied Physics, 2012, 45(18): 185302.

[68] Bretos I, Jiménez R, Wu A, et al. Activated solutions enabling low-temperature processing of functional ferroelectric oxides for flexible electronics[J]. Advanced Materials, 2014, 26(9): 1405-1409.

[69] Lee L P, Burns M J, Char K. Free-standing microstructures of $YBa_2Cu_3O_{7-\delta}$: A high-temperature superconducting air bridge[J]. Applied Physics Letters, 1992, 61(22): 2706-2708.

[70] Kim J H, Grishin A M. Free-standing epitaxial $La_{1-x}(Sr,Ca)_xMnO_3$ membrane on Si for uncooled infrared microbolometer[J]. Applied Physics Letters, 2005, 87(3): 033502.

[71] Tsakalakos L, Sands T. Epitaxial ferroelectric (Pb, La)(Zr, Ti) O_3 thin films on stainless steel by excimer laser liftoff[J]. Applied Physics Letters, 2000, 76(2): 227-229.

[72] Gan Q, Rao R A, Eom C B, et al. Direct measurement of strain effects on magnetic and electrical properties of epitaxial $SrRuO_3$ thin films[J]. Applied Physics Letters, 1998, 72(8): 978-980.

[73] Bakaul S R, Serrao C R, Lee M, et al. Single crystal functional oxides on silicon[J]. Nature Communications, 2016, 7(1): 1-5.

[74] Han S T, Zhou Y, Roy V A L. Towards the development of flexible non-volatile memories[J]. Advanced Materials, 2013, 25(38): 5425-5449.

[75] Ghoneim M T, Hussain M M. Review on physically flexible nonvolatile memory for

internet of everything electronics[J]. Electronics, 2015, 4(3): 424-479.

[76] Ghoneim M T, Zidan M A, Alnassar M Y, et al. Thin PZT-based ferroelectric capacitors on flexible silicon for nonvolatile memory applications[J]. Advanced Electronic Materials, 2015, 1(6): 1500045.

[77] Kim R H, Kim D H, Xiao J, et al. Waterproof AlInGaP optoelectronics on stretchable substrates with applications in biomedicine and robotics[J]. Nature Materials, 2010, 9(11): 929-937.

[78] Shen L, Wu L, Sheng Q, et al. Epitaxial lift-off of centimeter-scaled spinel ferrite oxide thin films for flexible electronics[J]. Advanced Materials, 2017, 29(33): 1702411.

[79] Alonso J A, Rasines I, Soubeyroux J L. Tristrontium dialuminum hexaoxide: an intricate superstructure of perovskite[J]. Inorganic Chemistry, 1990, 29(23): 4768-4771.

[80] Lu D, Baek D J, Hong S S, et al. Synthesis of freestanding single-crystal perovskite films and heterostructures by etching of sacrificial water-soluble layers[J]. Nature Materials, 2016, 15(12): 1255-1260.

[81] Baek D J, Lu D, Hikita Y, et al. Mapping cation diffusion through lattice defects in epitaxial oxide thin films on the water-soluble buffer layer $Sr_3Al_2O_6$ using atomic resolution electron microscopy[J]. APL Materials, 2017, 5(9): 096108.

[82] Hong S S, Yu J H, Lu D, et al. Two-dimensional limit of crystalline order in perovskite membrane films[J]. Science Advances, 2017, 3(11): eaao5173.

[83] Ji D, Cai S, Paudel T R, et al. Freestanding crystalline oxide perovskites down to the monolayer limit[J]. Nature, 2019, 570(7759): 87-90.

[84] Dong G, Li S, Yao M, et al. Super-elastic ferroelectric single-crystal membrane with continuous electric dipole rotation[J]. Science, 2019, 366(6464): 475-479.

[85] Lu D, Crossley S, Xu R, et al. Freestanding oxide ferroelectric tunnel junction memories transferred onto silicon[J]. Nano Letters, 2019, 19(6): 3999-4003.

[86] Luo Z D, Peters J J P, Sanchez A M, et al. Flexible memristors based on single-crystalline ferroelectric tunnel junctions[J]. ACS Applied Materials & Interfaces, 2019, 11(26): 23313-23319.

[87] Guo R, You L, Lin W, et al. Continuously controllable photoconductance in freestanding $BiFeO_3$ by the macroscopic flexoelectric effect[J]. Nature Communications, 2020, 11(1): 1-9.

[88] Zhao Z, Abdelsamie A, Guo R, et al. Flexible artificial synapse based on single-crystalline $BiFeO_3$ thin film[J]. Nano Research, 2022, 15(3): 2682-2688.

[89] Zhang Y, Ma C, Lu X, et al. Recent progress on flexible inorganic single-crystalline functional oxide films for advanced electronics[J]. Materials Horizons, 2019, 6(5): 911-930.

[90] Novoselov K S, Geim A K, Morozov S V, et al. Electric field effect in atomically thin carbon films[J]. Science, 2004, 306(5696): 666-669.

[91] Gupta A, Sakthivel T, Seal S. Recent development in 2D materials beyond graphene[J]. Progress in Materials Science, 2015, 73: 44-126.

[92] Yin S, Song C, Sun Y, et al. Electric and light dual-gate tunable MoS$_2$ memtransistor[J]. ACS Applied Materials & Interfaces, 2019, 11(46): 43344-43350.

[93] Lopez-Sanchez O, Lembke D, Kayci M, et al. Ultrasensitive photodetectors based on monolayer MoS$_2$[J]. Nature Nanotechnology, 2013, 8(7): 497-501.

[94] Lipp A, Schwetz K A, Hunold K. Hexagonal boron nitride: Fabrication, properties and applications[J]. Journal of the European Ceramic Society, 1989, 5(1): 3-9.

[95] De Padova P, Ottaviani C, Quaresima C, et al. 24 h stability of thick multilayer silicene in air[J]. 2D Materials, 2014, 1(2): 021003.

[96] Osada M, Sasaki T. Exfoliated oxide nanosheets: new solution to nanoelectronics[J]. Journal of Materials Chemistry, 2009, 19(17): 2503-2511.

[97] Ma R, Sasaki T. Nanosheets of oxides and hydroxides: Ultimate 2D charge-bearing functional crystallites[J]. Advanced Materials, 2010, 22(45): 5082-5104.

[98] Lee G H, Yu Y J, Cui X, et al. Flexible and transparent MoS$_2$ field-effect transistors on hexagonal boron nitride-graphene heterostructures[J]. ACS Nano, 2013, 7(9): 7931-7936.

[99] Chang H Y, Yang S, Lee J, et al. High-performance, highly bendable MoS$_2$ transistors with high-k dielectrics for flexible low-power systems[J]. ACS Nano, 2013, 7(6): 5446-5452.

[100] Cheng R, Jiang S, Chen Y, et al. Few-layer molybdenum disulfide transistors and circuits for high-speed flexible electronics[J]. Nature Communications, 2014, 5(1): 1-9.

[101] He Q, Zeng Z, Yin Z, et al. Fabrication of flexible MoS$_2$ thin-film transistor arrays for practical gas-sensing applications[J]. Small, 2012, 8(19): 2994-2999.

[102] Liu J, Yin Z, Cao X, et al. Fabrication of flexible, all-reduced graphene oxide nonvolatile memory devices[J]. Advanced Materials, 2013, 25(2): 233-238.

[103] Shirodkar S N, Waghmare U V. Emergence of ferroelectricity at a metal-semiconductor transition in a 1T monolayer of MoS$_2$[J]. Physical Review Letters, 2014, 112(15): 157601.

[104] Chang K, Liu J, Lin H, et al. Discovery of robust in-plane ferroelectricity in atomic-thick SnTe[J]. Science, 2016, 353(6296): 274-278.

[105] Fei R, Kang W, Yang L. Ferroelectricity and phase transitions in monolayer group-IV monochalcogenides[J]. Physical Review Letters, 2016, 117(9): 097601.

[106] Maisonneuve V, Reau J M, Dong M, et al. Ionic conductivity in ferroic CuInP$_2$S$_6$ and CuCrP$_2$S$_6$[J]. Ferroelectrics, 1997, 196(1): 257-260.

[107] Belianinov A, He Q, Dziaugys A, et al. CuInP$_2$S$_6$ room temperature layered ferroelectric[J]. Nano Letters, 2015, 15(6): 3808-3814.

[108] Liu F, You L, Seyler K L, et al. Room-temperature ferroelectricity in CuInP$_2$S$_6$ ultrathin flakes[J]. Nature Communications, 2016, 7(1): 1-6.

[109] You L, Zhang Y, Zhou S, et al. Origin of giant negative piezoelectricity in a layered van der Waals ferroelectric[J]. Science Advances, 2019, 5(4): eaav3780.

[110] Abrahams S C. Systematic prediction of new ferroelectrics on the basis of structure[J].

Ferroelectrics, 1990, 104(1): 37-50.

[111] Ding W, Zhu J, Wang Z, et al. Prediction of intrinsic two-dimensional ferroelectrics in In$_2$Se$_3$ and other III2-VI3 van der Waals materials[J]. Nature Communications, 2017, 8(1): 1-8.

[112] Zhou Y, Wu D, Zhu Y, et al. Out-of-plane piezoelectricity and ferroelectricity in layered α-In$_2$Se$_3$ nanoflakes[J]. Nano Letters, 2017, 17(9): 5508-5513.

[113] Osada M, Sasaki T. The rise of 2D dielectrics/ferroelectrics[J]. APL Materials, 2019, 7(12): 120902.

[114] Osamura K, Murakami Y, Tomiie Y. Crystal structures of α-and β-indium selenide, In$_2$Se$_3$[J]. Journal of the Physical Society of Japan, 1966, 21(9): 1848-1848.

[115] Popović S, Celustka B, Bidjin D. X-ray diffraction measurement of lattice parameters of In$_2$Se$_3$[J]. Physica Status Solidi, 2010, 6(1): 301-304.

[116] Popović S, Tonejc A, Gržeta-PlenkovićB, et al. Revised and new crystal data for indium selenides[J]. Journal of Applied Crystallography, 1979, 12(4): 416-420.

[117] Fei Z, Zhao W, Palomaki T A, et al. Ferroelectric switching of a two-dimensional metal[J]. Nature, 2018, 560(7718): 336-339.

[118] Xiao J, Wang Y, Wang H, et al. Berry curvature memory through electrically driven stacking transitions[J]. Nature Physics, 2020, 16(10): 1028-1034.

[119] Yasuda K, Wang X, Watanabe K, et al. Stacking-engineered ferroelectricity in bilayer boron nitride[J]. Science, 2021, 372(6549): 1458-1462.

[120] Stern M V, Waschitz Y, Cao W, et al. Interfacial ferroelectricity by van der Waals sliding[J]. Science, 2021, 372(6549): 1462-1466.

[121] Woods C R, Ares P, Nevison-Andrews H, et al. Charge-polarized interfacial superlattices in marginally twisted hexagonal boron nitride[J]. Nature Communications, 2021, 12(1): 1-7.

[122] Tsymbal E Y. Two-dimensional ferroelectricity by design[J]. Science, 2021, 372(6549): 1389-1390.

[123] Zheng Z, Ma Q, Bi Z, et al. Unconventional ferroelectricity in moiré heterostructures[J]. Nature, 2020, 588(7836): 71-76.

[124] Constantinescu G, Kuc A, Heine T. Stacking in bulk and bilayer hexagonal boron nitride[J]. Physical Review Letters, 2013, 111(3): 036104.

[125] Zhou S, Han J, Dai S, et al. van der Waals bilayer energetics: Generalized stacking-fault energy of graphene, boron nitride, and graphene/boron nitride bilayers[J]. Physical Review B, Condensed Matter and Materials Physics, 2015, 92(15): 155438.1-13.

[126] Warner J H, Rummeli M H, Bachmatiuk A, et al. Atomic resolution imaging and topography of boron nitride sheets produced by chemical exfoliation[J]. ACS Nano, 2010, 4(3): 1299-1304.

[127] Kim C J, Brown L, Graham M W, et al. Stacking order dependent second harmonic generation and topological defects in h-BN bilayers[J]. Nano Letters, 2013, 13(11): 5660-5665.

[128] Gilbert S M, Pham T, Dogan M, et al. Alternative stacking sequences in hexagonal boron nitride[J]. 2D Materials, 2019, 6(2): 021006.

[129] Park H J, Cha J, Choi M, et al. One-dimensional hexagonal boron nitride conducting channel[J]. Science Advances, 2020, 6(10): eaay4958.

[130] Li L, Wu M. Binary compound bilayer and multilayer with vertical polarizations: two-dimensional ferroelectrics, multiferroics, and nanogenerators[J]. ACS Nano, 2017, 11(6): 6382-6388.

[131] Yang Y, Iniguez J, Mao A J, et al. Prediction of a novel magnetoelectric switching mechanism in multiferroics[J]. Physical Review Letters, 2014, 112(5): 057202.

[132] Wu M, Dong S, Yao K, et al. Ferroelectricity in covalently functionalized two-dimensional materials: Integration of high-mobility semiconductors and nonvolatile memory[J]. Nano Letters, 2016, 16(11): 7309-7315.

[133] Zhao Y, Lin L, Zhou Q, et al. Surface vacancy-induced switchable electric polarization and enhanced ferromagnetism in monolayer metal trihalides[J]. Nano Letters, 2018, 18(5): 2943-2949.

[134] Xu T, Wang X, Mai J, et al. Strain engineering for 2D ferroelectricity in Lead chalcogenides[J]. Advanced Electronic Materials, 2020, 6(1): 1900932.

[135] Qi L, Ruan S, Zeng Y J. Review on recent developments in 2D ferroelectrics: Theories and applications[J]. Advanced Materials, 2021, 33(13): 2005098.

[136] Kang L, Jiang P, Hao H, et al. Giant tunneling electroresistance in two-dimensional ferroelectric tunnel junctions with out-of-plane ferroelectric polarization[J]. Physical Review B, 2020, 101(1): 014105.

[137] Xue F, Zhang J, Hu W, et al. Multidirection piezoelectricity in mono-and multilayered hexagonal α-In_2Se_3[J]. ACS Nano, 2018, 12(5): 4976-4983.

[138] Si M, Saha A K, Gao S, et al. A ferroelectric semiconductor field-effect transistor[J]. Nature Electronics, 2019, 2(12): 580-586.

[139] Dai M, Chen H, Wang F, et al. Robust piezo-phototronic effect in multilayer γ-InSe for high-performance self-powered flexible photodetectors[J]. ACS Nano, 2019, 13(6): 7291-7299.

[140] Dai M, Wang Z, Wang F, et al. Two-dimensional van der Waals materials with aligned in-plane polarization and large piezoelectric effect for self-powered piezoelectric sensors[J]. Nano Letters, 2019, 19(8): 5410-5416.

[141] Ding R, Lyu Y, Wu Z, et al. Effective piezo-phototronic enhancement of flexible photodetectors based on 2D hybrid perovskite ferroelectric single-crystalline thin-films[J]. Advanced Materials, 2021, 33(32): 2101263.

[142] Sun B, Zhang X, Zhou G, et al. A flexible nonvolatile resistive switching memory device based on ZnO film fabricated on a foldable PET substrate[J]. Journal of Colloid and Interface Science, 2018, 520: 19-24.

[143] Han S T, Zhou Y, Roy V A L. Towards the development of flexible non-volatile memories[J]. Advanced Materials, 2013, 25(38): 5424-5424.

[144] Casula G, Busby Y, Franquet A, et al. A flexible organic memory device with a clearly disclosed resistive switching mechanism[J]. Organic Electronics, 2019, 64: 209-215.

[145] Zhou L, Mao J, Ren Y, et al. Recent advances of flexible data storage devices based on organic nanoscaled materials[J]. Small, 2018, 14(10): 1703126.

[146] Lee S, Kim H, Yun D J, et al. Resistive switching characteristics of ZnO thin film grown on stainless steel for flexible nonvolatile memory devices[J]. Applied Physics Letters, 2009, 95(26): 262113.

[147] Jeong H Y, Kim Y I, Lee J Y, et al. A low-temperature-grown TiO_2-based device for the flexible stacked RRAM application[J]. Nanotechnology, 2010, 21(11): 115203.

[148] Lee M J, Ahn S E, Lee C B, et al. A simple device unit consisting of all NiO storage and switch elements for multilevel terabit nonvolatile random access memory[J]. ACS Applied Materials & Interfaces, 2011, 3(11): 4475-4479.

[149] Cai Y, Tan J, Liu Y, et al. A flexible organic resistance memory device for wearable biomedical applications[J]. Nanotechnology, 2016, 27(27): 275206.

[150] Kim S J, Lee J S. Flexible organic transistor memory devices[J]. Nano Letters, 2010, 10(8): 2884-2890.

[151] Kim R H, Kim H J, Bae I, et al. Non-volatile organic memory with sub-millimetre bending radius[J]. Nature Communications, 2014, 5(1): 1-12.

[152] Fu Y, Kong L, Chen Y, et al. Flexible neuromorphic architectures based on self-supported multiterminal organic transistors[J]. ACS Applied Materials & Interfaces, 2018, 10(31): 26443-26450.

[153] Wu C, Kim T W, Choi H Y, et al. Flexible three-dimensional artificial synapse networks with correlated learning and trainable memory capability[J]. Nature Communications, 2017, 8(1): 1-9.

[154] Fang M, Chen J H, Xu X L, et al. Antibacterial activities of inorganic agents on six bacteria associated with oral infections by two susceptibility tests[J]. International Journal of Antimicrobial Agents, 2006, 27(6): 513-517.

[155] Kalowekamo J, Baker E. Estimating the manufacturing cost of purely organic solar cells[J]. Solar Energy, 2009, 83(8): 1224-1231.

[156] Garcia V, Bibes M. Inside story of ferroelectric memories[J]. Nature, 2012, 483(7389): 279-280.

[157] Müller J, Yurchuk E, Schlösser T, et al. Ferroelectricity in HfO_2 enables nonvolatile data storage in 28 nm HKMG[C]//2012 Symposium on VLSI Technology (VLSIT). IEEE, 2012: 25-26.

[158] Pešić M, Schroeder U, Slesazeck S, et al. Comparative study of reliability of ferroelectric and anti-ferroelectric memories[J]. IEEE Transactions on Device and Materials Reliability, 2018, 18(2): 154-162.

[159] Hwang C S, Mikolajick T. 11 - Ferroelectric Memories[M]//Advances in Non-Volatile Memory and Storage Technology 2nd ed. Magyari-Kope B, Nishi Y. Woodhead Publishing, 2019: 393-441.

[160] Yan X, He H, Liu G, et al. Robust memristor based in epitaxy vertically aligned nanostructured baTiO$_3$-CeO$_2$ films on silicon[J]. Advanced Materials, 2022, 34(23): 2110343.

[161] Scott J F, Paz de Araujo C A. Ferroelectric memories[J]. Science, 1989, 246(4936): 1400-1405.

[162] Shkuratov S I, Baird J, Antipov V G, et al. Depolarization mechanisms of PbZr$_{0.52}$Ti$_{0.48}$O$_3$ and PbZr$_{0.95}$Ti$_{0.05}$O$_3$ poled ferroelectrics under high strain rate loading[J]. Applied Physics Letters, 2014, 104(21): 212901.

[163] Pintilie L, Dragoi C, Radu R, et al. Temperature induced change in the hysteretic behavior of the capacitance-voltage characteristics of Pt-ZnO-Pb(Zr$_{0.2}$Ti$_{0.8}$)O$_3$-Pt heterostructures[J]. Applied Physics Letters, 2010, 96(1): 012903.

[164] Setter N, Damjanovic D, Eng L, et al. Ferroelectric thin films: Review of materials, properties, and applications[J]. Journal of Applied Physics, 2006, 100(5): 051606.

[165] Böscke T S, Müller J, Bräuhaus D, et al. Ferroelectricity in hafnium oxide thin films[J]. Applied Physics Letters, 2011, 99(10): 102903.

[166] Sakai S, Ilangovan R. Metal-ferroelectric-insulator-semiconductor memory FET with long retention and high endurance[J]. IEEE Electron Device Letters, 2004, 25(6): 369-371.

[167] Mikolajick T, Schroeder U, Slesazeck S. The past, the present, and the future of ferroelectric memories[J]. IEEE Transactions on Electron Devices, 2020, 67(4): 1434-1443.

[168] Juan T P C, Chang C, Lee J Y M. A new metal-ferroelectric (PbZr$_{0.53}$Ti$_{0.47}$O$_3$)-insulator (Dy$_2$O$_3$)-semiconductor (MFIS) FET for nonvolatile memory applications[J]. IEEE Electron Device Letters, 2006, 27(4): 217-220.

[169] Müller J, Boscke T S, Schroder U, et al. Nanosecond polarization switching and long retention in a novel MFIS-FET based on ferroelectric HfO$_2$ [J]. IEEE Electron Device Letters, 2012, 33(2): 185-187.

[170] Florent K, Pesic M, Subirats A, et al. Vertical ferroelectric HfO$_2$ FET based on 3-D NAND architecture: Towards dense low-power memory[C]//2018 IEEE International Electron Devices Meeting (IEDM). IEEE, 2018: 2.5. 1-2.5. 4.

[171] Xiao W, Liu C, Peng Y, et al. Performance improvement of Hf$_{0.5}$Zr$_{0.5}$O$_2$-based ferroelectric-field-effect transistors with ZrO$_2$ seed layers[J]. IEEE Electron Device Letters, 2019, 40(5): 714-717.

[172] Chen L, Wang L, Peng Y, et al. A van der Waals synaptic transistor based on ferroelectric Hf$_{0.5}$Zr$_{0.5}$O$_2$ and 2D tungsten disulfide[J]. Advanced Electronic Materials, 2020, 6(6): 2000057.

[173] Zhuravlev M Y, Sabirianov R F, Jaswal S S, et al. Giant electroresistance in ferroelectric tunnel junctions[J]. Physical Review Letters, 2005, 94(24): 246802.

[174] Tsymbal E Y, Gruverman A. Beyond the barrier[J]. Nature Materials, 2013, 12(7): 602-604.

[175] Garcia V, Bibes M. Ferroelectric tunnel junctions for information storage and processing[J]. Nature Communications, 2014, 5(1): 1-12.

[176] Velev J P, Burton J D, Zhuravlev M Y, et al. Predictive modelling of ferroelectric tunnel junctions[J]. npj Computational Materials, 2016, 2(1): 1-13.

[177] Yin Y, Li Q. A review on all-perovskite multiferroic tunnel junctions[J]. Journal of Materiomics, 2017, 3(4): 245-254.

[178] Wang Z, Zhao W, Kang W, et al. A physics-based compact model of ferroelectric tunnel junction for memory and logic design[J]. Journal of Physics D: Applied Physics, 2013, 47(4): 045001.

[179] Chanthbouala A, Garcia V, Cherifi R O, et al. A ferroelectric memristor[J]. Nature Materials, 2012, 11(10): 860-864.

[180] Jin Hu W, Wang Z, Yu W, et al. Optically controlled electroresistance and electrically controlled photovoltage in ferroelectric tunnel junctions[J]. Nature Communications, 2016, 7(1): 1-9.

[181] Boyn S, Grollier J, Lecerf G, et al. Learning through ferroelectric domain dynamics in solid-state synapses[J]. Nature Communications, 2017, 8(1): 1-7.

[182] Yoong H Y, Wu H, Zhao J, et al. Epitaxial ferroelectric $Hf_{0.5}Zr_{0.5}O_2$ thin films and their implementations in memristors for brain-inspired computing[J]. Advanced Functional Materials, 2018, 28(50): 1806037.

[183] Guo R, Zhou Y, Wu L, et al. Control of synaptic plasticity learning of ferroelectric tunnel memristor by nanoscale interface engineering[J]. ACS Applied Materials & Interfaces, 2018, 10(15): 12862-12869.

[184] Guo R, Lin W, Yan X, et al. Ferroic tunnel junctions and their application in neuromorphic networks[J]. Applied Physics Reviews, 2020, 7(1): 011304.

[185] Ma C, Luo Z, Huang W, et al. Sub-nanosecond memristor based on ferroelectric tunnel junction[J]. Nature Communications, 2020, 11(1): 1-9.

[186] Boybat I, Le Gallo M, Nandakumar S R, et al. Neuromorphic computing with multi-memristive synapses[J]. Nature Communications, 2018, 9(1): 1-12.

[187] Chanthbouala A, Crassous A, Garcia V, et al. Solid-state memories based on ferroelectric tunnel junctions[J]. Nature Nanotechnology, 2012, 7(2): 101-104.

[188] Lee W, Kahya O, Toh C T, et al. Flexible graphene-PZT ferroelectric nonvolatile memory[J]. Nanotechnology, 2013, 24(47): 475202.

[189] Yu H, Chung C C, Shewmon N, et al. Flexible inorganic ferroelectric thin films for nonvolatile memory devices[J]. Advanced Functional Materials, 2017, 27(21): 1700461.

[190] Gao H, Yang Y, Wang Y, et al. Transparent, flexible, fatigue-free, optical-read, and nonvolatile ferroelectric memories[J]. ACS Applied Materials & Interfaces, 2019, 11(38): 35169-35176.

[191] Yang C, Han Y, Qian J, et al. Flexible, temperature-resistant, and fatigue-free ferroelectric memory based on Bi ($Fe_{0.93}Mn_{0.05}Ti_{0.02}$) O_3 thin film[J]. ACS Applied Materials & Interfaces, 2019, 11(13): 12647-12655.

[192] Gao W, You L, Wang Y, et al. Flexible PbZr$_{0.52}$Ti$_{0.48}$O$_3$ capacitors with giant piezoelectric response and dielectric tunability[J]. Advanced Electronic Materials, 2017, 3(8): 1600542.

[193] Chua L. Resistance Switching Memories are Memristors[B]. Handbook of Memristor Networks, Cham: Springer International Publishing, 2019: 197-203.

[194] Sun H, Luo Z, Liu C, et al. A flexible BiFeO$_3$-based ferroelectric tunnel junction memristor for neuromorphic computing[J]. Journal of Materiomics, 2022, 8(1): 144-149.

[195] Wang Z, Wang L, Nagai M, et al. Nanoionics-enabled memristive devices: strategies and materials for neuromorphic applications[J]. Advanced Electronic Materials, 2017, 3(7): 1600510.

[196] Yang R, Huang H M, Guo X. Memristive synapses and neurons for bioinspired computing[J]. Advanced Electronic Materials, 2019, 5(9): 1900287.

[197] Lee H E, Park J H, Kim T J, et al. Novel electronics for flexible and neuromorphic computing[J]. Advanced Functional Materials, 2018, 28(32): 1801690.

[198] Upadhyay N K, Jiang H, Wang Z, et al. Emerging memory devices for neuromorphic computing[J]. Advanced Materials Technologies, 2019, 4(4): 1800589.

[199] Yan X, Zhao J, Liu S, et al. Memristors: memristor with Ag-cluster-doped TiO$_2$ films as artificial synapse for neuroinspired computing[J]. Advanced Functional Materials, 2018, 28(1): 1870002.

[200] Fan Z, Chen J, Wang J. Ferroelectric HfO$_2$-based materials for next-generation ferroelectric memories[J]. Journal of Advanced Dielectrics, 2016, 6(02): 1630003.

[201] Zeng B, Xiao W, Liao J, et al. Compatibility of HfN metal gate electrodes with Hf$_{0.5}$Zr$_{0.5}$O$_2$ ferroelectric thin films for ferroelectric field-effect transistors[J]. IEEE Electron Device Letters, 2018, 39(10): 1508-1511.

[202] Kim S J, Mohan J, Summerfelt S R, et al. Ferroelectric Hf$_{0.5}$Zr$_{0.5}$O$_2$ thin films: a review of recent advances[J]. Journal of Operations Management, 2019, 71(1): 246-255.

[203] Xiao W, Liu C, Peng Y, et al. Thermally stable and radiation hard ferroelectric Hf$_{0.5}$Zr$_{0.5}$O$_2$ thin films on muscovite mica for flexible nonvolatile memory applications[J]. ACS Applied Electronic Materials, 2019, 1(6): 919-927.

[204] Wang T Y, Meng J L, He Z Y, et al. Atomic layer deposited Hf$_{0.5}$Zr$_{0.5}$O$_2$-based flexible memristor with short/long-term synaptic plasticity[J]. Nanoscale Research Letters, 2019, 14(1): 1-6.

[205] Tseng H J, Tian W C, Wu W J. Flexible PZT thin film tactile sensor for biomedical monitoring[J]. Sensors, 2013, 13(5): 5478-5492.

[206] Dagdeviren C, Su Y, Joe P, et al. Conformable amplified lead zirconate titanate sensors with enhanced piezoelectric response for cutaneous pressure monitoring[J]. Nature Communications, 2014, 5(1): 1-10.

[207] Lee Y, Park J, Cho S, et al. Flexible ferroelectric sensors with ultrahigh pressure sensitivity and linear response over exceptionally broad pressure range[J]. ACS Nano,

2018, 12(4): 4045-4054.

[208] Yang C, Guo M, Gao D, et al. A flexible strain sensor of Ba (Ti, Nb) O$_3$/mica with a broad working temperature range[J]. Advanced Materials Technologies, 2019, 4(11): 1900578.

[209] Stadlober B, Zirkl M, Irimia-Vladu M. Route towards sustainable smart sensors: ferroelectric polyvinylidene fluoride-based materials and their integration in flexible electronics[J]. Chemical Society Reviews, 2019, 48(6): 1787-1825.

[210] Yang M M, Kim D J, Alexe M. Flexo-photovoltaic effect[J]. Science, 2018, 360(6391): 904-907.

[211] Lee D, Yoon A, Jang S Y, et al. Giant flexoelectric effect in ferroelectric epitaxial thin films[J]. Physical Review Letters, 2011, 107(5): 057602.

[212] Park K I, Lee M, Liu Y, et al. Flexible nanocomposite generator made of BaTiO$_3$ nanoparticles and graphitic carbons[J]. Advanced Materials, 2012, 24(22): 2999-3004.

[213] Pozo B, Garate J I, Araujo J Á, et al. Energy harvesting technologies and equivalent electronic structural models[J]. Electronics, 2019, 8(5): 486.

[214] Shen D, Park J H, Noh J H, et al. Micromachined PZT cantilever based on SOI structure for low frequency vibration energy harvesting[J]. Sensors and Actuators A: Physical, 2009, 154(1): 103-108.

[215] Kim S G, Priya S, Kanno I. Piezoelectric MEMS for energy harvesting[J]. MRS Bulletin, 2012, 37(11): 1039-1050.

[216] Dagdeviren C, Yang B D, Su Y, et al. Conformal piezoelectric energy harvesting and storage from motions of the heart, lung, and diaphragm[J]. Proceedings of the National Academy of Sciences, 2014, 111(5): 1927-1932.

[217] Wang D, Yuan G, Hao G, et al. All-inorganic flexible piezoelectric energy harvester enabled by two-dimensional mica[J]. Nano Energy, 2018, 43: 351-358.

[218] Lee J H, Lee K Y, Kumar B, et al. Highly sensitive stretchable transparent piezoelectric nanogenerators[J]. Energy & Environmental Science, 2013, 6(1): 169-175.

[219] Koka A, Zhou Z, Sodano H A. Vertically aligned BaTiO$_3$ nanowire arrays for energy harvesting[J]. Energy & Environmental Science, 2014, 7(1): 288-296.

[220] Jeong C K, Park K I, Son J H, et al. Self-powered fully-flexible light-emitting system enabled by flexible energy harvester[J]. Energy & Environmental Science, 2014, 7(12): 4035-4043.

[221] Won S S, Seo H, Kawahara M, et al. Flexible vibrational energy harvesting devices using strain-engineered perovskite piezoelectric thin films[J]. Nano Energy, 2019, 55: 182-192.

[222] Berdan R, Marukame T, Ota K, et al. Low-power linear computation using nonlinear ferroelectric tunnel junction memristors[J]. Nature Electronics, 2020, 3(5): 259-266.

[223] Zhong G, Zi M, Ren C, et al. Flexible electronic synapse enabled by ferroelectric field effect transistor for robust neuromorphic computing[J]. Applied Physics Letters, 2020, 117(9): 092903.

[224] Yao P, Wu H, Gao B, et al. Face classification using electronic synapses[J]. Nature Communications, 2017, 8(1): 1-8.

[225] Yao P, Wu H, Gao B, et al. Fully hardware-implemented memristor convolutional neural network[J]. Nature, 2020, 577(7792): 641-646.

第 7 章 石墨烯场效应晶体管研究进展

7.1 石墨烯的发现、基本结构及性能

研究者们将从石墨中分离出来的碳原子层称为"石墨烯 (graphene)"。graphene 这种材料的定义可以追溯到 1986 年,Boehm 等定义 graphene 为:"The term graphene layer should be used for such a single carbon layer。" 到 1997 年,国际纯粹与应用化学联合会 (IUPAC) 将 "graphene" 正式纳入《化学技术纲要》中:"The term graphene should be used only when the reactions, structural relations or other properties of individual layers are discussed。"[1] 而中文的石墨烯定义还有待于进一步明确和完善。

从结构上看,石墨烯是由 sp^2 碳原子以蜂巢晶格构成的新型二维碳纳米材料,它与零维碳材料——富勒烯、一维碳纳米材料——碳纳米管、三维碳材料 (金刚石、石墨) 一起统称为碳的同素异形体。20 世纪 80~90 年代,由于零维碳材料和一维碳纳米材料的发现,碳的同素异形体受到了研究者们的广泛关注。与一维和三维碳材料相比,石墨烯的存在性曾是历史上备受争议的话题。直到 2004 年,Novoselov 等[2] 采用机械剥离方法制备出单层石墨烯,并观察到了石墨烯独有的电学性质,石墨烯的存在才真正得到认可。在这之后,石墨烯的其他制备方法,如外延生长法、化学气相沉积 (chemical vapour deposition, CVD) 法、机械剥离法等相继被报道,石墨烯及与其相关的材料、方法、器件等研究掀起了新的热潮。

1. 石墨烯的电子性能和能带结构

石墨烯中的每个碳原子有 4 个价电子,3 个价电子以 sp^2 杂化的形式构成 σ 键,另外 1 个 p 轨道的价电子在垂直方向上形成 π 键,π 电子即为石墨烯平面内自由移动的载流子。π 电子使石墨烯呈现理想的二维结构,即 π 和 π^* 状态下的单原子厚度互不影响。与传统的埋藏式半导体界面形成的二维结构不同,石墨烯的这种二维结构可以直接进行物理、化学修饰和测量。石墨烯的六个碳原子呈六边形排列,形成蜂窝状晶格,每个晶格单元含有两个碳原子 (图 7.1(a)),形成了一种相当独特的能带结构 (图 7.1(b)):π 状态形成价带,π^* 状态形成导带 (由于每个碳原子贡献一个 π 电子,因此石墨烯的价带刚好填满,而导带全空),价带和导带相交于第一布里渊区的六个点,称为狄拉克点或电荷中性点[3]。根据对称性

7.1 石墨烯的发现、基本结构及性能

可将这六个点简化成一对，即 K 和 K′，如果把能量限制在与电子输运相关的低能量状态下，能带就会产生线性色散，其结构可以看作在 E_{Dirac} 处接触的两个锥 (图 7.1(c)，(d))[3]。由于 π 和 π* 状态互不影响并可交叉，因此可将石墨烯看作带隙为零的半导体[3]。

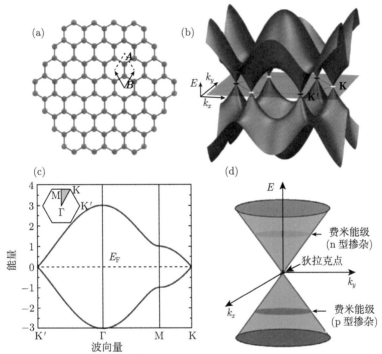

图 7.1　(a) 石墨烯的蜂窝状晶格[3]；(b) 石墨烯的三维能带结构图；(c) 石墨烯的线性色散；(d) 石墨烯的价带和导带呈锥形接触并相交于狄拉克点，费米能级的位置决定了掺杂的情况及载流子输运

2. 石墨烯的电荷输运特性

在石墨烯的分子结构中，每个 π 键相互共轭形成共轭大 π 键，电子和空穴载流子可以在如此大的共轭体系中以相当大的电子费米速率 ($v_{\text{F}} \approx 10^6$ m·s^{-1}) 移动，因此，石墨烯有非常高的载流子迁移率，且其载流子迁移率在 10~100 K 下几乎不受温度影响。研究表明，载流子密度是影响石墨烯载流子迁移率效率的重要因素。在单层石墨烯中，增加载流子密度通常会降低载流子迁移率，这主要取决于主导散射体的性质。在悬浮、剥离获得的石墨烯中，石墨烯与衬底的相互作用被消除，石墨烯的载流子迁移率高达 2×10^5 cm^2/(V·s)[4,5]。若衬底为绝缘体 (如非晶 SiO_2)，则载流子迁移率明显较低，根据绝缘体的性质和纯度，迁移率有所不

同，其值为几千到几万 $(cm^2/(V·s))$。在具有高表面声子频率的单晶衬底上，如氮化硼等，载流子迁移率更高[6]。目前，外延和化学气相沉积生长的石墨烯的载流子迁移率通常较低，为几千 $cm^2/(V·s)$[7]。

3. 石墨烯的化学性质

石墨烯的最基本化学键是 C=C，其最基本的结构单元是苯环，此外，石墨烯还含有边缘和平面缺陷，其化学性质主要体现在这些缺陷上。石墨烯的碳骨架结构非常稳定，一般的化学方法很难破坏其结构，因此，目前研究中的石墨烯的修饰方法主要有两种：①利用石墨烯的边界官能团，如羧基、羟基等基团和一些化学活性基团 (如乙二胺、戊二醛等) 之间的共价作用对石墨烯进行修饰；②利用带有化学活性基团的芳香分子和石墨烯之间的 π-π 作用非共价修饰石墨烯[8]。这两种修饰方法也为石墨烯的溶剂处理和生物功能化提供了实现手段。

7.2 石墨烯的表征方法

自 2004 年石墨烯的存在得到认可之后，科研人员采用多种表征手段验证获得到的石墨烯纳米片的质量。目前普遍采用的石墨烯表征方法主要有两大类：显微镜法及光谱法。下面从理论和应用角度分别对这两类方法进行介绍。

7.2.1 显微镜法

石墨烯的显微镜表征方法主要有光学显微镜法、扫描探针显微镜法、透射电子显微镜法、扫描隧道显微镜法。

1. 光学显微镜法

光学显微镜法是石墨烯材料被发现依靠的关键技术之一，也是目前石墨烯表征技术中最快速、简便、有效的方法。表征石墨烯的关键问题在于是否有合适的衬底，以使单层碳原子在具有最大灵敏度的波长范围内光学对比度达到最大。目前最常用的衬底为涂有 SiO_2 的硅片，沉积的石墨烯薄膜与硅上的 SiO_2 层形成的法布里-珀罗腔增加了一条小的光程，通过将 SiO_2 层厚度调节到 90 nm 或 300 nm，此时反射光强度在 550 nm 左右且达到最大，即人眼灵敏度的最大值。由于石墨烯与衬底的光学对比度高达 12%，因此单层石墨烯增加的短光程很容易被发现。这种方法极大地方便了对横向尺寸为几微米的石墨烯样品的表征 (图 7.2 (a))，自 2007 年来，研究者们已经发现了 SiO_2 的替代物，例如在 50 nm Si_3N_4 上使用蓝光，在硅片上沉积 72 nm 厚的 Al_2O_3，此外，在 90 nm 的聚甲基丙烯酸甲酯层上使用白光也可以观察到石墨烯[1]。虽然可以通过调整最大对比度下的单色光波长来简单分辨单层和多层石墨烯，但该方法仍不能作为辨别石墨烯样品层数的结论性证据[9]。

7.2 石墨烯的表征方法

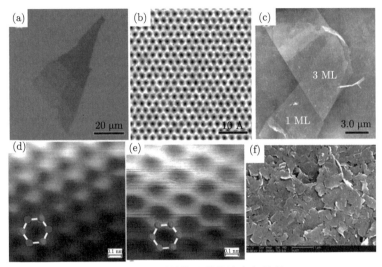

图 7.2 石墨烯的显微镜法表征结果

(a) 石墨烯的光学显微镜表征结果[1]；(b) 石墨烯的 TEM 表征结果[1]；(c) 石墨烯的 AFM 表征结果[1]；(d)、(e) 分别为石墨及石墨烯的 STM 表征结果[9]；(f) 石墨烯的 SEM 表征结果[10]

2. 透射电子显微镜

随着石墨烯纳米片的胶体制备方法的出现以及无支撑石墨烯器件特性的优化，近年来 TEM 被用作悬浮石墨烯的表征工具，该方法可以研究石墨烯的表面形态、晶格缺陷、边缘结构、吸附原子，并可分辨石墨烯层数 (图 7.2 (b))，是石墨烯高分辨分析研究的重要工具。

3. 扫描探针显微镜

扫描探针显微镜方法主要包括原子力显微镜 (atomic force microscope, AFM) 和扫描隧道显微镜 (scanning tunneling microscope, STM) 两种，是研究者们公认的表征材料厚度的最佳方法之一。AFM 是石墨烯表征技术史上最早的用于分辨石墨薄片是否为单原子层的技术之一，其扫描范围为几十纳米至 100 μm，它是利用样品和针尖之间的相互作用力分析样品表面形貌信息，可用来直接观察石墨烯样品的片层大小及厚度 (石墨烯的 AFM 表征结果如图 7.2 (c) 所示)。由于绝缘衬底、半金属石墨烯之间的尖端吸引/排斥差异以及薄层水在样品表面的吸附等问题的存在，对衬底–石墨烯高度剖面的准确解析具有一定的难度，因此目前报道的 AFM 观察到的衬底表面石墨烯的厚度与石墨烯片层间距 0.34 nm 相比具有较大差异，例如，在间歇接触模式下，结晶石墨上的单层石墨烯的厚度为 0.4 nm，氧化石墨烯薄片上的单层石墨烯厚度为 0.8~1.2 nm；在轻敲模式下，单层石墨烯的厚度为 0.9 nm[11]。尽管 AFM 存在测量速度慢、横向扫描尺寸有限、测量高度有误差的缺点，但是仍可以根据单层石墨烯的片层间距判断石墨烯样品的层数，并且可为石墨烯的折叠边缘提供一种更可靠、更准确的厚度测量方法，因此 AFM 是研究者们普遍认可的监测衬底表面石墨烯薄膜质量的最佳方法。

STM 具有极高的空间分辨能力 (平行方向的分辨率为 0.04 nm，垂直方向的分辨率达到 0.01 nm)，它可用于观察物质表面上的单个原子及其排列状态，并且还可以作为有效的工具来操纵表面上的单个原子。图 7.2 (d) 为石墨的 STM 图，由于石墨的 AB 堆积，六元环中只有三个碳原子清晰可见，而在单层石墨烯中 (图 7.2 (e))，六个碳原子完全等价，因此可呈现出相同的强度。STM 是石墨烯原子结构研究的重要工具。

4. 扫描电子显微镜

扫描电子显微镜 (SEM) 是石墨烯高分辨形貌表征中最常用的方法，图 7.2 (f) 为石墨烯的 SEM 表征结果。SEM 和其他显微镜表征方法相比有许多优点，例如，分辨率较高，最高可达 3 nm；能直接观察尺寸较大样品的原始表面，对试样形状无限制；样品室空间大，样品在样品室中的可移动范围大，因此可观察到不规则样品的各个区域；焦深大，比 TEM 大 10 倍，比光学显微镜大几百倍，立体感强；样品制备方法简单，可直接分析块状及粉末状样品，且对样品的损伤和污染程度小；在电视等装置的辅助下，能实现动态观察；除了能观察表面形貌外，还可进行表面元素分析等。因此，近年来，SEM 被研究者们广泛应用于石墨烯的形貌、元素组成等分析中。

7.2.2 光谱法

石墨烯的光谱表征方法主要有紫外可见光光谱、红外光谱、拉曼光谱、角分辨光电子谱、X 射线衍射、X 射线光电子能谱五种。

1. 紫外光可见光谱 (ultraviolet-visible spectrum, UV-Vis spectrum)

UV-Vis 光谱可用于石墨烯的定性分析。石墨烯的 UV-Vis 在 270 nm 左右有特征吸收峰，该峰对应于芳香 C—C 键的跃迁 (石墨烯的 UV-Vis 如图 7.3 (a) 所示)；对于氧化石墨烯来说，其 UV-Vis 在 230 nm 左右有最强吸收峰，该峰是由芳香环中 C=C 键的 π-π^* 跃迁导致的，并且在 300 nm 左右还有一较明显的吸收肩峰，该峰对应于 C=O 的 n-π^* 跃迁；将氧化石墨烯还原后，还原氧化石墨烯 300 nm 左右的肩峰消失，其最大吸收峰红移到 250~270 nm，该现象反映了 π 电子浓度的增加，即在还原过程中发生了局部重置或 sp^2 碳排序，C=O 相关的基团数量减少。此外，还可以通过 UV-Vis 进行基于朗伯–比尔定律的浓度测定，其关系式为 $A = \alpha CL$，其中 A 为石墨烯分散液的吸光度，C 为石墨烯浓度，L 为

比色皿宽度 (单位为 cm)，α 为摩尔吸收系数，可以根据绘制的校准曲线测定未知石墨烯样品的浓度。

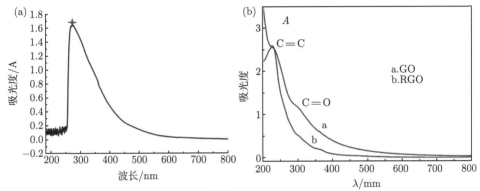

图 7.3 (a) 石墨烯的 UV-Vis 表征结果 [12]；(b) 氧化石墨烯 (GO) 及还原氧化石墨烯 (rGO) 的 UV-Vis 表征结果 [13]

2. 红外光谱 (infrared spectrum, IR spectrum)

IR 主要用于化合物鉴定及分子结构表征，是石墨烯表征中常用的定性分析方法。在化学方法制备石墨烯的过程中，总会发生天然石墨的氧化或氧化石墨的还原，因此总会伴随石墨烯 IR 中某些特征峰的减弱或消失。例如在氧化石墨烯的 IR 中，3430 cm^{-1} 处的峰为—OH 的伸缩振动峰，1725 cm^{-1} 处的峰为氧化石墨烯—COOH 上的 C=O 键，1630 cm^{-1} 处的峰对应于 C—OH 的弯曲振动，1110 cm^{-1} 处的峰为 C—O—C 的特征峰，当氧化石墨烯被还原后，—OH、C=O、C—OH、C—O—C 的吸收峰都相应减弱，说明石墨烯中的含氧量减少，氧化石墨烯被一定程度上还原 (氧化石墨烯及还原氧化石墨烯的 IR 表征结果如图 7.4 所示)。因此，IR 是石墨烯分子基团、分子结构研究中常用的分析方法。

3. 拉曼光谱

拉曼散射是一种快速、非破坏性的表征技术，它可直接检测电子-声子相互作用，对电子和晶体结构具有高度敏感性，被公认为碳材料结构的"指纹光谱"。石墨烯材料的拉曼光谱由 G 峰、D 峰及 G′ 峰组成 (图 7.5)：约 1560 cm^{-1} 处的 G 峰是石墨烯的主要特征峰，对应于布里渊区的声子的振动，是由 sp^2 碳原子的面内拉伸振动引起的，能反映石墨烯的层数信息；约 1360 cm^{-1} 处的 D 峰是由 sp^2 原子的对称伸缩振动造成的，该峰大小与石墨烯样品的缺陷程度有关，是表征石墨烯缺陷和杂质水平的特征峰，而纯石墨烯 (如机械剥离法制备的石墨烯) 的拉曼光谱 D 峰不存在或十分微弱；与其他碳材料相比，石墨烯的主要特征峰是约

2700 cm^{-1} 处的 2D 峰 (也称为 G′ 峰),该峰的形状、位置、强度与层数密切相关,例如,在单层石墨烯的拉曼光谱中可观察到以 2640 cm^{-1} 为中心的对称洛伦兹峰,而对于松散堆积的石墨烯层,该峰可以移动到 2655~2665 cm^{-1};此外,还可以通过 D 峰与 G 峰强度的比值 (I_D/I_G) 或 G 峰的半高宽表征石墨烯的缺陷密度。从以上介绍可以看出,拉曼光谱是表征石墨烯层数、缺陷程度等信息的有效手段。

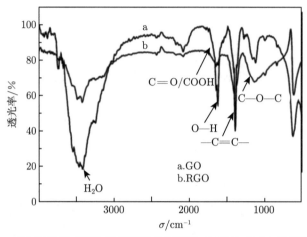

图 7.4　氧化石墨烯 (GO) 及还原氧化石墨烯 (RGO) 的 IR 表征结果[13]

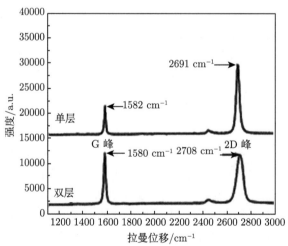

图 7.5　石墨烯的拉曼光谱[1]

4. 角分辨光电子谱

虽然角分辨光电子谱 (angular resolved photoemission spectroscopy，ARPES) 不属于标准的实验室表征技术，但它可为石墨烯和其他碳基材料的电子结构提供直接证据，因此在这里进行简单介绍。当用 10~300 eV 光子照射衬底时，光电子可从衬底表面被提取出来，其动量和能量可以用 15 meV 分辨率来分析，从而重建能带图 (图 7.6)。在石墨烯存在的情况下，可以直接观察到布里渊区 K 角附近的相对论类狄拉克线性色散和载流子的手性，以及由石墨烯层间或衬底–石墨烯相互作用而产生的小带隙。

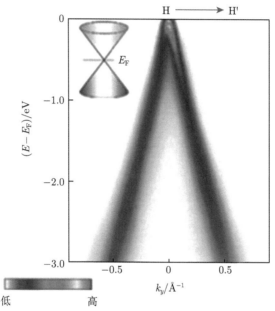

图 7.6　石墨烯的 ARPES 表征结果 [1] (彩图请扫封底二维码)

5. X 射线衍射

X 射线衍射 (X-ray diffraction，XRD) 表征技术可用于检测石墨烯的晶体结构和晶格尺寸。石墨烯的 XRD 图谱如图 7.7 所示：晶面 (001) 所对应的 26.48° 的峰为石墨烯的特征峰，而图谱上 45° 和 55° 的石墨特征峰的缺失也证明了石墨烯的成功制备 [14]。此外，将图谱上获得的衍射峰偏转角度代入布拉格方程中，还可获得石墨烯晶体晶格的相关参数，从而分析实验成果。

6. X 射线光电子能谱

X 射线光电子能谱 (X-ray photo-electron spectroscopy，XPS) 可以用于石墨烯化学结构、化学组分的定性及定量研究。在测试获得的 XPS 全谱中 (图 7.8(a))，可获得样品中各元素 (如 C、O) 的含量信息,对各元素的核心谱进行分峰后,可获得样品含有的分子基团信息,例如,在 C 1s 拟合峰 (图 7.8(b)) 中,284.6 eV 处的峰对应于 sp^2 碳 (C=C),285.4 eV 处的峰对应于 sp^3 碳 (C—C),286.1 eV 处的峰对应于 C—O;在 O 1s 拟合峰 (图 7.8(c)) 中,531.7 eV 处的峰对应于 O—H,533.3 eV 处的峰对应于 O=C—O 等。此外,还可以根据石墨烯修饰前后谱图中信号峰的改变来监控石墨烯的改性效果。

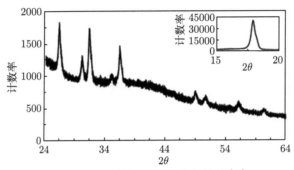

图 7.7 石墨烯的 XRD 表征结果[14]

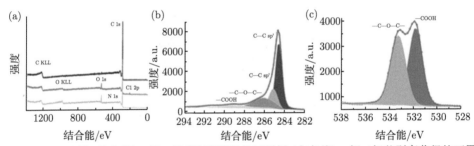

图 7.8 (a) 石墨 (黑色线)、邻二氯苯剥离获得的石墨烯 (灰色线)、邻二氯苯剥离获得的石墨烯 (浅灰色线) 的 XPS 全谱;(b) 和 (c) 分别为 C 1s 和 O 1s 核心谱[15]

7.3 石墨烯的制备方法

自 1960 年以来,研究者们一直致力于少层石墨烯纳米片的制备,但都以失败告终。直到 2004 年,Geim 的研究小组采用胶带反复粘揭石墨块的方法成功制备了单层石墨烯,石墨烯的存在才得到认可。之后,微机械剥离法外延生长法、化学气相沉积法、化学合成法及液相剥离法制备石墨烯被发展出来,而化学合成制备法的出现又带来了新的问题,虽然该方法可以实现大规模石墨烯的制备,但如何更充分地把 sp^3 型氧化石墨烯还原为还原氧化石墨烯,是一项新的挑战。下面对这四种石墨烯制备方法的最新概况进行介绍,并对每种方法的优点和不足进行评述。

1. 微机械剥离法

层间范德瓦耳斯作用力约为 $2~\text{eV}\cdot\text{nm}^{-2}$，剥离石墨所需力约为 $300~\text{nN}\cdot\mu\text{m}^{-2}$，这种非常弱的作用力可以通过胶带剥离实现。最初采用这种方法的是 Novoselov 和 Geim，他们使用普通胶带反复粘贴、剥离石墨块十几次，从而在 $1~\mu\text{m}$ 厚的石墨块上分离单层石墨片，之后轻轻按压胶带，将石墨烯和石墨片转移到清洁的 Si/SiO_2 衬底上。胶带剥离法的一个缺点是胶水的残留，从而影响石墨烯样品的载流子迁移率，因此，还需要在氢气/氩气气氛下热处理来去除残留物。

为了避免胶水残留的问题，一些研究者报道了通过施加高压来促进石墨烯在绝缘衬底的粘附的方法，他们将一千伏到几十千伏电压持续作用到衬底上新剥离的石墨片上几秒钟，然后将石墨片从衬底上移除，可在衬底上观察到单层和少层石墨烯样品，但该方法制备的石墨烯仍会受到轻微污染。这之后，Huc 的研究小组报道了一种基于 "scotch-tape" 的石墨烯制备技术，其过程如图 7.9 所示，主要包括两步：① 使用薄层环氧树脂胶键合高定向热解石墨 (highly oriented pyrolytic graphite, HOPG)，然后在压力下固化；② 采用防粘胶带对 HOPG 进行反复粘

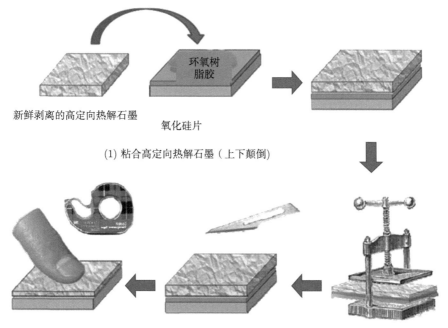

图 7.9 石墨烯的反向剥落制备流程图 [16]

贴剥离。与以前报道的机械剥离石墨烯制备方法相比，该方法中石墨烯产品被留在固化的环氧树脂一面 (也称为反向剥离)，螺旋压力机施加的压力可使石墨烯片面积更大、褶皱更少。

微机械剥离法虽然可获得高质量的石墨烯产品，但其产率低、难以实现大规模生产的缺点限制了该方法的广泛应用。

2. 化学气相沉积法

化学气相沉积 (CVD) 法制备石墨烯的机理可概括为：在高温 (650~1000℃) 及金属催化剂 (如 Cu 和 Ni 薄膜) 作用下，甲烷、乙炔、乙烯、己烷等碳氢化合物气体可发生裂解而形成碳原子和氢原子，然后，碳原子通过金属催化剂的表面和主体扩散，并在达到碳溶解度极限时在金属表面形成少层石墨烯片。2006 年，Somani 等 [17] 首次报道了 CVD 法成功制备平面少层石墨烯，他们以樟脑为原料，在 Ni 箔上制备了少层石墨烯，这项研究虽然未解决石墨烯制备中存在的层数控制、最小化褶皱等问题，但仍为平面少层石墨烯的合成开辟了一条新的工艺路线。2007 年，Obraztsov 等 [18] 以比例为 92:8 的 H_2 和 CH_4 混合物为前驱体，在气体气压为 80Torr (1Torr = 133.322Pa) 的条件下制备了厚度为 1~2 nm 的少层石墨烯，俄歇电子能谱表征结果表明，制备的石墨烯膜具有原子般光滑的由脊状物隔开的微米尺寸区域，并把脊状物的形成归因于镍和石墨热膨胀系数的差异。之后，Yu 等 [19] 以 CH_4、H_2 和 Ar 组成的气体混合物 (比例为 0.15:1:2) 为前驱体，多晶镍箔为衬底，在温度为 1000℃、时间为 20 min 的条件下成功制备了厚度为 3~4 层的石墨烯薄膜；TEM 和拉曼光谱分析表明，中等冷却速率有助于石墨烯的形成，而高、低冷却速率对石墨烯的合成过程不利，并将不同冷却速率下石墨烯形成的差异归因于碳在镍中的溶解度和碳的偏析动力学；他们的工作为利用 CVD 法生长石墨烯提供了重要的理论支持。为了实现石墨烯的大规模生长，Wang 等 [20] 提出了一种无衬底生长少层石墨烯的新方法，他们以 MgO 负载的 Co 为催化剂，在 CH_4 和 Ar 混合气氛 (体积比为 1:4) 中生长石墨烯，反应产物经浓盐酸和蒸馏水洗涤后，在 70℃ 下干燥；实验结果表明，500 mg 催化剂粉末可制备出 50 mg 石墨烯；SEM、TEM 和拉曼光谱分析表明，该方法至少可制备厚度为 5 层的少层石墨烯。

为了将制备的石墨烯应用到石墨烯场效应晶体管、太阳能电池等电子器件的制备中，就需将石墨烯转移到玻璃、Si、PDMS 等基底上。最常用的转移方法为基于聚甲基丙烯酸甲酯 (polymethyl methacrylate, PMMA) 的支撑法，首先在 CVD 生长的石墨烯薄膜表面旋涂 PMMA，之后将 PMMA/石墨烯/Cu 结构放在 $Fe(NO_3)_3$ 溶液中，使 PMMA 面朝上以刻蚀掉铜箔，并在去离子水中漂洗几次以去除残留的 $Fe(NO_3)_3$ 溶液。将 SiO_2/Si 基板浸入去离子水后，在范德瓦耳斯力的

7.3 石墨烯的制备方法

作用下，石墨烯薄膜可粘附到 SiO_2/Si 衬底上，最后用丙酮和异丙醇去除 PMMA 载体。CVD 法可以生长出低缺陷、大表面积的高质量石墨烯，但其高成本、复杂的工艺及转移过程等问题仍限制了这种方法的广泛应用[21]。为了解决这些问题，许多研究人员致力于在较低的温度和环境压力下优化合成条件，但该方法又会对石墨烯的质量造成影响。例如 Kalita 等[22] 报道了一种基于表面波等离子体增强 CVD(surface wave plasma enhanced CVD, PECVD) 的石墨烯制备方法，该方法显著降低了生长温度 (450°C) 和沉积时间 (小于 5min)，从而提高了整个生长过程的综合性和可扩展性 (图 7.10)。总而言之，基于 CVD 的石墨烯制备方法尚不成熟，其生产成本、产量及合成工艺仍有待于改进。

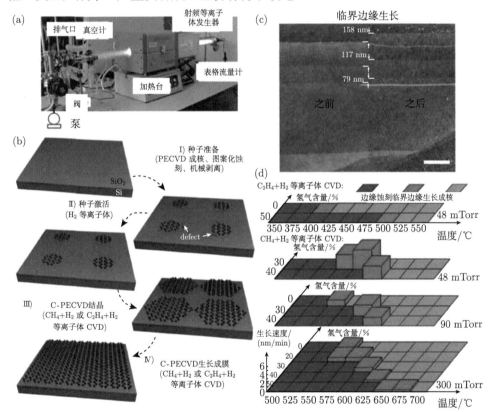

图 7.10 (a) PECVD 系统示意图；(b) C-PECVD 程序示意图；(c) C-PECVD 生长之前和之后的 AFM 图像；(d) 48 mTorr、90 mTorr 和 300 mTorr 时，实验结果随温度 (T) 和 H_2 含量的变化曲线[23]

3. SiC 外延生长法

如图 7.11 所示，在真空或惰性条件下，石墨烯也可以通过 SiC 衬底的高温热分解 (1200~1600°C) 来制备：高温处理会导致 Si(Si 的熔点为 1100°C) 的升华，留在衬底上的 C 原子会聚集并形成 sp^2 杂化网络,从而诱导石墨烯的生长,此过程称为 SiC 上石墨烯的外延生长[24]。1975 年,Bommel 等首次报道了超高压及高温 (1000~1500℃) 处理下,在 SiC 的两个极性平面上生长少层石墨的方法[25]。2004 年,de Heer 等采用 SiC 外延生长法在单晶 6H-SiC 的 Si(0001) 面上成功制备了厚度为 1~2 层的少层石墨烯[26],尽管该方法可以制备出高质量的石墨烯,但仍存在制备的石墨烯无法转移到其他衬底上的问题。2009 年,Juang 等的研究成果取得了突破性进展,他们在沉积有 Ni 催化剂薄膜的 SiC 衬底上生长石墨烯,用该方法制备的石墨烯可以很容易地转移到其他衬底上[27]。尽管该方法可以制备出高质量的石墨烯并通过调节加热温度控制石墨烯产品的层数,但由于目前合成条件中能源密集型工艺和商用 SiC 衬底尺寸的局限性,石墨烯的外延生长价格十分昂贵。此外,外延生长产生的不同极性面 (如 Si 面或 C 面) 会严重影响石墨烯产品的质量,由于硅面可以确保石墨烯的均匀生长,因此成为了最佳选择。到目前为止,由于石墨烯的生长机理以及石墨烯与底物之间相互作用的有关数据仍然很少,因此这种合成方法仍处于进一步研究中。

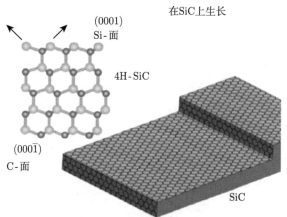

图 7.11 SiC 外延生长石墨烯[28](彩图请扫封底二维码)

金色和灰色球分别代表 Si 和 C 原子,高温下,Si 原子蒸发 (如箭头所示) 后,留下富含 C 的表面,形成石墨烯片

4. 化学合成法

氧化石墨 (graphite oxide, GO) 的化学还原是制备石墨烯的常规方法之一。GO 的合成可采用三种方法:Brodie 法、Staudenmaier 法和 Hummers 法,其中 Hummers 法包括通过强酸 (例如浓 H_2SO_4、HNO_3 和 $KMnO_4$) 和氧化剂来氧化石墨的过程,而石墨的氧化程度会随着化学计量、反应条件 (压力、温度等) 和石墨类型 (高定向热解石墨、纯石墨等) 的不同而变化[29]。1859 年,Brodie 等首

次报道了在发烟硝酸存在下,向石墨浆中加入 KClO₃ 来合成 GO 的方法,但是整个过程具有危险程度高、耗时、操作烦琐等问题。1898 年,Staudenmaier 等改善了上述方法,他们在石墨浆中加入浓 H₂SO₄ 并通过一步法制备了高氧化程度的 GO。1958 年,Hummers 等报道了至今仍流行的方法,该方法包括在不使用 HNO₃ 的情况下将石墨浆液混入由 NaNO₃、浓 H₂SO₄ 和 KMnO₄ 组成的混合溶液中来制备氧化石墨。图 7.12 为上述三种方法制备 GO 的示意图。

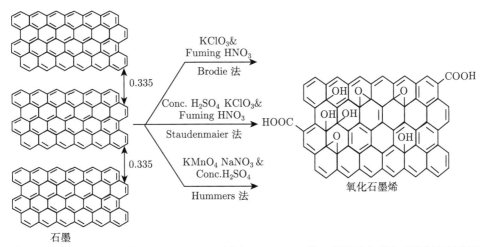

图 7.12　使用 Brodie 法、Staudenmaier 法和 Hummers 法三种通过化学合成法制备氧化石墨烯的示意图[29]

当石墨变成 GO 时,随着氧化时间的延长,石墨的层间距增加到原来的 2~3 倍。氧化 1 h 后,原始石墨的层间距由 3.34 Å 增加到 5.62 Å,随着氧化时间达到 24 h,层间距增大到 7.35 Å。在 N,N-二甲基甲酰胺/水 (9:1) 混合溶液中超声处理后,GO 的层间距进一步扩大,形成了单层 GO 悬浮液。水合肼处理后,GO 被还原为石墨烯或还原氧化石墨烯 (reduced graphene oxide, rGO),整个石墨化学合成石墨烯的过程如图 7.13 所示。除水合肼外,硼氢化钠、羟胺、对苯二酚、抗坏血酸等也可作为还原剂实现 GO 的还原。

H₂SO₄、HNO₃ 等有毒化合物的使用促进了环境友好性更好的氧化石墨烯制备方法的开发,但在众多化学合成法中 Hummers 法仍被认为是一种更快、更安全、更有效的氧化石墨烯制备方法。2010 年,Marcano 等[30]对 Hummers 法进行了改进,改进的 Hummers 法使用了大量的 KMnO₄ 及 H₂SO₄/H₃PO₄(体积比为 9:1) 的反应混合物,去掉了 NaNO₃:首先对 H₂SO₄/H₃PO₄ 混合溶液进行搅拌,在搅拌过程中,缓慢加入石墨粉和 KMnO₄;连续搅拌 6 h 后,溶液颜色变为深绿色;之后加入少量 H₂O₂,将含 H₂O₂ 的混合溶液搅拌一段时间后,冰浴冷却以降低反应体系温度；在混合溶液中添加 HCl 和去离子水以去除金属离子，并对溶液进行离心处理；生成的产物经盐酸和去离子水洗涤及干燥后，形成粉末状的氧化石墨烯；将氧化石墨烯进一步与水混合，并在规定的时间内进行超声处理，在高温和低搅拌速度下，将少量的硫黄系还原剂 ($NaHSO_3$, $Na_2S·9H_2O$, SO_2) 添加到混合物中，经过滤、去离子水洗涤及冷冻干燥后可获得粉状黑色石墨烯。与 Hummers 法相比，改进的 Hummers 法具有更高的氧化效率，并产生了亲水性更好的氧化石墨烯产品。

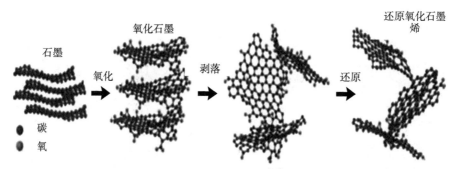

图 7.13　石墨化学合成石墨烯的流程图[29](彩图请扫封底二维码)

制备好氧化石墨烯后，许多研究者报道了通过氧化石墨烯的还原来制备石墨烯的报道，如热还原、电化学还原、光化学还原、水热还原、微波辅助还原，以及基于化学试剂的化学还原法[29]。在上述还原方法中，化学还原法被认为是最佳方法，因为该方法可去除石墨烯氧化物中存在的大多数含氧官能团，并可恢复还原后石墨烯的表面结构。然而，由于存在残留的含氧官能团 (如—OH、—COOH 等)，制备的石墨烯仍存在含氧缺陷，与无缺陷的原始石墨烯相比，该方法制备的石墨烯导电性能较差。石墨烯和 rGO 上残留的含氧官能团可用于调节石墨烯产品本身的电学和化学性质，以适应不同的应用。

5. 液相剥离法

液相剥离法是应用最广泛的石墨烯制备方法之一。2008 年，Coleman 的研究小组首次采用液相剥离法成功制备了少层石墨烯，他们发现，在特定溶剂的作用下，通过超声剥离可直接制备无结构缺陷的少层石墨烯，原理为将少量石墨粉分散在特定的剥离试剂中，在超声波或湍流辅助作用下破坏石墨层与层之间的范德瓦耳斯力，然后利用剥离试剂和石墨之间的插层作用将石墨层层剥离[31,32]。石墨烯的液相剥离制备过程主要包含三个步骤 (图 7.14)：① 将石墨分散在合适的剥离溶剂中；② 石墨的层层剥离；③ 离心纯化。

7.3 石墨烯的制备方法

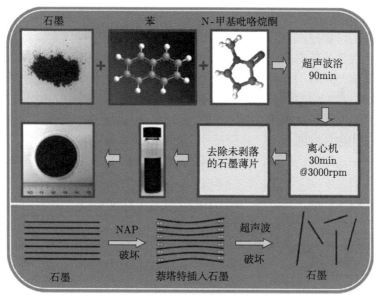

图 7.14 为以萘为剥离溶剂的石墨烯的液相剥离制备过程 (上)，包括混合、剥离、离心和真空过滤步骤；以及采用 π-π 堆叠作用剥离石墨烯的原理图 (下)[33]

液相剥离石墨烯制备的核心问题为剥离溶剂的选择、外力作用 (超声/剪切混合) 及离心纯化过程[31]。

1) 剥离溶剂的选择

由于液相剥离法是通过克服石墨层和层之间的范德瓦耳斯力来实现石墨的剥落的，因此剥离溶剂的选择主要取决于表面能、表面张力、Hildebrand 溶解度和 Hansen 溶解度参数等液体特性。Hernandez 等[32] 的研究表明，剥离溶剂和石墨之间的作用力是范德瓦耳斯力而不是共价作用力，理想溶剂应该能有效克服石墨层之间的范德瓦耳斯力，即使石墨层与层之间的 π-π 堆积距离保持在 3.35~3.4 Å，他们发现剥离溶剂的表面张力达到 40~50 mJ·m^{-2} 时能使克服范德瓦耳斯力的能量输出最小化。目前报道的剥离溶剂主要包括有机溶剂 (如 N-甲基吡咯烷酮 (N-methyl pyrrolidone, NMP)、N, N'-二甲基甲酰胺 (N, N-dimethyl formamide, DMF)、邻二氯苯等)、有机溶剂/辅助剂 (如 NMP/NaOH、NMP/有机盐、环己酮/萘等)、有机溶剂/稳定剂 (NMP/二十二烷直链烷烃、NMP/n-辛基苯、NMP/卟啉等)、水/表面活性剂溶液 (如水/十二烷基苯磺酸钠等)、水/表面稳定剂溶液 (如水/芘甲基胺盐酸盐、水/1, 3, 6, 8-芘四磺酸、水/纳米微晶纤维素等)、共溶剂 (如水/DMF、水/乙醇、水/异丙醇等)、离子液体 (1-丁基-3-甲基咪唑双 (三氟甲磺酰) 亚胺、1-己基-3-甲基咪唑六氟磷酸盐、1-苄基-3-甲基咪唑双 (三氟甲基磺酰基) 酰胺等) 六种。

2) 外力作用

外力作用,如超声或剪切混合是石墨烯剥离过程的关键因素。超声波会对石墨片层产生强烈的压缩及疏松作用,由此产生的真空腔及高压射流会导致石墨片层的层层剥离。例如,Khan 的研究小组 [34] 以 NMP 为剥离溶剂,在低超声功率下 (23W) 通过增加超声时间至 460h 获得了浓度高达 1.2 mg·mL^{-1} 的石墨烯溶液,其中单层石墨烯产率为 4 wt%,多数为少于 10 层的石墨烯,拉曼光谱分析表明,过长的超声时间减小了片层尺寸,虽然片层内部结构未受损伤,但在片层边缘引入了一定程度的缺陷。Arao 等以 NMP 和水/表面活性剂为剥离溶剂,通过优化初始石墨浓度、表面活性剂浓度、溶液体积等,获得了 1.17 g/h(NMP) 和 0.277 g/h(水/表面活性剂) 的产率;超声时间为 150 min 时,石墨烯/表面活性剂比值为 1.7,达到了目前报道的最高水平 [35]。

超声剥离技术具有简单、成本低廉的优点,但由于其产率较低 (据文献报道,产率最高为 1.17 g/h[35]),因此难以实现石墨烯的规模化制备。为此,Paton 等 [36] 报道了一种石墨剪切混合新技术,他们利用安装有转子/定子的可产生高速剪切力的搅拌器,以 NMP 和水/表面活性剂溶液为剥离溶剂,成功制备出无缺陷、高浓度石墨烯分散液,产率高达 5.3 g/h,他们发现,快速旋转叶片产生的局部湍流高剪切速率 (大于 10^4 s^{-1}) 是石墨烯剥落的主要原因,也就是说,任何可达到 10^4 s^{-1} 剪切速率的搅拌机都可实现石墨烯的剥离。文献报道,剪切混合技术制备的石墨烯的平均尺寸大小为 300~800 nm,平均厚度为 7 层 [36],而这些参数可能无法满足石墨烯的某些应用 (如聚合物复合材料的加固)。为了实现剥落的石墨烯片尺寸的控制,Arao 等 [37] 利用压力均化器产生的高速层流实现了石墨烯的剥离,产率高达 3.6 g/h:压力均化器可以在狭窄的流路上产生强流,在活塞泵驱动下,柱塞将溶液挤压到阀间隙中,当流体离开阀间隙时,形成径向射流,从而产生空化效应,最后,射流撞击撞击圈;高速层流可以对石墨施加均匀的剪切应力,从而导致石墨烯的剥落,而撞击圈中的碰撞过程是保持石墨烯片尺寸的主要原因。

3) 离心纯化作用

由于超声后的分散液中主要含有小的石墨烯薄片及石墨微晶,因此必须通过离心作用将石墨微晶除去,从而获得稳定的石墨烯分散液。分散液中的石墨烯薄片和石墨微晶的形状、大小和浮力密度是决定沉降速率的主要因素,当采用高的离心速度时,大的薄片沉降速度较快,因此,离心完成后,较小的薄片位于离心管顶部,而较大的薄片位于离心管底部。液相剥离法一般可获得尺寸为 1μm 左右的石墨烯片,但一些应用 (如石墨烯-有机玻璃复合材料的制备) 可能需要更大尺寸的薄片。离心会优先去除较大的薄片,从而导致片层平均尺寸的减小,因此可以通过调节离心速度来控制片层的尺寸大小。Lotya 等的实验结果很好地证明了这一点,他们发现,较高的离心速度会导致较低的分散液浓度:当离心速度由 500 rpm 增加到 5000 rpm 时,片层的平均厚度由 5nm 下降到 3.5nm、平均长度由 1.2μm 下降到 0.5μm、平均宽度由 600nm 下降到 300 nm[38]。

尽管液相剥离法是一种相当有前途的石墨烯合成方法,但多项研究表明,超声处理可能会在石墨烯的片层边缘和基面产生缺陷。通过优化超声处理时间、水浴温度和超声功率等参数,可以避免石墨烯片层上缺陷的产生,但是目前这方面的研究较少。尽管液相剥离法表现出某些缺点,但该方法在产品合成和功能化方面具有很高的通用性,并可以制备出高质量的少层石墨烯。此外,和化学气相沉积、SiC 外延生长等石墨烯制备方法相比,液相剥离法具有制备过程简单、成本低廉等优点,在生物传感等领域具有潜在的应用价值。

7.4 石墨烯场效应晶体管基本结构及原理

石墨烯场效应晶体管是一种以石墨烯为导电沟道的场效应器件,根据栅介质的不同可将其分为背栅、顶栅及液栅三种,下面分别进行介绍。

7.4.1 背栅石墨烯场效应晶体管

2004 年,Novoselov 等以含有 300 nm SiO_2 层的高度掺杂的 n 型 Si 为衬底,以 n 型 Si 为背栅制备了首个石墨烯背栅场效应晶体管 (graphene based back-gated field effect transistor, G-BgFET)[39]。在 G-BgFET 的制备中,一般选用沉积 300 nm SiO_2 层的高掺杂 Si 为衬底,并通过机械剥离或 CVD 方法在 SiO_2 层表面生长石墨烯膜,采用光刻和电子束蚀刻技术将电极图案转移到光刻胶表面后,通过电子束蒸发和剥离工艺制作出金属电极[40]。G-BgFET 的载流子密度及电导率由施加到硅衬底的栅压诱导的电场调制,典型的 G-BgFET 由源极 (source, S)、漏极 (drain, D)、栅极 (gate, G) 及沉积在 SiO_2 层顶部的石墨烯导电沟道组成,G-BgFET 的测量过程为:在 S 和 D 之间施加恒定的偏置电压 V_{sd},并监测源漏电流 I_{sd} 的变化 (图 7.15 (a))。当向石墨烯导电沟道施加栅压 V_g 时,由于石墨烯的高电导率,石墨烯导电沟道中可感生出电子或空穴,在 V_{sd} 恒定的情况下,I_{sd} 可由 V_g 调制;此外,在 V_g 的调制作用下,石墨烯导电沟道内的多数载流子可由空穴 (图 7.15 (b) 左侧的红色曲线) 转变为电子 (图 7.15 (b) 右侧的灰色曲线),并产生所谓的"双极效应",通常将 I_{sd} 最小的点称为电荷中性点 (charge neutral point, CNP)[41,42]。在 G-BgFET 中,电场的变化既可以采用背栅电压来实现,也可以通过 DNA、细胞等靶分子的物理或化学吸附来诱导。当背栅电压一定时,I_{sd} 的变化可归因于吸附在石墨烯表面的分子,这是由 2007 年 Schedin 等的研究结果证实的,他们制备的 G-bgFET 可实现对化学吸附在石墨烯表面的 NO_2 分子的检测。

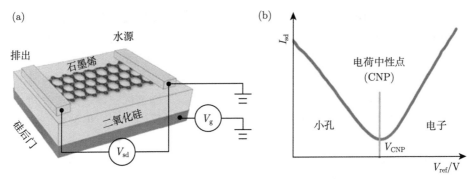

图 7.15 (a) 背栅石墨烯场效应晶体管的结构示意图和 (b) 双极性转移特性 [42]

7.4.2 顶栅石墨烯场效应晶体管

2007 年，Lemme 等制备了首个顶栅石墨烯场效应晶体管 (graphene based top-gated field effect transistor, G-TgFET) [43]，此后，研究者们通常以机械剥离法制备的石墨烯，CVD 法生长在镍、铜等金属上的石墨烯，或 SiC 外延生长制备的石墨烯为 (TgFET) 的导电沟道，SiO_2、HfO_2 及 Al_2O_3 等氧化物为顶栅电介质，n 掺杂的 Si 为衬底来制备 G-TgFET。典型的 G-TgFET 结构图如图 7.16 所示，由源极、漏极、顶栅 (top-gate, TG) 以及沉积在 SiO_2 层顶部的石墨烯导电沟道组成，顶栅和 SiO_2 层之间的 Y_2O_3 为顶栅绝缘层。与 G-BgFET，G-TgFET 在电路应用中显示出更高的灵活性，但 G-TgFET 的载流子迁移率比 G-BgFET 低得多，这主要是由于 SiO_2 等顶栅电介质会引起更多的散射源，且石墨烯薄膜在制备过程中容易被破坏。因此，石墨烯高载流子迁移率的保持对于性能良好的 G-TgFET 至关重要。例如 Farmer 等采用石墨烯和 HfO_2 栅介质之间的有机聚合物缓冲层 (聚合物 NFC 1400-3CP) 实现了 G-TgFET 的高载流子迁移率 [44]。

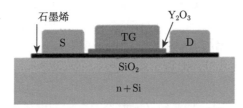

图 7.16 G-TgFET 的结构示意图 [45]

7.4.3 液栅石墨烯场效应晶体管

2008 年，Ang 的研究小组 [46] 首次提出了液栅石墨烯场效应晶体管 (graphene based solution-gated field effect transistor, G-SgFET)，并将其应用于 pH 的传感

中。G-SgFET 由源极、漏极及栅级组成，石墨烯材料可在源极和漏极之间形成导电沟道，栅级悬挂在石墨烯导电通道顶部 (G-SgFET 的结构图如图 7.17 所示)。在 G-SgFET 中，栅电压 V_g 通过参比电极施加到电解质溶液中，而参比电极通过界面电容 C (C 由石墨烯的量子电容 (C_Q) 及电解质溶液的双电层电容 (C_{DL}) 组成) 将 V_g 耦合到石墨烯导电沟道中。根据泊松–玻尔兹曼 (Poisson-Boltzmann) 方程，双电层电容是由位于固体和溶液界面处的分离电荷形成的虚拟电容。G-SgFET 属于离子敏场效应晶体管 (ion sensitive field effective transistor, ISFET) 大家族的一员 [42]，而 ISFET 是一种基于场效应的离子选择性敏感器件。尽管通道材料、参比电极、操作模式和器件封装因情况而异，但任何 ISFET 的核心都位于电解质和固体材料之间的界面上。G-SgFET 通常在低 V_g (小于 1 V) 下工作，因此任何电化学过程和交换离子电流都可以忽略不计，即该界面被认为是惰性并且是纯电容性的。在 G-SgFET 中，石墨烯导电沟道及栅极可与电解质直接接触，因此，栅极可灵敏地调制电解质溶液中的化学势 [47]；和 G-BgFET 和 G-TgFET 相比，G-SgFET 中较低的栅极操作电位 (小于 1V) 避免了不必要的氧化还原反应或水分解过程，使得场效应管在水环境中的应用成为可能，近年来，G-SgFET 在生化样品的分析中发挥着越来越重要的作用 [41]。

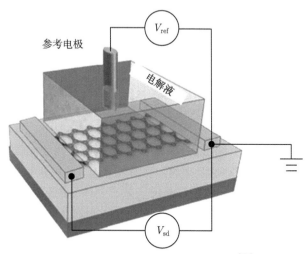

图 7.17　G-SgFET 的结构示意图 [42]

7.5　石墨烯场效应晶体管的制备工艺

石墨烯场效应晶体管的制备过程非常复杂，主要包括衬底上石墨烯薄膜的沉积及通过光刻和剥离工艺制备源极、漏极、栅极三个电极两个步骤。当将机械剥离法、CVD、SiC 外延生长等方法制备的石墨烯薄膜转移到衬底上后，下一个步骤即为三个电极的制备。石墨烯场效应晶体管的制备通常需要用到标准光刻技术。源极和漏极可以直接沉积在石墨烯薄膜上，也可以位于石墨烯薄膜的底部 (即先沉积源极和漏极，后沉积石墨烯薄膜)。石墨烯导电沟道及源极和漏极形成后，通常采用氧等离子体刻蚀技术使石墨烯薄膜图案化，从而形成多个单独的器件。为了防止电解以及其他与电极相关的有害反应的发生，则采用介电材料或聚合物覆盖导电电极与水溶液接触的部分 (即封装步骤) 是十分必要的。光刻过程的常见问题为光刻胶的残留。光刻胶一般都含有芳香树脂，而芳香树脂可以通过 π-π 堆积作用粘附在石墨烯表面，为了解决该问题，可以在石墨烯薄膜和光刻胶之间沉积一层 PMMA 作为牺牲层，从而防止光刻胶和石墨烯薄膜的直接接触，但 PMMA 也会导致残留物的形成。

喷墨打印作为一种简单、廉价、快速的器件制备方法，在石墨烯场效应晶体管的大规模生产制备中具有良好的应用潜力，因此成为了光刻技术的替代方法之一。尽管该方法会与诸如光刻之类的标准硅制造方法相抵触，但它在不适合硅制造方法的领域 (如纺织品、聚合物基和可循环利用的基底) 中具有重大意义。此外，喷墨打印法还有助于减少有毒化学品的使用。例如 Dua 的研究小组以剥离获得的氧化石墨烯溶液为墨水，采用喷墨打印法成功制备了石墨烯场效应晶体管型传感器[48]。近年来，喷墨打印已经成为了一种可替代传统半导体器件制备工艺的廉价且简单的方法。该方法首先采用市售的铬基油墨及喷墨打印机将所需的传感器图案反面印在铜箔上，然后使用标准的 CVD 工艺在图案化的箔片上 (未印刷区域) 生长石墨烯，将生长的石墨烯转移到任意衬底上后，用银纳米颗粒墨水将接触引线喷墨印刷到石墨烯上，最后在 180℃ 的条件下烧结以提高银纳米颗粒的导电性[49]。

对于 G-SgFETL 来说，栅极为悬挂的形式，因此不需要在器件结构上添加栅极。对于 G-BgFET 及 G-TgFET 来说，石墨烯薄膜、源极和漏极制备好后，最后一步为即栅极的沉积。最早报道的石墨烯场效应晶体管的制备方法是将机械剥离法制备的石墨烯薄膜转移到沉积有 300 nmSiO$_2$ 的硅片上，并以高掺杂的 p 型硅为背栅，由于氧化物层较厚，因此需要一个较大的 V_g 才能实现石墨烯导电沟道电导率的控制[39]。尽管 G-BgFET 结构验证了石墨烯场效应晶体管这种器件的存在，但实际应用并不多。为了实现高性能石墨烯场效应晶体管的制备，则必须采用低泄漏、薄而均匀、界面陷阱密度低的高 k 电介质为顶栅。一般采用热沉积或原子层沉积 (atomic layer deposition, ALD) 技术沉积栅极电介质。由于 ALD 法可精确控制薄膜的厚度和均匀性，因此该方法为栅极电介质制备时的最优方法。但由于 ALD 是一种基于水的方法，而石墨烯的表面呈疏水特性，因此直接沉积氧化物层 (如 Al_2O_3、HfO_2 等) 具有挑战性。此外，石墨烯表面的化学惰性进一步加剧了这个问题，为了促进成核和生长，一般通过化学功能化方法增强石墨烯表面的反应活性，如基于 NO_2 或 O_3 的功能化方法，但这种方法会破坏石墨烯的 sp^2 晶格，并导致载流子迁移率的降低。另一种方法是在 ALD 高 k 沉积之前引入聚合物的缓冲层，但由于聚合物一般为低 k 材料，因此电介质堆叠将导致较小的有效电容，而上述其他方法可在直接沉积或化学功能化过程引入石墨烯晶格上的缺陷，从而导致载流子迁移率的显著降低。化学功能化或引入缓冲层的替代方法为电子束蒸发沉积金属薄层，形成的金属薄层可迅速氧化并充当 ALD 氧化物的成核层，该方法可实现 $8000\ cm^2/(V·s)$ 的高载流子迁移率。

7.6 石墨烯场效应晶体管中石墨烯的功能化方法

石墨烯具有大的芳香 sp^2 碳晶格，但无悬挂键，因此其本质上是化学惰性的。在石墨烯场效应晶体管的石墨烯导电沟道中引入无机基团、有机或有机金属分子、DNA、蛋白质、二维异质结构等敏化剂后，这些分子可以对其附近的环境产生物理或化学响应，并将这些响应转化为石墨烯电导率的变化，从而实现石墨烯场效应晶体管的传感能力。在石墨烯表面或片层边缘引入上述分子的过程称为石墨烯的功能化。功能化后的石墨烯含有生化传感所需的特定识别部分，但在很大程度上仍具有与未功能化石墨烯相同的碳蜂窝骨架以及场效应等电学性能，因此可根据靶分子识别后石墨烯场效应晶体管的响应实现靶分子的检测。根据石墨烯场效应晶体管在生化物质传感中的应用，石墨烯的功能化方法主要包括两种：共价功能化方法及非共价功能化方法，下面分别进行介绍。

7.6.1 共价功能化方法

共价功能化方法是指将功能性生化分子与芳香晶格中的 sp^2 碳反应，从而在反应位点引入 sp^3 碳并将功能性生化分子修饰在石墨烯表面。石墨烯的共价功能化可以很大程度上改善石墨烯的性能，特别是在带隙工程、表面改性和生物界面等领域。Manchester 大学的研究小组首次报道了在石墨烯的蜂窝状结构中引入氢或氟原子的方法，他们的研究成果证明了将石墨烯这种高导电性的零带隙半金属转变为石墨烷 (图 7.18) 或二维 teflon 绝缘体的可能性。关于共价功能化在生化物质传感中的应用，理论计算结果表明，氢化石墨烯 (部分) 对 NO_2 具有高亲和性；而掺杂有 Li 原子的石墨烷对 H_2S 和 NH_3 敏感。此外，高度氢化后石墨烯的载流子迁移率仍可以满足传感器应用的需求。

对于以氧化石墨烯为导电沟道的石墨烯场效应晶体管来说，由于氧化石墨烯表面含有羧基、羟基、环氧基等含氧缺陷，因此该功能化方法最容易实现，具体操作方法为：在化学还原氧化石墨烯之前，将乙二胺连接到氧化石墨烯上的环氧基上,氧化石墨烯上连接的胺不仅可以在还原过程中保留下来,还可以用于进一步的生物功能化[50]。除此之外,将 CVD 生长的石墨烯暴露于低能氨等离子中,也可以形成适用于生物功能化的氨基基团[51]。其他共价功能化方法还包括石墨烯与重氮盐的反应,例如 Kasry 等利用该方法将生物素共价结合到石墨烯表面,并实现了对靶分子链霉亲和素的检测[52]。另外一种共价功能化方法称为表面引发光接枝和光聚合,通过紫外光照射甲基丙烯酸酯单体从而诱导石墨烯边缘和基面缺陷处的氢原子发生共价反应,这种方法不会显著影响石墨烯的电学特性[53,54]。虽然共价功能化方法利用的是石墨烯中已有的缺陷,并且不会对石墨烯片层造成额外的损伤,但该方法会降低石墨烯薄膜的平整度、破坏石墨烯的 sp^2 碳晶格,从而导致石墨烯载流子迁移率的降低[42,49]。

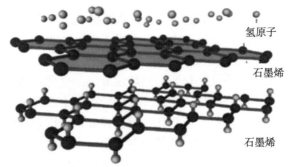

图 7.18 石墨烯层 (绿色) 与冷等离子体氢原子反应生成石墨烷[55](彩图请扫封底二维码)

7.6.2 非共价功能化方法

非共价功能化方法可在不破坏石墨烯固有芳香特性的情况下功能化石墨烯,此外,和原始石墨烯相比,该方法还可提高功能化后石墨烯的载流子迁移率。因此,非共价功能化方法对于高性能生化传感器的实现非常有吸引力。一般来说,石墨烯的非共价功能化方法可根据相应的分子间相互作用力进行分类,包括 π-π 堆叠、疏水堆叠、静电相互作用及范德瓦耳斯相互作用四种。基于 π-π 作用的非共价功能化方法一般是指将带有化学活性基团的芳香分子 (如 1-吡咯丁酸、苯酚等) 通过 π-π 作用堆叠到石墨烯表面,利用芳香分子上的化学活性基团共价连接功能性生物分子,从而实现对靶分子的特异性识别。值得注意的是,在 π-π 堆积的情况下,单链 DNA(single-stranded DNA,ssDNA) 也可以通过其含有的五碳糖和磷酸碱基 π-π 固定到石墨烯表面,因此非共价功能化方法还必须考虑非特异性结合及沾污造成的假阳性结果。1-吡咯丁酸、苯酚、牛血清蛋白等分子在石墨烯表面上的自组装过程可以得到有效控制,并可以通过 STM 等技术表征自组装的效果,从而有助于生化传感器的设计。图 7.19 (a) 为芳香苝-3,4,9,10-四羧酸-3,

4,9,10-二酐分子 π-π 非共价功能化后的石墨烯的 STM 表征结果,从图中可以看出,即使暴露在环境条件下,石墨烯表面形成的基于苝的单层膜也是稳定、坚固的[56]。芳香分子和核酸之间的 π-π 或疏水相互作用也可以促进 ssDNA 在石墨烯表面的修饰 (图 7.19 (b) 的右图)[57]。为了避免 ssDNA 在石墨烯表面的非特异性吸附,可以采用在石墨烯表面首先自组装单层苝-乙二醇的方式,由于苝-乙二醇自组装后的石墨烯表面具有亲水特性,因此可通过 ssDNA 的疏水作用防止 ssDNA 的吸附 (图 7.19 (b) 的左图)。除 DNA 之外,含有芳香环的蛋白质或肽也可以通过 π-π 作用自组装在石墨烯表面,如图 7.19 (c) 所示,肽的孵化导致石墨烯表面均匀的网状膜的形成,而氧化硅表面未受影响,这表明吸附只发生在石墨烯上[58]。另外一种常用的非共价功能化方法是利用吸附到石墨烯表面的纳米粒子 (如金纳米粒子、铂纳米粒子等) 作为 ssDNA 探针分子附着的位点,例如将金纳米粒子自组装到石墨烯表面后,利用金纳米粒子上的金硫键共价固定巯巯键修饰的功能性生物分子,从而特异性识别靶分子[49]。

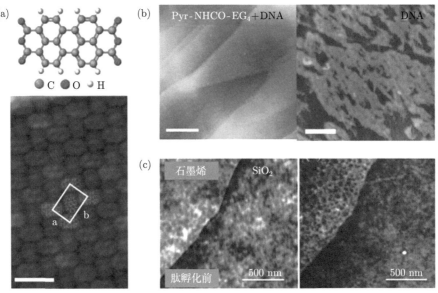

图 7.19 (a) 芳香苝-3,4,9,10-四羧酸-3,4,9,10-二酐分子 π-π 非共价功能化后的石墨烯的 STM 表征结果[56];(b) 左图为 3 M KCl 和 8 M 尿素溶液孵化 5 min 并用超纯水冲洗后的高定向热解石墨的 AFM 表征结果图,右图为单链 M13 DNA(10 ng·μL^{-1}) 孵化后的高定向热解石墨的 AFM 图[57];(c) 肽孵化前 (左图) 和孵化后 (右图) 的石墨烯的 AFM 表征结果[58](彩图请扫封底二维码)

尽管非共价功能化方法可在不破坏石墨烯键合结构的情况下实现石墨烯的功能化,但仍然存在以下两点缺陷:① 通过物理吸附作用吸附到石墨烯表面的分子作为某些生物分子探针的附着层,会使生物特异性识别事件远离石墨烯表面,从而导致德拜屏蔽效应的增强及靶分子检测灵敏度的降低;② 非共价功能化方法独特的物理吸附方式也造成了探针分子解析的可能,从而导致靶分子检测灵敏度的降低及 ssDNA 等生物分子的非特异性吸附性 [49]。

7.7 石墨烯场效应晶体管在生化物质传感中的应用

7.7.1 石墨烯场效应晶体管生化传感基本理论

在 G-BgFET、G-TgFET 及 G-SgFET 三种结构的石墨烯场效应晶体管中,研究者们多采用 G-SgFET 实现生化物质的传感,因此这里主要介绍 G-SgFET 在生化物质传感中的应用。G-SgFET 的生物传感原理如图 7.20 (a)~(c) 所示。一般可以将 G-SgFET 集成到微流体系统中:微流体通道对溶液的限制作用有助于将分析物带到传感器表面。在典型的测量中,受体分子被固定在石墨烯表面以选择性识别靶生物分子 (图 7.20 (b) 的上图)。G-SgFET 获得的 I_{sd} 与参比电压 (V_{ref}) 的关系曲线如图 7.20 (b) 的中图所示,其转移特性与 G-BgFET 获得的结果相似 (图 7.15 (b))。图 7.20 (b) 下图显示了 V_{ref} 一定时 I_{sd} 的变化 (如灰色虚线所示):当多数载流子为空穴 (图中 "h" 代表空穴) 或电子 (图中 "e" 代表电子) 时,由于场效应,带正电的靶分子的结合 (图 7.20 (c) 的上图) 会导致空穴载流子的耗尽 (或电子载流子的累积),这种掺杂效应会导致 $I_{sd}(V_{ref})$ 曲线的负移,如图 7.20 (c) 中图的红色箭头所示。在随时间变化的测量中 (图 7.20 (c) 的下图),带正电荷的靶分子与受体分子的结合导致空穴状态下的 I_{sd} 减小,而电子状态下的 I_{sd} 增大。相反,带负电荷的分子 (图 7.20 (a)) 与受体分子的结合会导致 $I_{sd}(V_{ref})$ 曲线的正移,并导致空穴状态下的 I_{sd} 增加,而在电子状态下,同样的反应会引起 $I_{sd}(V_{ref})$ 曲线的负移及 I_{sd} 的减小。石墨烯导电沟道的这种电流调制作用可以表示为载流子密度 (Δn) 变化的函数,即 Δn 与吸附在石墨烯表面的带电生物分子的总量 N 成正比:

$$\Delta I_{ds} = \frac{w}{l} V_{ds} e \mu \Delta n \infty N \tag{7.1}$$

其中,w 和 l 分别代表石墨烯导电沟道的宽度和长度;e 为电子电荷量;μ 为电荷载流子迁移率。根据公式 (7.1) 可以得出结论,G-SgFET 的响应信号与石墨烯表面吸附的带电生物分子的总量 N 成正比。但是,确定每种生物分子带的电荷数量,控制石墨烯导电沟道的化学功能化效果,以及准确识别不同状态下石墨烯表面发生的传感反应并非易事,因此对生物分子进行定量监测具有很大的挑战性。此外,还应指出,原理上,G-SgFET 对不带电荷的分子不会产生场效应感应,除非这些分子会引起电荷的变化 (例如通过细微的偶极波动或分子工程)。为了推导公式

(7.1)，我们假设石墨烯在吸附生物分子时具有恒定的载流子迁移率 μ。在大多数吸附的生物分子与受体结合后并与石墨烯晶格之间存在弱相互作用的情况下，这种假设是正确的。但是，直接结合在石墨烯表面的生物分子会形成其他散射中心，从而导致载流子迁移率的改变。此外，在传感器的实际设计中还应考虑生物分子吸附后界面电容的变化。

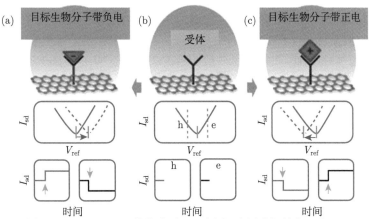

图 7.20 G-SgFET 的传感原理示意图 (彩图请扫封底二维码)

在 (b) 的上图中，受体分子被固定在石墨烯表面。I_{sd} 和 V_{ref} 的关系曲线以及 I_{sd} 和时间 t 的关系曲线分别显示在 (b) 的中图和下图。红色字母 "h" 和黑色字母 "e" 分别代表在空穴和电子状态下进行测量。(c)(或 (a)) 表示带正电 (或负电) 的靶生物分子与受体结合引起的场效应 (如 $I_{sd}(t)$ 曲线中的灰色箭头所示)，红色 (或蓝色) 箭头表示带电生物分子与受体的结合导致 I_{sd} 与 V_{ref} 的关系曲线的偏移

微弱的场效应诱导产生的电流变化 ΔI_{sd} 定义了器件对生物分子的传感响应 $S = \Delta I_{sd}/N$。根据公式 (7.1)，S 与石墨烯的载流子迁移率 μ 成比例。在其他参数相同的情况下，较大的 S 代表了较好的传感器性能。由于石墨烯场效应晶体管型传感器的传感性能取决于载流子迁移率 μ，因此使用高质量的石墨烯为导电沟道对于器件的传感性能非常重要，例如，基于机械剥离法 ($\mu \approx 3000 \sim 15000$ cm^2/(V·s)) 及 CVD 法 ($\mu \approx 1000 \sim 10000$ cm^2/(V·s)) 制备的石墨烯，其场效应晶体管都具有较好的传感性能 [42]。

7.7.2 G-SgFET 在生化物质传感领域的应用

G-SgFET 是一种以石墨烯为导电沟道的表面电荷型器件，通过导电沟道顶部的参比电极将栅电压施加到电解液并作用于导电沟道，在栅电压的作用下，石墨烯/电解液界面会产生载流子的堆积，从而引起表面电荷密度的改变，并对氧化还原物质的电子转移、带电分子的吸附等过程产生响应 [46]。石墨烯的高载流子迁移率、大的比表面积及良好的生物相容性赋予了 G-SgFET 响应速度快、检测面积大、对外界环境的变化反应灵敏等突出性能 [40]，近年来被广泛应用于 pH、重

金属离子、芳香剂 (丁酸戊酯)、细胞、生物大分子 (如蛋白质、核酸、疾病相关生物标志物等)、生物小分子 (葡萄糖、谷氨酸等)、细菌等生化物质的检测中，下面分别进行介绍。

1. pH 检测

2008 年，在 Ang 等[46]的研究基础上，Ohno 的研究小组[59]以机械剥离法制备的非功能化单层石墨烯为导电沟道，采用 G-SgFET 实现了对水溶液中 pH 的传感，他们发现，G-SgFET 的跨导比在真空中高 30 倍，且电导率随着电解液 pH 的增加而线性增加。此外，蛋白质的吸附实验也证明了 G-SgFET 对蛋白质的传感能力。

2010 年，Ohno 等[60]采用机械剥离法制备的石墨烯成功构建了 G-SgFET，并实现了水溶液中 pH 的灵敏检测。实验结果证明，当水溶液中 pH 增加时，G-SgFET 转移曲线的狄拉克点向正方向移动，方法检出限为 0.025。此外，根据生物分子的等电点，构建的 G-SgFET 还可以准确检测吸附生物分子的不同电荷类型。

Ristein 等[61]采用 SiC 外延生长法制备的石墨烯成功构建了 G-SgFET，并系统地研究了电解质 pH 对器件输出特性和转移特性的影响。实验结果证明，器件的转移特性具有 (19 ±1) mV/pH 的对溶液 pH 的依赖性。

为了研究 G-SgFET 对 pH 的敏感机制，Fu 等[62]采用 CVD 法制备的石墨烯成功构建了 G-SgFET，并系统地研究了 G-SgFET 对不同 pH 溶液的响应，实验结果证明，制备的石墨烯对 pH 的敏感性较弱，为 6mV/pH；若采用疏水氟苯分子钝化石墨烯表面，可进一步降低其对的 pH 敏感性；若在表面覆盖一层薄的无机氧化物层，则能显著增加石墨烯对 pH 的敏感性。这是因为，无机氧化物层可提供能被质子化和去质子化的末端羟基，从而形成一个表面电荷层，其电荷密度取决于溶液中的质子浓度。因此可以得出结论，理想的无缺陷石墨烯对 pH 不响应，有缺陷的石墨烯才会有响应。

非共价功能化方法以其独特的无损特性被广泛应用于石墨烯、碳纳米管等碳纳米材料的功能化中。例如 Fu 等[63]采用 CVD 生长的石墨烯成功制备了 G-SgFET，以具有 pH 活性的羟基基团的苯酚分子 π-π 共轭功能化石墨烯。实验结果证明，不同 pH 的缓冲液会造成电荷中性点 V_{CNP} 的不同程度偏移，这种非共价功能化的方法不仅保证了未受干扰的 sp^2 石墨烯晶格卓越的电子质量，同时获得了比最先进的 pH 计高两个数量级的检出限。

为了研究不同基底以及 CVD 石墨烯制备和转移过程中的残留物对 G-SgFET pH 传感的影响，Mailly-Giacchetti 等[64]分别以 SiO_2 和聚 (乙烯 2,6-萘二甲酸酯) 柔性材料为基底，利用 CVD 法生长的石墨烯制备了 G-SgFET，通过比较两种基底以及石墨烯表面不同程度的残留物对 G-SgFET pH 传感表现的影响,结果证明,这两个因素都不会显著影响石墨烯对 pH 变化的电响应,羟基和水合氢离子在石墨烯表面的吸附是 G-SgFET pH 传感的主要机理。

和 CVD 法制备的石墨烯相比,氧化还原法不仅成本低、易于操作,rGO 上丰富的活性位点还有助于提高 G-SgFET 的 pH 传感响应。Sohn 等[65] 成功制备了基于 rGO 的 G-SgFET,以 1-吡啶丁酸丁二酰酯 π-π 共轭功能化 rGO,将功能化的器件用于 pH 传感实验,H^+ 可与 rGO 表面的 OH^- 相互作用,导致表面电位的变化及电荷中性点的改变,最终获得了 29 mV/pH 的高灵敏度。

溶液中羟基和水合氢离子在石墨烯表面的吸附是 G-SgFET pH 传感的主要机理,而溶液中的其他离子可能会对 pH 的传感结果造成影响。为了了解 G-SgFET 对溶液中其他离子的响应,Lee 等[66] 设计了一系列 pH 缓冲液 (pH = 6~8,在一定浓度范围内加入各种离子),pH 传感结果表明,与 pH 相关的狄拉克电压 (V_{Dirac}) 的移动方向强烈依赖于缓冲液的浓度及离子组合。因此 G-SgFET 的 pH 传感过程必须充分考虑缓冲液中其他离子的影响。Falina 等[67] 利用基于 CVD 生长的石墨烯的 G-SgFET 研究了羧基化石墨烯的 pH 传感机理。导电沟道石墨烯的羧基化是在 pH 为 7 的缓冲溶液中连续阳极氧化实现的。由于羧基化可导致石墨烯中缺陷的增加,因此在低 pH 区域内,羧基化后的 G-SgFET 较非羧基化 G-SgFET 的 pH 敏感性增加,检出限为 32.6 mV/pH;由于羧基化石墨烯导电沟道表面的负电荷和碱性溶液中羟基离子的排斥作用,因此羧基化 G-SgFET 在高 pH 区域的 pH 敏感性较弱。

2. 重金属离子的检测

Sudibya 等[68] 成功制备了基于 rGO 的 G-SgFET,通过 1-吡啶丁酸丁二酰酯与 rGO 之间的 π-π 堆叠作用将连接分子固定在 rGO 表面,3-氨基丙基三乙氧基硅烷 (3-aminopropyl triethoxysilane, APTES) 的另一端可共价连接 Ca^{2+} 结合蛋白钙调蛋白 (CaM) 或金属硫蛋白 II 型蛋白 (MT-II),CaM 可选择性识别 Ca^{2+},而 MT-II 与生理重金属 (如锌、铜、硒) 和环境重金属 (如镉、汞) 都具有高亲和力。实验结果证明,该方法可高灵敏度、高选择性地检测各种离子。

Tu 等[69] 构建了基于 CVD 生长的石墨烯的 G-SgFET 阵列 (6 × 6),用于环境污染物中 Hg^{2+} 的超灵敏检测。首先通过 π-π 堆叠作用将连接分子 1, 5-二氨基萘 (1, 5-diaminonaphthalene, DAN) 固定在石墨烯表面,之后利用戊二醛 (GA) 耦合 DAN 上的氨基,当加入氨基修饰的适体 ssDNA 时,戊二醛一端的醛基可以和适体 ssDNA 上的氨基发生席夫碱反应,从而将适体 DNA 固定在石墨烯表面。Hg^{2+} 孵化后,由于 T-Hg^{2+}-T 反应,与 ssDNA 适配体结合的 Hg^{2+} 会在石墨烯界面发生静电变化,从而引起 G-SgFET 电信号的变化。方法检测范围为 100 pM ~ 100 nM，检出限低至 40 pM。

3. 芳香剂（丁酸戊酯）的检测

2012 年，Park 等[70] 首次报道了一种在柔性 PET 基底上制作的基于 CVD 生长的双层石墨烯的 G-SgFET 型嗅觉系统（即电子鼻），该系统可实现芳香剂丁酸戊酯 (amyl butyrate, AB) 的超灵敏检测。首先，采用 O_2 或 NH_3 等离子体处理双层石墨烯，使其具有 p 型和 n 型行为；其次，采用 DANπ-π 共轭石墨烯，并将与 AB 特异性结合的人类嗅觉受体 2AG1(hOR2AG1: OR) 化学连接到 DAN 上；最后，进行戊二醛 (glutaraldehyde, GA) 处理，通过席夫碱反应形成 GA 耦合的 DAN/石墨烯。GA 上的醛基和 OR 上的氨基可形成共价键，从而实现 OR 的特异性捕获。方法检出限为 0.04 fM，OR-O_2 等离子体处理的石墨烯及 OR-NH_3 等离子体处理的石墨烯的平衡常数分别为 $3.44×10^{14}M^{-1}$ 及 $1.47×10^{14}M^{-1}$。

4. 细胞信号的检测

2010 年，Cohen-Karni 等[71] 首次报道了基于石墨烯的 G-SgFET 用于记录鸡胚胎心肌细胞跳动所产生的电导信号的研究，他们将制备的 G-SgFET 与自发跳动的鸡胚胎心肌细胞直接接触，实验结果表明，G-SgFET 产生的电信号可清楚地记录鸡胚胎心肌细胞跳动所产生的电导信号，且随着 G-SgFET 面积的增加，记录的峰值信号宽度也随之增大，这些信号表明了自发跳动细胞的细胞膜外不同位置发出的平均信号。

2011 年，Hess 等[72] 采用 CVD 生长的石墨烯制备了 G-SgFET 阵列，并将与心肌细胞类似的 HL-1 细胞孵化在阵列上，当在液栅介质下对器件进行检测时，HL-1 细胞自发跳动所产生的动作电位可得到有效的分辨和跟踪。

5. 生物大分子的检测

1) DNA 或 RNA

2010 年，Dong 等[73] 采用 CVD 生长的单层石墨烯和少层石墨烯成功制备了 G-SgFET，并将其应用于 DNA 杂交的传感中。他们首先将探针 DNA 通过饱和吸附固定到石墨烯表面，互补 DNA 和错配 DNA 孵化后，互补 DNA 和探针 DNA 的杂交会造成器件的电子掺杂，石墨烯电导率的最小值 ($V_{g,min}$) 随着固定的 DNA 浓度的增加而左移，而错配 DNA 只造成了 $V_{g,min}$ 的较小程度偏移；在石墨烯表面加入金纳米粒子后，由于 DNA 负载量的增加，DNA 的检测上限由 10 nM 增加到 500 nM。

Ohno 等[74] 利用基于 CVD 石墨烯的 G-SgFET 及简单的平板电容器模型，实现了 DNA 杂交的检测及 DNA 数量的估算。他们通过 PASE 与石墨烯之间的 π-π 堆叠作用将连接分子 PASE 固定在石墨烯表面，然后利用共价固定在 PASE

上的氨基化探针 DNA 来特异性捕获靶 DNA，实验结果证明，探针 DNA 上的负电荷诱导了 G-SgFET 转移特性的改变，互补 DNA 的杂交可引起漏极电流的改变，非互补 DNA 孵化后未观察到变化，说明制备的 G-SgFET 可实现 DNA 杂交的检测。此外，他们利用平板电容器模型，估计出每 (10×10) nm^2 石墨烯通道上吸附一个探针 DNA，石墨烯通道上 30% 的探针 DNA 与 200 nM 互补 DNA 杂交，只有 5% 的探针 DNA 与非互补 DNA 结合。

Cai 等[75] 利用基于 rGO 的 G-SgFET 实现了 DNA 的免标记超灵敏检测，通过 PASE 与 rGO 之间的 π-π 堆叠作用将连接分子 PASE 固定在石墨烯表面，然后将共价固定在石墨烯表面的肽核酸 (PNA) 作为捕获探针来识别互补 DNA，检出限为 100 fM。他们研制的 rGO FET 型 DNA 生物传感器具有高灵敏度、高特异性，在疾病诊断中具有潜在的应用价值。

Xu 等[76] 采用 CVD 法制备了厘米级单晶石墨烯结构域，并将其图案化为 6 个 G-SgFET，用于 DNA 结合动力学和亲和性的多通道分析。他们将石墨烯导电通道用 1-芘丁酸丁二酰酯连接分子 π-π 共轭功能化，1-芘丁酸丁二酰酯上的琥珀酰亚胺部分可共价固定氨基修饰的探针 DNA，从而实现对靶 DNA 的特异性捕获。实验结果证明：该方法可多通道检测具有反应时间和浓度依赖性的 DNA 杂交动力学和亲和性，检出限为 10 pM；可实现单碱基突变的实时定量区分；并可根据建立的分析模型估计 DNA 探针密度、DNA 杂交效率和传感器的最大响应。

Tian 等[77] 利用基于 CVD 生长的石墨烯的 G-SgFET 实现了 RNA 的免标记检测，通过 π-π 堆叠的 PBASE 共价连接探针 DNA 来识别互补 RNA，方法检出限低至 0.1 fM。与传统的 RNA 检测方法比较，G-SgFET RNA 生物传感器灵敏度高、成本低、分析速度快、操作简单，并可实现 RNA 测量的小型化。

2) 蛋白质

2010 年，He 等[78] 在包括柔性 PET 薄膜在内的各种基底上，利用毛细管微塑法制备了厘米级、超薄 (13 nm)、具有可连续导电微图案的均匀平行的 rGO 薄膜阵列 (即 G-SgFET)，并且基于 "多巴胺可以通过 π-π 相互作用与 rGO 耦合，从而导致 rGO 的掺杂" 的机理，实现了对多巴胺分子及神经内分泌 PC12 活细胞中的多巴胺动态分泌的免标记检测。

此外，Ohno 等[79] 报道了一种基于 G-SgFET 的免标记生物传感器，通过在石墨烯表面固定免疫球蛋白 E(IgE) 适体来特异性识别 IgE，原子力显微镜的分析结果证明了 IgE 的成功固定。实验结果证明，该方法能实现对靶蛋白 IgE 的选择性检测，并且根据源漏电流的净变化量与孵化的 IgE 浓度之间的依赖关系估算出了 IgE 的解离常数 (47 nM)。

2011 年，He 等[80] 报道了一种在 PET 基板上利用 rGO 制备柔性、透明 G-SgFET 的方法，并且首次实现了缓冲液中纤连蛋白非特异性吸附的实时检测。

他们利用 1-吡啶丁酸丁二酰酯连接分子 π-π 共轭功能化 rGO 导电沟道，PASE 连接分子一端的羧基可实现对缓冲液中纤连蛋白的特异性捕获，检出限为 0.5 nM；此外，利用生物素化的 G-SgFET 还实现了对抗生物素蛋白的特异性检测。

胰岛素是一种促进血糖吸收的内分泌肽激素，是糖尿病诊断中的重要参考指标。Hao 等 [81] 报道了一种基于 G-SgFET 的可实时、免标记、高特异性、高灵敏度地监测胰岛素水平的纳米传感器。他们首先通过 π-π 作用将连接分子 PASE 固定在石墨烯表面，然后利用 1-吡啶丁酸丁二酰酯上的二氢丁二胺酯将氨基修饰的 IGA3 适配体共价固定，带负电荷的 IGA3 适配体与胰岛素结合后会折叠成紧凑而稳定的反平行或平行的 G-四链体结构，从而导致石墨烯上载流子密度及电导率的改变。通过测定电导率的变化可实时检测胰岛素水平，该方法检测范围为 100 pM~1 μM，检出限低至 35 pM。

2018 年，Kawata 等 [82] 报道了一种基于 G-SgFET 的 IgE 生物传感器，他们分别以四 (4-羧苯基) 卟啉 (TCPP) 和 PASE 为连接剂对石墨烯进行 π-π 共轭功能化，并比较了两种连接剂功能化的 G-SgFET 传感 IgE 的表现。和 PASE 相比，TCPP 有更大的 π 共轭系统，因此可更稳定地与石墨烯结合，除此之外，TCPP 上含有的 4 个羧基也可作为 IgE 适配体的结合位点，从而提高适配体的结合密度。PASE 功能化的 G-SgFET 的 IgE 检出限为 13 nM，而 TCPP 功能化的 G-SgFET 的 IgE 检出限提高到了 2.2 nM，证明了 TCPP 作为 G-SgFET 功能化材料的潜力。

3) 疾病相关生物标志物

人类表皮生长因子受体 2(HER2) 和表皮生长因子受体 (EGFR) 是乳腺癌的重要生物标志物。2011 年，Myung 等 [83] 报道了一种基于 rGO 封装的纳米颗粒 (NP) 的 G-SgFET，并实现了对 HER2 和 EGFR 的高灵敏度、高选择性检测。他们首先在 rGO 上涂敷 3-氨丙基三乙氧基硅烷 (APTES) 功能化的氧化硅纳米颗粒，该方法能防止 rGO 的团聚、提高 rGO 的导电性，更重要的是，rGO 封装的纳米颗粒 (rGO-NP) 具有高表面体积比，可生成三维电子表面，并能显著提高生物标志物的检出限。之后，rGO-NP 被 4-(1-芘基) 甲基丁醛 π-π 共轭功能化，4-(1-芘基) 甲基丁醛上的醛基可共价固定 HER2 抗体和 EGFR 抗体，从而实现了对 HER2 和 EGFR 的选择性捕获。该方法检出限为 1 pM(HER2) 和 100 pM(EGFR)。

前列腺特异抗原/α1-抗胰凝乳蛋白酶 (PSA/ACT) 复合物是前列腺癌的重要生物标志物。Kim 等 [84] 报道了一种基于 rGO 的 G-SgFET 超灵敏检测前列腺癌生物标志物 (PSA/ACT) 的方法。G-SgFET 的导电沟道首先被 PASE π-π 共轭功能化，之后共价固定 PSA 单克隆抗体 (PSA mAb)，PSA/ACT 复合物与 PSA mAb 的免疫反应会导致转移曲线中狄拉克点的线性改变，检出限可达飞摩尔级。

Gao 等 [85] 利用基于 CVD 生长的石墨烯的 G-SgFET 实现了对 PSA 的实时可逆检测。他们通过 π-π 堆叠的芘丁酸 (PYCOOH) 共价固定带有氨基末端的聚乙烯乙二醇 (PEG)，以间隔分子乙醇胺 (ETA) 或 DNA 适体作为特异性蛋白受体，PEG/ETA 修饰的石墨烯 FET 实现了对磷酸盐缓冲液中 1~1000 nM 的 PSA 的实时可逆检测，PEG/DNA 适体修饰的石墨烯场效应管实现了对 PSA 的不可逆特异性结合及检测。

基质金属蛋白酶 (MMP-7) 有助于肿瘤和癌的侵袭性生长和转移，是肺癌及唾腺癌诊断及预后的重要指标。2016 年，Chen 等 [86] 报道了一种基于多肽 (JR2EC) 功能化的 rGO-SgFET 的快速、灵敏地检测 MMP-7 的新方法。MMP-7 可特异性消化了固定在 rGO 上的带负电荷的 JR2EC，从而调节 rGO-SgFET 的电导率。当检测不同临床浓度 MMP-7 时，检出限为 10 ng·mL^{-1}；人血浆中 MMP-7 的检出限为 40 ng·mL^{-1}。

脑钠素 (BNP) 是心力衰竭诊断中重要的生物标志物，对于心血管系统疾病的诊断及预后具有很大的应用价值。2017 年，Lei 等 [87] 利用铂纳米粒子 (PtNP) 修饰的基于 rGO 的 G-SgFET 实现了对全血中 BNP 的灵敏检测，通过在铂纳米粒子表面共价固定 BNP 抗体来免疫识别 BNP，检出限为 100 fM，该方法被成功地应用于人全血样品中 BNP 的检测中，表明了 rGO 场效应管在复杂样本中的传感能力。

癌胚抗原 (CEA) 是一种癌症相关的生物标志物，可用于评估癌症预后和监测癌症的发展。Zhou 等 [88] 利用基于 CVD 生长的石墨烯的 G-SgFET 实现了对 CEA 的免标记检测，通过非共价 π 堆积的 1-芘丁酸琥珀酰亚胺酯双功能分子共价固定靶向癌胚抗原的抗体 (Anti-CEA) 来免疫识别 CEA，检出限为 100 pg·mL^{-1}。

人绒毛膜促性腺激素 (hCG) 是一种糖蛋白激素，其浓度水平与睾丸癌、胰腺癌、前列腺癌等癌症的发生密切相关。Haslam 等 [89] 利用基于 CVD 生长的石墨烯的 G-SgFET 免疫传感器实现了对癌症风险生物标志物 hCG 的免标记检测，通过 π-π 堆叠的 PASE 共价连接抗-hCG 单克隆抗体来免疫识别 hCG，检出限为 1 pg·mL^{-1}。所开发的 GFET 生物传感器还可用于阿尔茨海默病、帕金森病和心血管疾病等的医学诊断中。

6. G-SgFET 用于生物小分子的检测

2010 年，Huang 等 [90] 利用 CVD 生长的石墨烯成功制备了 G-SgFET，通过与石墨烯 π-π 堆叠的 PASE 实现葡萄糖氧化酶 (GOD) 和谷氨酸脱氢酶 (GluD) 的共价固定。当加入葡萄糖和谷氨酸时，葡萄糖和 GOD 或谷氨酸和 GluD 之间的催化氧化反应所生成的产物会导致石墨烯电导率的增加，从而诱导源漏电流的增加。实验获得的检出限为 0.1 mM(葡萄糖) 和 5 mM(谷氨酸)，和电化学检测方法获得的结果相当。

2012 年，Kwak 等[91] 以 PET 为基板，成功制备了基于 CVD 生长的石墨烯的柔性 G-SgFET，通过 π-π 堆叠的 1-芘丁酸琥珀酰亚胺酯酸 (PSE) 共价连接 GOD，从而实现对葡萄糖的催化氧化。由于葡萄糖的催化氧化过程产生的双氧水会造成石墨烯的 n 型掺杂，因此加入的葡萄糖浓度的增加会导致转移曲线的狄拉克点的移动及源漏电流的变化，检测范围为 3.3~10.9 mM。

2017 年，Siddique 等[92] 利用基于 CVD 生长的石墨烯的 G-SgFET 实现了对胆固醇的检测，随着胆固醇浓度的增加，狄拉克点位置逐渐向负栅极电压移动，石墨烯载流子迁移率从 ~2000 $cm^2/(V·s)$ 提高到 ~3900 $cm^2/(V·s)$。真实样品应用表明，基于石墨烯的传感器能够以快速、高效的方式高精度地定量测定血清样品中的胆固醇浓度。

7. G-SgFET 用于细菌的检测

饮用水和地表水中的大肠杆菌污染严重威胁着公众健康。Huang 等[93] 采用 CVD 生长的石墨烯成功制备了 G-SgFET，并实现了对大肠杆菌 (E. coli) 的高灵敏度、高选择性检测。他们首先通过 PASE 与石墨烯之间的 π-π 作用非共价功能化石墨烯，之后将抗-E. coli O & K 抗体共价固定在 PASE 上，从而实现对 E. coli 的特异性识别。实验结果证明：E. coli 的孵化可导致 G-SgFET 的电导率升高，检出限为 10 $cfu·mL^{-1}$；此外，该方法还可实现对葡萄糖诱导的大肠杆菌代谢活性的实时检测。

Wu 等[94] 报道了一种基于 G-SgFET 及芘标记的 DNA 适配体 (PTDA) 的 E. coli 检测方法。他们首先将 PTDA 固定在石墨烯表面，E. coli 孵化后，PTDA 可与 E. coli 的外膜蛋白或脂多糖特异性结合，并可导致 PTDA 构向的改变，从而诱导带负电荷的大肠杆菌靠近石墨烯表面，有效提高了石墨烯中的空穴载流子密度。实验结果表明，G-SgFET 的源漏电流与大肠杆菌的浓度有强烈依赖关系，检出限为 100 $cfu·mL^{-1}$。

Thakur 等[95] 构建了基于 rGO 的 G-SgFET，利用锚定在金纳米颗粒上的大肠杆菌特异性抗体作为选择性捕获大肠杆菌的探针，检出限为 10^3~10^5 $cfu·mL^{-1}$，报道的 rGO 场效应管提供了一种低成本且无标签地检测水或其他液体介质中病原体的方法。

除此之外，G-SgFET 还被应用于对环境污染物 2,4-二氯代酚[96]、双酚 A(检出限为 10 $ng·mL^{-1}$)[97]、自由氯 (检出限为 100 nM)[98] 等物质的检测中。

参 考 文 献

[1] Soldano C, Mahmood A, Dujardin E. Production, properties and potential of graphene [J]. Carbon, 2010, 48(8): 2127-2150.

[2] Novoselov K S, Geim A K, Morozov S V, et al. Electric field effect in atomically thin carbon films[J]. Science, 2004, 306(5696): 666-669.

[3] Avouris P. Graphene: electronic and photonic properties and devices[J]. Nano Letters, 2010, 10(11): 4285-4294.

[4] Bolotin K I, Sikes K J, Jiang Z, et al. Ultrahigh electron mobility in suspended graphene[J]. Solid State Communications, 2008, 146(9-10): 351-355.

[5] Du X, Skachko I, Barker A, et al. Approaching ballistic transport in suspended graphene[J]. Nature Nanotechnology, 2008, 3(8): 491.

[6] Mayorov A S, Gorbachev R V, Morozov S V, et al. Micrometer-scale ballistic transport in encapsulated graphene at room temperature[J]. Nano Letters, 2011, 11(6): 2396-2399.

[7] Dimitrakopoulos C, Lin Y M, Grill A, et al. Wafer-scale epitaxial graphene growth on the Si-face of hexagonal SiC (0001) for high frequency transistors[J]. Journal of Vacuum Science & Technology B, Nanotechnology and Microelectronics: Materials, Processing, Measurement, and Phenomena, 2010, 28(5): 985-992.

[8] Mao H Y, Lu Y H, Lin J D, et al. Manipulating the electronic and chemical properties of graphene via molecular functionalization[J]. Progress in Surface Science, 2013, 88(2): 132-159.

[9] Allen M J, Tung V C, Kaner R B. Honeycomb carbon: a review of graphene[J]. Chemical Reviews, 2010, 110(1): 132-145.

[10] Hu L, Cheng Q, Chen D, et al. Liquid-phase exfoliated graphene as highly-sensitive sensor for simultaneous determination of endocrine disruptors: diethylstilbestrol and estradiol[J]. Journal of Hazardous Materials, 2015, 283: 157-163.

[11] Arao Y, Mori F, Kubouchi M. Efficient solvent systems for improving production of few-layer graphene in liquid phase exfoliation[J]. Carbon, 2017, 118: 18-24.

[12] Wang Z R, Jia Y F. Graphene solution-gated field effect transistor DNA sensor fabricated by liquid exfoliation and double glutaraldehyde cross-linking[J]. Carbon, 2018, 130: 758-767.

[13] 单云, 张红琳, 张凤. 氧化石墨烯水热还原前后的发光光谱 [J]. 应用化学, 2015, 32(7): 837-842.

[14] Bari R, Tamas G, Irin F, et al. Direct exfoliation of graphene in ionic liquids with aromatic groups[J]. Colloids and Surfaces A-physicochemical and Engineering Aspects, 2014, 463: 63-69.

[15] Skaltsas T, Ke X X, Bittencourt C, et al. Ultrasonication induces oxygenated species and defects onto exfoliated graphene[J]. Journal of Physical Chemistry C, 2013, 117(44): 23272-23278.

[16] Huc V, Bendiab N, Rosman N, et al. Large and flat graphene flakes produced by epoxy bonding and reverse exfoliation of highly oriented pyrolytic graphite[J]. Nanotechnology, 2008, 19(45): 455601.

[17] Somani P R, Somani S P, Umeno M. Planer nano-graphenes from camphor by CVD[J]. Chemical Physics Letters, 2006, 430(1-3): 56-59.

[18] Obraztsov A N, Obraztsova E A, Tyurnina A V, et al. Chemical vapor deposition of thin graphite films of nanometer thickness[J]. Carbon, 2007, 45(10): 2017-2021.

[19] Yu Q, Lian J, Siriponglert S, et al. Graphene segregated on Ni surfaces and transferred to insulators[J]. Applied Physics Letters, 2008, 93(11): 113103.

[20] Wang X, You H, Liu F, et al. Large-scale synthesis of few-layered graphene using CVD[J]. Chemical Vapor Deposition, 2009, 15(1-3): 53-56.

[21] Lee J, Zheng X, Roberts R C, et al. Scanning electron microscopy characterization of structural features in suspended and non-suspended graphene by customized CVD growth[J]. Diamond and Related Materials, 2015, 54: 64-73.

[22] Kalita G, Ayhan M E, Sharma S, et al. Low temperature deposited graphene by surface wave plasma CVD as effective oxidation resistive barrier[J]. Corrosion Science, 2014, 78: 183-187.

[23] Wei D, Lu Y, Han C, et al. Critical crystal growth of graphene on dielectric substrates at low temperature for electronic devices[J]. Angew. Chem. Int. Ed. Engl., 2013, 52(52): 14121-14126.

[24] Lee X J, Hiew B Y Z, Lai K C, et al. Review on graphene and its derivatives: Synthesis methods and potential industrial implementation[J]. Journal of the Taiwan Institute of Chemical Engineers, 2019, 98: 163-180.

[25] Van Bommel A J, Crombeen J E, van Tooren A. LEED and Auger electron observations of the SiC(0001) surface[J]. Surface Science, 1975, 48(2): 463-472.

[26] de Heer W A, Berger C, Wu X, et al. Epitaxial graphene[J]. Solid State Communications, 2007, 143(1-2): 92-100.

[27] Juang Z Y, Wu C Y, Lo C W, et al. Synthesis of graphene on silicon carbide substrates at low temperature[J]. Carbon, 2009, 47(8): 2026-2031.

[28] Bonaccorso F, Lombardo A, Hasan T, et al. Production and processing of graphene and 2d crystals[J]. Materials Today, 2012, 15(12): 564-589.

[29] Adetayo A, Runsewe D. Synthesis and fabrication of graphene and graphene oxide: a review[J]. Open Journal of Composite Materials, 2019, 09(02): 207-229.

[30] Marcano D C, Kosynkin D V, Berlin J M, et al. Improved synthesis of graphene oxide[J]. ACS Nano, 2010, 4(8): 4806-4814.

[31] Narayan R, Kim S O. Surfactant mediated liquid phase exfoliation of graphene[J]. Nano Converg, 2015, 2(1): 20.

[32] Hernandez Y, Nicolosi V, Lotya M, et al. High-yield production of graphene by liquid-phase exfoliation of graphite[J]. Nature Nanotechnology, 2008, 3(9): 563-568.

[33] Xu J, Dang D K, Tran V T, et al. Liquid-phase exfoliation of graphene in organic

solvents with addition of naphthalene[J]. Journal Of Colloid and Interface Science, 2014, 418: 37-42.

[34] Khan U, O'Neill A, Lotya M, et al. High-concentration solvent exfoliation of graphene [J]. Small, 2010, 6(7): 864-871.

[35] Arao Y, Kubouchi M. High-rate production of few-layer graphene by high-power probe sonication[J]. Carbon, 2015, 95: 802-808.

[36] Paton K R, Varrla E, Backes C, et al. Scalable production of large quantities of defect-free few-layer graphene by shear exfoliation in liquids[J]. Nature Materials, 2014, 13(6): 624-630.

[37] Arao Y, Mizuno Y, Araki K, et al. Mass production of high-aspect-ratio few-layer-graphene by high-speed laminar flow[J]. Carbon, 2016, 102: 330-338.

[38] Lotya M, King P J, Khan U, et al. High-concentration, surfactant-stabilized graphene dispersions[J]. ACS Nano, 2010, 4(6): 3155-3162.

[39] Novoselov K S, Geim A K, Morozov S V, et al. Electric field effect in atomically thin carbon films[J]. Science, 2004, 306(5696): 666-669.

[40] Zhan B, Li C, Yang J, et al. Graphene field-effect transistor and its application for electronic sensing[J]. Small, 2014, 10(20): 4042-4065.

[41] Andronescu C, Schuhmann W. Graphene-based field effect transistors as biosensors[J]. Current Opinion in Electrochemistry, 2017, 3(1): 11-17.

[42] Fu W, Jiang L, van Geest E P, et al. Sensing at the surface of graphene field-effect transistors[J]. Advanced Materials, 2017, 29(6): 1603610.

[43] Lemme M C, Echtermeyer T J, Baus M, et al. A graphene field-effect device[J]. IEEE Electron Device Letters, 2007, 28(4): 282-284.

[44] Farmer D B, Chiu H Y, Lin Y M, et al. Utilization of a buffered dielectric to achieve high field-effect carrier mobility in graphene transistors[J]. Nano Lett., 2009, 9(12): 4474-4478.

[45] Xu H, Zhang Z, Xu H, et al. Top-gated graphene field-effect transistors with high normalized transconductance and designable dirac point voltage[J]. ACS Nano, 2011, 5(6): 5031-5037.

[46] Ang P K, Chen W, Wee A T, et al. Solution-gated epitaxial graphene as pH sensor[J]. Journal of the American Chemical Society, 2008, 130(44): 14392-14393.

[47] Wang D, Noël V, Piro B. Electrolytic gated organic field-effect transistors for application in biosensors-a review[J]. Electronics, 2016, 5(1): 9.

[48] Dua V, Surwade S P, Ammu S, et al. All-organic vapor sensor using inkjet-printed reduced graphene oxide[J]. Angewandte Chemie, 2010, 49(12): 2154-2157.

[49] Tamanaha C R. A survey of graphene-based field effect transistors for bio-sensing[J]. Carbon-Based Nanosensor Technology, 2017, 17: 165-200.

[50] Stine R, Robinson J T, Sheehan P E, et al. Real-time DNA detection using reduced graphene oxide field effect transistors[J]. Advanced Materials, 2010, 22(46): 5297-5300.

[51] Baraket M, Stine R, Lee W K, et al. Aminated graphene for DNA attachment produced via plasma functionalization[J]. Applied Physics Letters, 2012, 100(23): 233123.

[52] Kasry A, Afzali A A, Oida S, et al. Detection of biomolecules via benign surface modification of graphene[J]. Chemistry of Materials, 2011, 23(22): 4879-4881.

[53] Steenackers M, Gigler A M, Zhang N, et al. Polymer brushes on graphene[J]. Journal of the American Chemical Society, 2011, 133(27): 10490-10498.

[54] Hess L H, Lyuleeva A, Blaschke B M, et al. Graphene transistors with multifunctional polymer brushes for biosensing applications[J]. ACS Applied Materials & Interfaces, 2014, 6(12): 9705-9710.

[55] Elias D C, Nair R, Mohiuddin T, et al. Control of graphene's properties by reversible hydrogenation: evidence for graphane[J]. Science, 2009, 323(5914): 610-613.

[56] Wang Q H, Hersam M C. Room-temperature molecular-resolution characterization of self-assembled organic monolayers on epitaxial graphene[J]. Nature Chemistry, 2009, 1(3): 206-211.

[57] Schneider G F, Xu Q, Hage S, et al. Tailoring the hydrophobicity of graphene for its use as nanopores for DNA translocation[J]. Nature Communications, 2013, 4(1): 2619-2619.

[58] Katoch J, Kim S N, Kuang Z, et al. Structure of a peptide adsorbed on graphene and graphite[J]. Nano Letters, 2012, 12(5): 2342-2346.

[59] Ohno Y, Maehashi K, Yamashiro Y, et al. Electrolyte-gated graphene field-effect transistors for detecting pH and protein adsorption[J]. Nano Letters, 2009, 9(9): 3318-3322.

[60] Ohno Y, Maehashi K, Matsumoto K. Chemical and biological sensing applications based on graphene field-effect transistors[J]. Biosensors & Bioelectronics, 2010, 26(4): 1727-1730.

[61] Ristein J, Zhang W, Speck F, et al. Characteristics of solution gated field effect transistors on the basis of epitaxial graphene on silicon carbide[J]. Journal of Physics D: Applied Physics, 2010, 43(34): 345303.

[62] Fu W, Nef C, Knopfmacher O, et al. Graphene transistors are insensitive to pH changes in solution[J]. Nano Letters, 2011, 11(9): 3597-3600.

[63] Fu W, Nef C, Tarasov A, et al. High mobility graphene ion-sensitive field-effect transistors by noncovalent functionalization[J]. Nanoscale, 2013, 5(24): 12104-12110.

[64] Mailly-Giacchetti B, Hsu A, Wang H, et al. pH sensing properties of graphene solution-gated field-effect transistors[J]. Journal of Applied Physics, 2013, 114(8): 084505.

[65] Sohn I Y, Kim D J, Jung J H, et al. pH sensing characteristics and biosensing application of solution-gated reduced graphene oxide field-effect transistors[J]. Biosensors & Bioelectronics, 2013, 45: 70-76.

[66] Lee M H, Kim B J, Lee K H, et al. Apparent pH sensitivity of solution-gated graphene transistors[J]. Nanoscale, 2015, 7(17): 7540-7544.

[67] Falina S, Syamsul M, Iyama Y, et al. Carboxyl-functionalized graphene SGFET: pH sensing mechanism and reliability of anodization[J]. Diamond and Related Materials, 2019, 91: 15-21.

[68] Sudibya H G, He Q Y, Zhang H, et al. Electrical detection of metal ions using field-effect transistors based on micropatterned reduced graphene oxide films[J]. ACS Nano, 2011, 5(3): 1990-1994.

[69] Tu J, Gan Y, Liang T, et al. Graphene FET array biosensor based on ssDNA aptamer for ultrasensitive Hg^{2+} detection in environmental pollutants[J]. Frontiers in Chemistry, 2018, 6: 333.

[70] Park S J, Kwon O S, Lee S H, et al. Ultrasensitive flexible graphene based field-effect transistor (FET)-type bioelectronic nose[J]. Nano Letters, 2012, 12(10): 5082-5090.

[71] Cohen-Karni T, Qing Q, Li Q, et al. Graphene and nanowire transistors for cellular interfaces and electrical recording[J]. Nano Letters, 2010, 10(3): 1098-1102.

[72] Hess L H, Jansen M, Maybeck V, et al. Graphene transistor arrays for recording action potentials from electrogenic cells[J]. Advanced Materials, 2011, 23(43): 5045-5049.

[73] Dong X, Shi Y, Huang W, et al. Electrical detection of DNA hybridization with single-base specificity using transistors based on CVD-grown graphene sheets[J]. Advanced Materials, 2010, 22(14): 1649-1653.

[74] Ohno Y, Okamoto S, Maehashi K, et al. Direct electrical detection of DNA hybridization based on electrolyte-gated graphene field-effect transistor[J]. Japanese Journal of Applied Physics, 2013, 52(11R): 110107.

[75] Cai B, Wang S, Huang L, et al. Ultrasensitive label-free detection of PNA-DNA hybridization by reduced graphene oxide field-effect transistor biosensor[J]. ACS Nano, 2014, 8(3): 2632-2638.

[76] Xu S, Zhan J, Man B, et al. Real-time reliable determination of binding kinetics of DNA hybridization using a multi-channel graphene biosensor[J]. Nature Communications, 2017, 8: 14902.

[77] Tian M, Xu S C, Zhang J Y, et al. RNA detection based on graphene field-effect transistor biosensor[J]. Advances in Condensed Matter Physics, 2018, 2018(1): 8146765.

[78] He Q, Sudibya H G, Yin Z, et al. Centimeter-long and large-scale micropatterns of reduced graphene oxide films: fabrication and sensing applications[J]. ACS Nano, 2010, 4(6): 3201-3208.

[79] Ohno Y, Maehashi K, Matsumoto K. Label-free biosensors based on aptamer-modified graphene field-effect transistors[J]. Journal of the American Chemical Society, 2010, 132(51): 18012-18013.

[80] He Q, Wu S, Gao S, et al. Transparent, flexible, all-reduced graphene oxide thin film transistors[J]. ACS Nano, 2011, 5(6): 5038-5044.

[81] Hao Z, Zhu Y, Wang X, et al. Real-time monitoring of insulin using a graphene field-effect transistor aptameric nanosensor[J]. ACS Applied Material & Interfaces, 2017,

9(33): 27504-27511.

[82] Kawata T, Ono T, Kanai Y, et al. Improved sensitivity of a graphene FET biosensor using porphyrin linkers[J]. Japanese Journal of Applied Physics, 2018, 57(6): 065103.

[83] Myung S, Solanki A, Kim C, et al. Graphene-encapsulated nanoparticle-based biosensor for the selective detection of cancer biomarkers[J]. Advanced Materials, 2011, 23(19): 2221-2225.

[84] Kim D J, Sohn I Y, Jung J H, et al. Reduced graphene oxide field-effect transistor for label-free femtomolar protein detection[J]. Biosensors & Bioelectronics, 2013, 41: 621-626.

[85] Gao N, Gao T, Yang X, et al. Specific detection of biomolecules in physiological solutions using graphene transistor biosensors[J]. Proceedings of the National Academy of Sciences of the United States of America, 2016, 113(51): 14633-14638.

[86] Chen H, Chen P, Huang J, et al. Detection of matrilysin activity using polypeptide functionalized reduced graphene oxide field-effect transistor sensor[J]. Analytical Chemistry, 2016, 88(6): 2994-2998.

[87] Lei Y M, Xiao M M, Li Y T, et al. Detection of heart failure-related biomarker in whole blood with graphene field effect transistor biosensor[J]. Biosensors & Bioelectronics, 2017, 91: 1-7.

[88] Zhou L, Mao H, Wu C, et al. Label-free graphene biosensor targeting cancer molecules based on non-covalent modification[J]. Biosensors & Bioelectronics, 2017, 87: 701-707.

[89] Haslam C, Damiati S, Whitley T, et al. Label-free sensors based on graphene field-effect transistors for the detection of human chorionic gonadotropin cancer risk biomarker[J]. Diagnostics (Basel), 2018, 8(1): 5.

[90] Huang Y, Dong X, Shi Y, et al. Nanoelectronic biosensors based on CVD grown graphene[J]. Nanoscale, 2010, 2(8): 1485-1488.

[91] Kwak Y H, Choi D S, Kim Y N, et al. Flexible glucose sensor using CVD-grown graphene-based field effect transistor[J]. Biosensors & Bioelectronics, 2012, 37(1): 82-87.

[92] Siddique S, Iqbal M Z, Mukhtar H. Cholesterol immobilization on chemical vapor deposition grown graphene nanosheets for biosensors and bioFETs with enhanced electrical performance[J]. Sensors and Actuators B-chemical, 2017, 253: 559-565.

[93] Huang Y, Dong X, Liu Y, et al. Graphene-based biosensors for detection of bacteria and their metabolic activities[J]. Journal of Materials Chemistry, 2011, 21(33): 12358-12362.

[94] Wu G, Dai Z, Tang X, et al. Graphene field-effect transistors for the sensitive and selective detection of escherichia coli using pyrene-tagged DNA aptamer[J]. Advanced Healthcare Materials, 2017, 6(19): 1700736.

[95] Thakur B, Zhou G, Chang J, et al. Rapid detection of single E. coli bacteria using a graphene-based field-effect transistor device[J]. Biosensors & Bioelectronics, 2018, 110: 16-22.

[96] Li Y, Li X, Dong C, et al. A graphene oxide-based molecularly imprinted polymer platform for detecting endocrine disrupting chemicals[J]. Carbon, 2010, 48(12): 3427-3433.

[97] Liu S, Fu Y, Xiong C, et al. Detection of bisphenol a using DNA-functionalized graphene field effect transistors integrated in microfluidic systems[J]. ACS Applied Materials & Interfaces, 2018, 10(28): 23522-23528.

[98] Xiong C, Zhang T, Wang D, et al. Highly sensitive solution-gated graphene transistor based sensor for continuous and real-time detection of free chlorine[J]. Analytica Chimica Acta, 2018, 1033: 65-72.

第 8 章 氮化物激光器技术新进展

8.1 引　言

8.1.1 氮化物激光器的应用

氮化镓 (GaN) 基材料包括氮化镓、氮化铝、氮化铟和它们的合金，是继硅、砷化镓之后的第三代半导体，其禁带宽度范围为 $0.7 \sim 6.2$ eV，能够实现从紫外到可见光波段的连续可调，弥补了砷化镓基和磷化铟基等第二代半导体激光器短波输出的空白，同时其还具有效率高、寿命长、体积小和响应速度快等优点，使其在激光显示、激光照明、水下通信等领域具有广阔的应用价值和研究意义[1]。

近年来氮化物激光器的波长已经扩展到了紫外和绿光波段[2-4]，使得以红、绿、蓝半导体激光器作为光源的激光显示成为一种新兴的显示技术[5]。激光二极管相比于发光二极管 (LED)、液晶显示 (LCD)、阴极射线管 (CRT) 具有更高的色纯度，在色度图上形成的色度三角形面积最大，因而激光显示的图像更加真实客观，如图 8.1 所示[6]。另外，半导体激光器体积小、亮度高、光束易调控，可以大大降低激光显示的成本，并给终端设计 (比如激光电视、微型投影等) 带来更大的自由度。氮化物激光器由于是受激辐射发光机制，可以有效地避免传统 LED 在大电流密度下的 droop 现象[7-10]，因此可以通过提高单芯片的输出光功率来降低光源成本；氮化物蓝光激光器光谱半高宽小，易于与荧光材料的最佳激发波长相匹配而实现较高的能量转换效率；相比于传统 LED，氮化物蓝光激光器的工作电流密度高，可以实现高功率照明[11]，例如，欧司朗 (OSRAM) 公司研究的 GaN 基蓝光激光器已配备在奥迪 "R8 LMX" 和宝马 "i8" 上。

氮化物蓝、绿光激光器本身还具有高调制带宽的特性，可以同时实现照明与通信应用——可见光通信技术[12-14]。另外，470~540 nm 波段的蓝、绿光在海水中具有较低的吸收系数，因而氮化物蓝、绿光激光器可以作为深海探测和对潜通信等应用的光源[15,16]，如图 8.2 所示[17]；400 nm 波段附近的紫光在大气中具有较低的吸收系数，因而氮化物紫光激光器可以作为大气探测、新型激光引信等应用的光源[18]。此外，氮化物紫光激光器具有较小的衍射极限光斑尺寸，可用于光盘存储[19]、激光打印[1] 等。

8.1 引 言

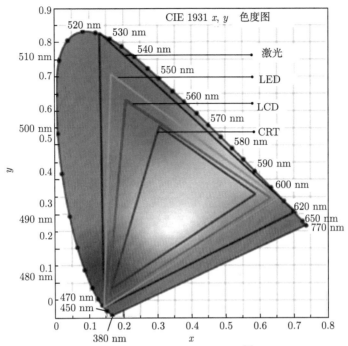

图 8.1 激光、LED、LCD 和 CRT 显示器的色域[6] (彩图请扫封底二维码)

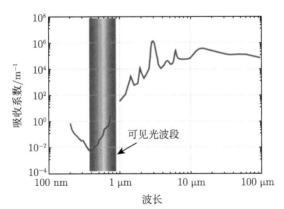

图 8.2 可见光在纯水中的吸收系数[17] (彩图请扫封底二维码)

8.1.2 氮化物激光器发展历程

20 世纪 70~80 年代，人们已经开始研究 GaN 基材料及其光电器件。1971 年，Dingle 等在低温的条件下观测到 GaN 材料中的激射现象[20]，由于蓝宝石衬底与 GaN 材料具有较大的晶格失配，生长的 GaN 晶体质量较差，没能够实现室温下激射。因此生长高质量 GaN 材料的方法成为当时的研究热点。日本名古屋大学的 Amano 等和日本日亚公司的 Nakamura 等分别于 1986 年和 1991 年提出低温 AlN 缓冲层和低温 GaN 缓冲层技术，解决了异质外延过程中由晶格失配和热失配产生高密度位错的问题[21]。在解决高质量 GaN 外延问题后，还需解决生长高质量 p-GaN 的问题。由于受到镁杂质电离能高、杂质及缺陷补偿等限制，很难实现 GaN 材料的 p 型掺杂。20 世纪 80 年代末，Amano 和 Nakamura 等利用低能电子束辐照和高温退火方法解决了 GaN 材料体系的 p 型掺杂问题[22,23]。在得到高质量的 GaN 材料以及 p-GaN 后，要实现 GaN 基激光器，还需要制备高质量的 InGaN 材料，以作为激光器的有源区。但是由于 InN 生长的最优温度远低于 GaN 生长的最优温度，且 In 存在相分凝，很难生长高质量的 InGaN 材料。1993 年，Nakamura 和 Mukai 利用改进的金属有机化合物化学气相沉积 (MOCVD) 方法成功生长出了高质量的 InGaN 材料，并在室温下获得了高效的带间发射[24]。

GaN 基激光器的研制成功主要归因于以上三个关键问题的解决，一是 GaN 材料的高质量生长，二是 GaN 材料的 p 型掺杂，三是 InGaN 材料的高质量生长。

1. InGaN 基激光器

通过调控激光器有源区 InGaN 中的 In 组分可以实现紫光、蓝光和绿光的输出。当 In 组分含量较低时，输出的为紫光和蓝光波段；当 In 组分含量增加至约 33% 时，输出的为绿光波段。然而，随着有源区中 In 组分的增加，将导致 InGaN 量子阱内的 In 组分分布不均、热稳定性变差、极化电场增强等问题，这就会限制 InGaN 量子阱的发光效率。

1996 年，日本日亚公司的 Nakamura 等研制出首支 GaN 基紫光激光器，其激射波长为 417 nm(脉冲工作模式)[25]；1998 年，Nakamura 等将 GaN 基紫光激光器的最大输出功率提升至 420 mW[26]；1999 年，日亚公司首次报道了单量子阱蓝光激光器，其激射波长 450 nm，室温下输出功率为 5mW[27]。随着外延材料技术水平、器件结构设计和散热封装技术的不断提升，激光器的输出功率不断增加。目前，日亚公司商品化的紫光激光器主要性能参数为：单模紫光激光器 (370~380 nm、390~425 nm) 的输出功率已达到 300 mW，工作电压为 5.3V；多模紫光激光器 (370~380 nm、400~405 nm) 的输出功率已达到 1200 mW，工作电压为 4.1V。商品化的蓝光激光器主要性能参数为：单模蓝光激光器 (440~460 nm) 的输出功率已达到 100 mW，工作电压为 5.3 V；多模蓝光激光器 (440~455 nm) 的输出功率已达到 3500 mW，工作电压为 4.3 V[28]。此外，德国欧司朗公司相比于日本日亚公司在蓝紫光激光器的研究起步虽然相对较晚，但是发展迅速，该公司在 2013 年报道了输出功率高达 4 W 的蓝光激光器[1]，并且，目前已推出输出功率达 60W 的蓝光激光器产品[29]。

由于 GaN 基绿光激光器的制备难度大，国际上的相关研究直到 2009 年才取得了突破。2009 年,德国欧司朗公司的 Avramescu 等在 c 面 GaN 衬底上首次实现绿光的脉冲激射,激射波长为 519 nm,最大输出功率为 50 mW[30]。同年,日本日亚公司的 Miyoshi 等同样在 c 面 GaN 衬底上首次实现绿光的连续激射,激射波长为 515 nm,最大输出功率为 5mW[31]。自此之后,InGaN 基绿光激光器的波长及效率得到了大幅提升[32-34]。目前,日亚公司 GaN 基绿光激光器 (503~507 nm、510~520 nm) 已经商品化,部分性能参数如下:单模绿光激光器的输出功率达 150 mW,工作电压为 5.1 V;多模绿光激光器输出功率达 1000 mW,工作电压 4.7 V[28]。欧司朗公司 GaN 基绿光激光器 (510~520 nm) 已经商品化,单模绿光激光器的输出功率达 140 mW,工作电压为 6 V[35]。

2. AlGaN 基激光器

随着激光器的波长向短波方向发展,激光器光学限制层中的 Al 组分需相应增加以保证足够的光学限制。然而,一方面,高质量高 Al 组分的 AlGaN 材料难以生长;另一方面,在 p-AlGaN 材料中,Mg 受主的激活能较大,并且随 Al 组分的增加而增大[36-38],因此 p-AlGaN 的电阻率很大,AlGaN 基紫外激光器的工作电压非常高,这使得紫外光激光器很难实现激射。

在 AlGaN 基紫外光电器件中,深紫外 (200~320 nm) 发光二极管已经有产品上市,而深紫外激光二极管 (LD) 始终停留在光泵浦的水平。尽管早在 2006 年就已经实现了 AlN 材料在 214 nm 的光泵浦激射[39],但在常规的法布里–珀罗 (F-P) 激光器中电泵浦激射的最短波长目前只能达到 2008 年滨松光子学株式会社报道的 336 nm[40]。为了实现深紫外激光二极管在电注入条件下的激射,就要寄希望于核心外延材料技术水平和新型器件结构设计的突破进展。

8.2 氮化物激光器基本特性

8.2.1 氮化物激光器工作原理

激光器一般由激励源、工作物质和光学谐振腔三个基本要素构成。激励源也叫泵浦系统,把能量供给低能级上的电子,激发使其跃迁到高能级上,能量供给的方式有载流子注入、光场泵浦激发、化学作用等。工作物质是激光器的核心,以半导体激光器为例,其利用正向偏置的 pn 结提供载流子注入,使得有源区内导带底部的电子数目远大于价带顶部的空穴数目,实现粒子数反转。工作物质的物理特性决定了所产生激光的波长等特性。光学谐振腔一般为两面互相平行的镜子或是有很强光限制作用的腔体,光在反射镜间或是增益腔体内来回反射,将光场限制在有增益介质的区域内,目的是使被激发的光多次经过增益介质以得到足够的放大。要获得连续而稳定的激光振荡输出,要求激励源注入足够大,使得光增益大于或等于受激辐射在谐振腔来回反射时的能量损耗（包括载流子吸收、缺陷散射、端面透射损耗等），即满足阈值条件，这在半导体激光器中一般是通过重掺杂或是高注入来实现的。光学谐振腔两镜面之间的距离也对输出的激光波长有着选择作用，只有在两镜间的距离能产生共振的波长才能产生对应波长的激光。

半导体激光器主要包括边发射激光二极管和面发射激光二极管两种结构，如图 8.3 所示。边发射激光器 (edge-emitting laser, EEL) 也称为法布里–珀罗激光器，而面发射激光器也称为垂直腔面发射激光器 (vertical-cavity surface-emitting laser, VCSEL)。两者主要是出光方向不同，VCSEL 是垂直于衬底表面出光，输出光束具有较低的发散角，远场光斑呈圆形，与光纤的耦合效率更高；而 EEL 是侧面出光，横向尺寸与激光器波长相同量级，远小于侧向尺寸，导致了横向上较强的衍射作用，激光器的垂直结平面方向上的发散角远大于平行于结平面的发散角，则远场光斑往往呈椭圆形。

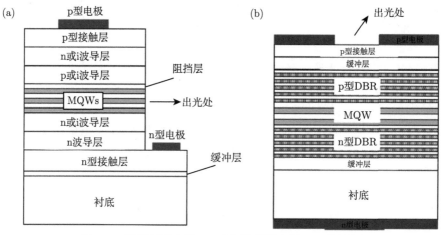

图 8.3　边发射激光器 (a) 和面发射激光器 (b) 的结构

半导体激光器的材料通常采用 MOCVD 方法制备外延层，外延层包括缓冲层、限制层、有源层、顶层和帽层，其中有源层又包含上波导层、量子阱和下波导层。有源层的带隙比限制层小，折射率比它们大，因此由 p 面和 n 面限制层注入的空穴和电子会被限制在有源层中，载流子复合产生的光又能被很好地限制在波导层中，使辐射效率得到了提高。帽层采用高掺杂，目的是与 p 面金属电极形成更好的欧姆接触，降低欧姆接触电阻。GaN 基半导体激光器主要工作原理如下：随着外加电流的注入，电子和空穴分别从 n 电极和 p 电极注入，它们经扩散和漂移进入多量子阱有源区，随后被量子阱捕获。被量子阱捕获的载流子占据量子阱各分立能级，同时电子和空穴在量子阱中发生辐射复合，如图 8.4 所示。由于大

8.2 氮化物激光器基本特性

电流注入时在量子阱中注入的大量载流子，激光器获得较强的粒子数反转，使受激光子在谐振腔来回反射的过程中增益大于损耗，最终满足受激发射条件，在器件腔面发射出稳定的激光。

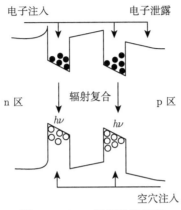

图 8.4 GaN 量子阱能带图

8.2.2 氮化物激光器工作特性

氮化物激光器工作特性主要包括发射波长、阈值特性、光谱特性、转换效率、温度特性等。

1. 发射波长

半导体激光器的发射波长是由导带的电子跃迁到价带时所释放的能量决定的，这个能量近似等于禁带宽度 $E_g(\text{eV})$：

$$h\nu = E_g \tag{8.1}$$

$$\nu = \frac{c}{\lambda} \tag{8.2}$$

式中，ν 和 λ 分别为发射光的频率 (Hz) 和波长 (μm)；c 为真空中的光速，大小为 $3\times10^8 \text{ m}\cdot\text{s}^{-1}$；$h$ 为普朗克常量，大小为 6.628×10^{-34} J·s；1 eV=1.60×10^{-19} J。代入式 (8.1) 和式 (8.2) 得

$$\lambda = \frac{1.24}{E_g(\text{eV})} \tag{8.3}$$

由于禁带宽度与半导体材料的成分及含量有关，因此根据这个原理可以制成不同波长的激光器。

2. 阈值特性

对于 LD,当外加正向电流达到某一数值时,输出光功率急剧增加,这时将产生激光振荡,这个电流称为阈值电流,用 I_{th} 表示。当驱动电流小于阈值时,有源区无法达到粒子数反转,也无法达到谐振条件,此时半导体激光器表现为发光二极管模式,即自发辐射;当驱动电流大于阈值时,有源区不仅有粒子数反转,而且达到了谐振条件,此时半导体激光器表现为激光模式,即受激辐射,如图 8.5(a) 所示。有无阈值是区分半导体激光器和半导体发光二极管的依据。从图 8.5(b) 和 (c) 中可以看出,$I\text{-}V$ 是典型的二极管伏-安特性。在达到阈值电流之前,其电流与电压关系表现为指数函数,激光器的 $V\text{-}I$ 特性为[41]

$$I = I_0[\exp[\alpha_j(V_a - Ir_s) - 1]] \tag{8.4}$$

式中,I_0 为饱和电流;α_j 为二极管参数;V_a 为正向电压;r_s 为 LD 的串联电阻。

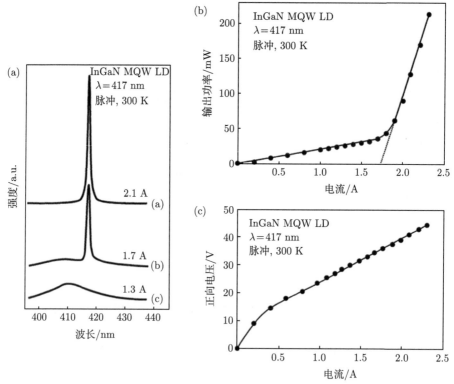

图 8.5　InGaN 多量子阱 (MQW)LD[25]
(a) 光谱图;(b) 输出功率-电流特性曲线;(c) 电压-电流特性曲线

8.2 氮化物激光器基本特性

当激光器激射以后，其电流与电压的关系表现为线性，即 [41]

$$V \approx \frac{E_g}{e} + I_{r_s} \qquad (8.5)$$

式中，E_g 为禁带宽度；e 为电子的电荷。V-I 曲线斜率越大，则意味着在 r_s 上消耗的电功率较大而使电功率下降。

3. 光谱特性

当注入电流大于阈值电流时，辐射光在腔内建立起来的电磁场模式称为激光的模式。半导体激光器的模式分为纵模和横模，通常纵模表示谐振传播方向上的驻波振荡特性，横模表示谐振腔截面上的场型分布，如图 8.6 所示。纵模决定半导体激光器的频谱特性，横模决定半导体激光器的光场的空间特性，即横模决定近场特性和远场特性。

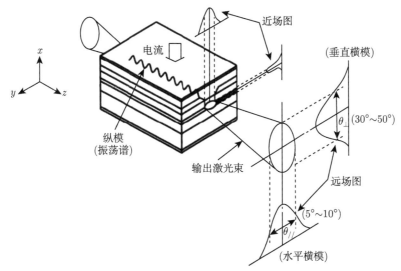

图 8.6　半导体激光器的模式类型

横模分为水平横模和垂直横模两种类型。水平横模反映了有源区中平行于 pn 结方向光场的空间分布，主要取决于谐振腔宽度、边壁材料及其制作工艺；垂直横模则表示与 pn 结垂直方向上电磁场的空间分布。激光器的横模直接影响到器件与光纤的耦合效率，通常用近场图和远场图来表示横向光场的分布规律。半导体激光器输出的近场图为解理面上的光强分布，远场图是光束输出腔面后的在自由空间的光强分布。由于激光器有源区厚度相当薄，数量级与激光器波长相当，光束在垂直于 pn 结平面的方向上发生较强的衍射作用，因此垂直于结平面方向半功率点所张的角 $\theta_\perp$ 较大。而在平行于 pn 结平面的方向上，条宽通常为几微米，这比激光器的波长大几倍到十几倍，光束在平行于 pn 结平面的方向上发生较弱的衍射作用，因此平行于结平面方向半功率点所张的角 $\theta_{//}$ 较小。因此，半导体激光器发出的是一束 $\theta_\perp \times \theta_{//}$ 的椭圆形光束，如图 8.6 所示。

纵模反映光谱性质，它表示激光器所发射光束的功率在不同频率 (或波长) 分量上的分布。对于半导体激光器，当注入电流低于阈值时，发射光谱是导带和价带的自发发射谱，谱线较宽；只有当激光器的注入电流大于阈值后，谐振腔里的增益才大于损耗，自发发射谱线中满足驻波条件的光频率才能在谐振腔里振荡并建立起场强，这个场强使粒子数反转分布的能级间产生受激辐射，而其他频率的光受到抑制，使激光器的输出光谱呈现出一个或几个模式振荡，这种振荡称为激光器的纵模，如图 8.7 所示。两个相邻纵模之间的距离为[42]

$$\Delta\lambda_0 = \frac{\lambda_0^2}{2nL} \tag{8.6}$$

式中，$\Delta\lambda_0$ 为纵模间隔；λ_0 为空气中测得的激光峰值波长；n 为有源区材料的折射率；L 为激光器的腔长。由式 (8.6) 可以看出，半导体激光器腔长越大，纵模间隔越小。

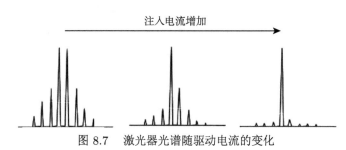

图 8.7　激光器光谱随驱动电流的变化

4. 转换效率

功率效率指的是半导体激光器的输出光功率与注入电功率之比，其表达式为

$$\eta_P = \frac{\text{激光器输出的光功率}}{\text{激光器消耗的电功率}} = \frac{P_{\text{out}}}{IV + I_{r_s}^2} \tag{8.7}$$

式中，P_{out} 为激光器输出的光功率；I 为工作电流；V 为激光器 pn 结正向电压；r_s 为串联电阻 (包括半导体材料的体电阻和电极接触电阻)。提高功率效率的关键是金属电极与半导体材料之间形成良好的欧姆接触，从而减少 r_s。

内量子效率指的是半导体激光器有源区内电子与空穴复合产生的光子数与注入的电子空穴对数之比，其表达式为

$$\eta_i = \frac{\text{有源区内单位时间产生的光子数}}{\text{有源区内单位时间内注入的电子--空穴对数}} \tag{8.8}$$

由于界面态、杂质、缺陷和俄歇复合的存在，有源区内注入的载流子不能够完全以辐射复合的形式产生光子，部分载流子还会发生非辐射复合，使得半导体激光器的内量子效率小于 1。由此可见，提高内量子效率的关键是提高半导体材料本身的晶体质量。

外量子效率指的是半导体激光器向体外辐射的光子数与有源区内注入的电子–空穴对数之比，其表达式为

$$\eta_{\text{out}} = \frac{\text{有源区内单位时间向体外辐射的光子数}}{\text{有得区内单位时间内注入的电子–空穴对致}} \quad (8.9)$$

与内量子效率不同的是，外量子效率不仅考虑有源区内的非辐射复合损耗，还包括光子在腔内产生的散射、衍射和吸收等损耗。

外微分量子效率指的是半导体激光器达到阈值之后，输出光子数的增量与注入电子数的增量之比，其表达式为

$$\eta_{\text{d}} = \frac{(P - P_{\text{th}})/hf}{(I - I_{\text{th}})/e} = \frac{P - P_{\text{th}}}{I - I_{\text{th}}} \cdot \frac{e}{hf} \quad (8.10)$$

式中，P 为激光器的输出光功率；I 为激光器的驱动电流；P_{th} 为激光器的阈值功率；I_{th} 为激光器的阈值电流；h 为普朗克常量；f 为光子频率；e 为电子电荷。

5. 温度特性

温度升高对半导体激光器阈值电流密度的影响主要取决于增益系数随温度的变化。随着温度的不断上升，损耗系数会随之变大，内量子效率会不断减小，上述因素都会使得阈值电流密度变大。阈值电流随温度 T 的变化关系可以用以下方程式描述[43]：

$$I_{\text{th}} = I_0 \exp\left(\frac{T}{T_0}\right) \quad (8.11)$$

式中，I_{th} 为激光器的阈值电流；I_0 为常数；T 为结区的热力学温度；T_0 为特征温度，它只与激光器的结构和材料有关，是判定半导体激光器温度是否稳定的一个关键值。

大量的实验和理论研究证明，随着温度的升高[44,45]，半导体材料的带隙宽度将变窄，从而导致半导体激光器发射波长的红移。带隙宽度随温度增加而减小，主要是由于温度增加时伴随着键长增大，键能减小。

温度对半导体激光器的输出功率也有很大影响，随着半导体激光器有源区温度的不断上升，激光器的输出功率会不断地变小。激光器输出功率减小了，能量就被更多地用来产生热量，导致温度不断上升而以此恶性循环。由式 (8.10) 可以

得到，达到阈值之后半导体激光器的输出功率表达式为

$$P = P_{\text{th}} + \eta_{\text{d}} \cdot \frac{hf(I - I_{\text{th}})}{e} \tag{8.12}$$

通过分析式 (8.12) 得出，随着温度的不断增大，阈值电流不断增加，外微分量子效率不断减小，最终使得半导体激光器输出的光功率不断减小。

由于半导体激光器各层所用的材料与材料之间的热膨胀系数不同，随着温度的上升，半导体激光器内部会出现热应力。而应力会在各材料层之间扩散，甚至造成器件撕裂更加剧烈，最终导致半导体激光器的性能减弱或直接失效。此外，随着半导体激光器温度升高，器件端面复合吸收的过程可能会使得谐振腔的端面毁坏。因此半导体激光器对工作温度有较高的要求。

8.2.3 氮化物激光器调制特性

在半导体激光器上加直流偏置和连续正弦波调制信号，在激光器偏置于阈值电流之上时，输出光功率将随调制信号幅度的变化而变化，此时输出的光功率不再是驱动电流的线性函数，因此当激光器偏置在不同电流时，激光器将有不同的响应特性。图 8.8 为半导体激光器的小信号电流调制原理图。

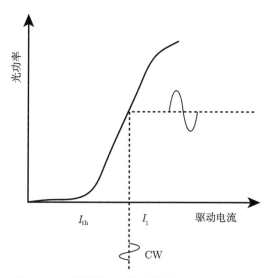

图 8.8　半导体激光器的小信号电流调制原理图

假设直流电流大小为 I_1 (I_1 在激光器线性工作范围)，在只有直流偏置下激光器的输出光功率为 P_1，叠加的正弦信号大小为 $\Delta I_{\text{m}} \sin(2\pi ft)$，其中 ΔI_{m} 为交流信号的最大振幅，f 为调制频率。则 LD 的输出光功率会在直流光功率 P_1 附近浮动，变化值为 $\Delta P_{\text{m}} \sin(2\pi ft + \Phi)$，其中 ΔP_{m} 为调制输出光功率的最大振幅，Φ

8.3 氮化物白光激光技术

为输入信号 (调制电流) 与输出信号 (调制光功率) 的相位差。相位差的产生是由于激光器在激射之前通常有几个纳秒的延迟，延迟时间与少数载流子寿命有关。

当 ΔI_m 保持不变时，$\Delta P_m(f)$ 随调制频率的变化曲线即为激光器的频率响应曲线。假设直流偏置时激光器的输出光功率为 $\Delta P_m(0)$，则 $\Delta P_m(f)/\Delta P_m(0)$ 随调制频率的变化曲线如图 8.9 所示。可以看出，当调制频率较低时，$\Delta P_m(f)/\Delta P_m(0)$ 相对平坦，但当调制频率上升到一定值，即弛豫振荡频率 f_r 时，$\Delta P_m(f)/\Delta P_m(0)$ 获得最大值。f_r 取决于激光器特性，大小可以表示为[46]

$$f_r = \frac{1}{2\pi}\sqrt{\frac{1}{\tau\tau_{ph}}\left(\frac{I_1}{I_{th}} - 1\right)} \tag{8.13}$$

式中，τ 为载流子寿命，即非平衡载流子的平均生存时间，包含辐射复合和非辐射复合两个过程，一般为纳秒量级；τ_{ph} 为光子寿命，即光子离开谐振腔所花的平均时间，包含腔内光子吸收和端面发射两个过程，一般为皮秒量级；I_1 为激光器的偏置电流；I_{th} 为激光器的阈值电流。但是在弛豫振荡频率附近对激光器调制是不稳定的，容易引起调制非线性。为了避免这种现象，一般取对应于弛豫振荡频率以下 3dB 作为直接调制允许的带宽[47-50]。因此激光器的调制带宽与自身的弛豫振荡频率成正比。

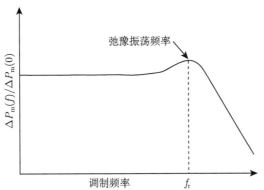

图 8.9 $\Delta P_m(f)/\Delta P_m(0)$ 随调制频率的变化曲线

除弛豫振荡频率外，半导体激光器的调制带宽还受其本身的寄生电容、串联电阻和键合引线电感的限制。因此在设计用于光通信的 LD 时，需要从多方面来研究优化器件的调制带宽。

8.3 氮化物白光激光技术

III 族氮化物激光二极管 (laser diode, LD) 近年来成为固态照明 (solid-state lighting, SSL) 的研究热点，在未来可能是发光二极管 (light emitting diode, LED)

的良好"补充者"甚至是"替代者"。LD 相对于 LED 的潜在优势如下所述。第一，与 LED 相比，LD 在更高的电流下具有更高的光电转换效率。这是因为引起效率下降的俄歇复合在激光器达到阈值之后不再增长。第二，与 LED 相同的荧光体转换方法也可以与 LD 一起使用，以产生具有相同显色指数和色温的白光。第三，对于 LED 和 LD 而言，从彩色混合发射器发出白光同样具有挑战性，这两种光源都不具有直接优势。第四，LD 的发射是定向的，可以更容易地被捕获和聚焦，从而有可能提供新颖、紧凑的照明设备。第五，LD 的更小面积和更高的电流密度操作为它提供了优于 LED 的潜在成本优势。2014 年，"蓝光 LED 之父"日本科学家中村修二在瑞典举行的诺贝尔演讲上表示，激光照明在未来将取代 LED 照明[51]。激光照明是新一代的具有优异性能的固态照明光源。此外，LD 的调制带宽也远高于 LED，这就使得 LD 在光通信领域也具有巨大的应用潜力。

8.3.1 荧光转化激光白光技术

荧光转换激光白光技术 (phosphor-conversion laser white light, pc-LD)：指的是蓝光或紫光 LD 激发荧光材料使剩余蓝光与下转换荧光形成的复合白光。以蓝光 LD 为例，其结构和光谱如图 8.10 所示[52]。

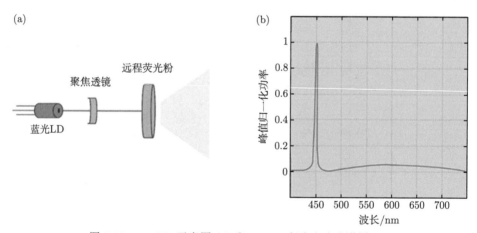

图 8.10　pc-LD 示意图 (a) 和 pc-LD 复合白光光谱图 (b)

2005 年，日本日亚 (Nichia) 公司最早提出以半导体激光器实现白光输出，将 GaN 基蓝光 LD 经光纤耦合输出来激发荧光粉，这是世界上首个半导体激光白光光源[53]。2007 年，日亚公司又成功制备出了 500 mW 的 GaN 基 LD，并在此基础上提高了白光的输出参数，光效达到 40 lm/W，但是当时的研究仅测试了白光光强的角度均匀性，而未对激光的光束进行匀化，光品质也有待提升[54]。随后的研究热点开始关注于激光白光的光品质，2008 年，中国科学院半导体研究所

8.3 氮化物白光激光技术

徐云小组采用 405 nm 蓝紫光半导体激光器来激发荧光粉，制备了高稳定性的固态白光光源[55]；随后该团队又采用 445 nm 的 GaN 基蓝光 LD 作为激发光源激发黄色和红色混合荧光粉，当注入电流为 350mA 时，实现了光通量为 52 lm，光效为 31 lm/W，显色指数为 72.7，色温为 5225 K 的固态白光，这是国内最早有关 pc-LD 的研究[56]。2010 年该团队将混合荧光粉改为单一的黄色荧光粉，445 nm 的 GaN 基蓝光 LD 仍作为激发源，当注入电流为 500mA 时，获得了光通量为 113 lm，光效为 44 lm/W，显色指数为 60 的固态白光，进一步提高了 pc-LD 复合白光的光效[57]。2010 年，韩国仁荷大学采用功率为 50 mW 的 445 nm 蓝色激光激发黄色荧光粉，实现白光输出，当注入电流为 100 mA 时，白光光源的光通量为 5 lm，发光效率为 10 lm/W，其研究还表明，基于 LD 的白光光源具有很高的光色稳定性[58]。与 LED 相比，激光器的工作电流密度更大，因此具有更大的光功率密度，导致其对荧光材料的要求更高。LD 和荧光材料两者之间的匹配至关重要，影响着复合光源的光品质及稳定性。2013 年，美国加州大学圣巴巴拉分校 (UCSB) 分别采用 442 nm 蓝光 LD 和 402 nm 蓝紫光 LD，激发置于反射衬底上的 YAG:Ce 黄色荧光粉和红、绿、蓝混合荧光粉得到了激光白光，其光通量分别为 252 lm 和 53 lm，发光效率分别为 76 lm/W 和 19 lm/W，显色指数分别为 57 和 95，色温分别为 4400 K 和 2700 K。但是其在实验中发现荧光粉由于长时间受到高功率密度激光的直接辐射会发生淬灭现象[59]。为改善这种情况，2015 年，UCSB 采用单晶体荧光材料作为被激发层，得到了光通量为 1100 lm、发光效率为 86.7 lm/W、显色指数为 62 的白光，但是由于光谱成分没有优化，其色温较高，为 7300 K[60]。2016 年，韩国成均馆大学通过在荧光陶瓷片上制作纳米结构，将 pc-LD 复合激光白光的光效提高到 218 lm/W[61]。2018 年，厦门大学将荧光粉/玻璃复合材料 (phosphor in glass, PiG) 膜直接烧结在由一维光子晶体覆盖的高导热蓝宝石衬底上，这种具有复合结构的彩色转换器具有接近原始荧光粉的内量子效率，以及高达 90% 的出色封装效率，如图 8.11 所示。被 450 nm LD 激发后产生均匀的白光，具有 845 Mcd/m^2 的高亮度 (光通量为 1839 lm)，210 lm/W 的发光效率以及色温 6504 K；通过添加橙色或红色荧光粉层，可以将显色指数提高到 74[62]。

表 8.1 总结了使用蓝紫光 LD 激发 RGB 荧光粉，以及蓝色 LD 激发黄色荧光粉合成的激光白光的最新进展。通常，前者具有较高的显色指数，但是由于多次吸收 (斯托克斯损耗严重) 而具有相对较低的发光效率。后者是最常用的方式，具有发光效率高、成本低和尺寸紧凑的优点。

目前限制荧光转换激光白光技术进一步发展的关键因素主要有如下三个问题。

(1) 激光器的散热问题：目前蓝色 LD 的光电转化效率相对较低，比如以功率密度 15 kW/cm^2 驱动，光电转化效率约为 30 %，而此时 LD 芯片的热通量约为

10^8W/m^2[72]。由于需要高输出功率，因此需要增加输入功率，并且会产生大量的热量，从而芯片的结温非常高。较高结温会降低载流子的限制作用并提高非辐射复合率，导致较高的阈值电流和较低的转换效率，从而限制了 LD 的最大输出功率。同样，由于结温升高，LD 的可靠性和使用寿命也会降低。

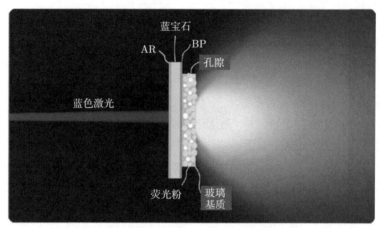

图 8.11　激光复合白光示意图

表 8.1　pc-LD 复合白光的最新进展

λ_{ex}/nm	荧光粉类型	CCT/K	Ra	Φ_ν/lm	η_ν/(lm/W)	年份	参考文献
445	黄色荧光粉	—	—	96	40	2007	[54]
405	蓝、黄荧光粉	5200	70	5.7	13	2008	[55]
445	黄、红荧光粉	5225	72.7	52	31	2008	[56]
445	YAG	—	60	113	44	2010	[57]
442	YAG:Ce	4400	57	252	76	2013	[59]
402	RGB	2700	95	53	19	2013	
442	YAG:Ce 单晶	7300	62	1100	86.7	2015	[60]
450	黄色荧光粉	7045	—	250	45	2015	[63]
450	YAG	6314	—	1093	48.9	2016	[64]
450	YAG 陶瓷	6403	68	—	40	2016	[65]
445	YAG 陶瓷	5994	54.2	1424.6	218	2016	[61]
445	荧光微晶玻璃	5649	—	—	110	2017	[66]
455	荧光微晶玻璃	6990	56	850	70	2017	[67]
450	荧光微晶玻璃	6504	74	1839	—	2018	[62]
450	荧光微晶玻璃	5666	60.8	—	23.8	2018	[68]
450	荧光微晶玻璃	6230	62.5	—	26.5	2019	[69]
450	荧光微晶玻璃	6593	—	369	—	2019	[70]
445	荧光微晶玻璃	3646	57	651	—	2019	[71]

(2) 激光光束匀化问题：对于边发射激光器而言，激光光束在垂直方向和平行方向上的发散角是不同的，呈椭圆状。如果直接用来照射荧光样品，则输出的

白光在这两个正交方向的强度和光斑直径差别会非常大，不利于后续的二次光学设计。此外，激光器的照度呈高斯分布，当荧光样品被激光光束辐照时，荧光样品只有中心区域能够被有效激发，能量利用率相对较低。图 8.12(a) 为由非均匀和均匀激光光斑激发的荧光样品发出的光通量随输入功率的变化[73]。可以看到，均匀的激光光束可以获得更高的光通量。总体而言，较差的激光光束质量是限制激光复合白光进一步应用的瓶颈。

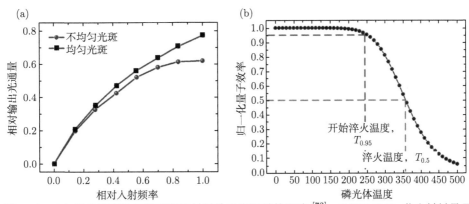

图 8.12　(a) 激光光斑均匀性对荧光材料输出光通量的影响[73]；(b)YAG:Ce 荧光材料量子效率与温度的关系[74,75]

(3) 荧光材料热淬灭问题：为了获得高亮度的激光照明，激光光束通常会被准直或聚焦在荧光材料表面，导致非常高的激发强度。当激光透过荧光材料时，由于斯托克斯位移损失、量子效率损失、吸收损失等，部分光能转换为热能。传统的封装方式由于有机黏合剂的使用导致相对较低的热导率，从而进一步提升荧光材料的温度。图 8.12(b) 为商用 YAG:Ce 荧光材料的量子效率随温度的变化关系[74,75]，可以看出，随着温度的升高，量子效率首先保持稳定，然后开始迅速下降至峰值量子效率的 95% 以上。量子效率的降低导致荧光材料中产生的热量增加，使荧光材料的温度进一步升高，反过来导致量子效率的进一步降低。这种热失控效应最终导致荧光材料的量子效率下降到峰值的 50% 以下，即荧光材料发生热淬灭[75]，大大缩短了激光复合白光光源的寿命。

8.3.2　RGB 三基色复合激光白光技术

RGB 三基色复合激光白光技术指的是将红、绿、蓝半导体激光器通过合束的方法形成复合白光，其基本结构和光谱如图 8.13 所示[52]。这种方式得到的白光不需要荧光材料，因此电光转换效率比较高，而且可以通过控制激光器的驱动电流调节白光的色温，从而满足不同场合。但是也存在以下缺点：RGB 激光器衰减速率不同，在使用后可能引起色温、色坐标变化，影响照明效果[76]；RGB 激光器的光谱不连续, 会降低复合白光的显色指数; 激光光束的辐照度呈高斯分布, 会造成复合白光的光色不均匀; 由于激光具有较强的相干性, 直接用来照明会产生散斑现象[77]。因此, 如果采用这种方式实现照明, 就需要对激光光束进行复杂的光学设计。

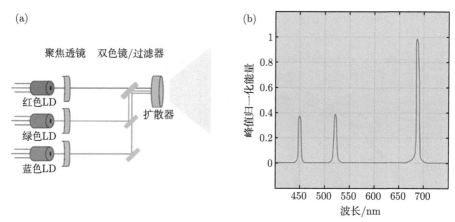

图 8.13　RGB LD 示意图 (a) 和复合白光光谱图 (b)

早在 2011 年, 新墨西哥大学的 Neumann 等就采用红光半导体激光器和倍频的蓝、绿、黄光固态激光器通过合束得到了复合激光白光, 并证明了这种四色复合激光白光可以与传统光源显色指数相媲美[76]。同年, 韩国电子与电信研究所提出了一种 RGB 三基色复合激光白光合成器, 其包含红色、绿色和蓝色 LD 以及半导体制冷器 (TEC), 结构如图 8.14 所示, 总体积为 3.84cc[78]。2015 年, 英国爱丁堡大学将红、绿、蓝光半导体激光器 (RGBLD) 通过合束实现了复合激光白光, 照度达到 971.28lx, 满足室内照明水平[79]。2015 年, 阿卜杜拉国王科

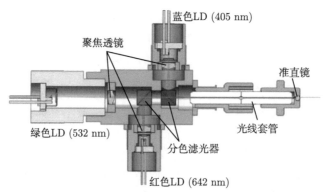

图 8.14　RGB 三基色复合激光白光合成器结构示意图

技大学 (KAUST) 以及我国台湾省的台湾大学和交通大学利用商业扩散器元件对 RGB LD 复合白光进行了简单匀化, 发现蓝光 LD 的驱动电流需要有一个折中来平衡白光的色温以及蓝光 LD 的调制带宽[80]。2017 年台湾大学的林恭如小组采用磨砂玻璃作为扩散器对 RGB LD 复合激光白光进行了匀化, 其色温为 8382 K, 但是显色指数仅为 54.4[81]。2018 年, 该小组又对比了不同材质扩散器对 RGB LD 复合激光白光的匀化作用, 但均是以牺牲光强为代价, 极大降低了光能利用率[82]。2018 年, 中国科学院半导体研究所利用微透镜阵列对 RGB LD 复合激光白光进行了匀化, 照度均匀度高达 90%, 并且获得的白光显色指数高达 64.8[83]。

表 8.2 总结了 RGB LD 复合激光白光的最新进展。由于激光光谱较窄, 无法像 pc-LD 那样覆盖整个可见光谱, 所以采用三基色激光器直接合成的激光白光的显色指数普遍偏低, 但是由于 RGB LD 是单独驱动的, 所以可以通过调节三者的输入电流以获得不同色温的白光。此外, 由于没有荧光材料的使用, RGB LD 与 pc-LD 相比前者合成的复合激光白光往往具有较高的光电转化效率。

表 8.2 RGB LD 复合激光白光的最新进展

λ/nm			CCT/K	Ra	Φ_ν/lm	η_ν/(lm/W)	年份	参考文献
R	G	B						
642	532	405	—	—	—	—	2011	[78]
658	520	450	8000	—	—	—	2015	[79]
642	520	450	5835	—	—	—	2015	[80]
659	516	452	8382	54.4			2017	[81]
650	520	450	6500				2018	[82]
638	520	450	5171	64.8			2018	[83]

目前限制 RGB LD 激光白光技术进一发展的关键因素主要有如下三个。

(1) 激光器的散热问题: 目前 LD 的光电转化效率相对较低, 在工作过程中会产生大量的热, 将三种激光器封装在一起更不利于散热, 如果没有良好的散热系统, 那么将导致三种激光器芯片的结温非常高。较高结温会降低载流子的限制作用并提高非辐射复合率, 导致较高的阈值电流和较低的转换效率, 从而限制了 LD 的最大输出功率。同样, 由于结温升高, LD 的可靠性和使用寿命也会降低。

(2) 光谱不连续问题: 激光器的光谱半高宽很窄, 通常只有 1~2 nm, 显然将 RGB 三种激光器进行合束是无法覆盖整个可见光谱的。现在很多研究机构开始在 RGB LD 的基础上添加黄色 LED 或者黄色荧光粉来扩展其光谱, 显色指数确实有了较大的提升。2016 年, Janjua 等采用 RGB LD 和黄色纳米线 LED 实现了显色指数高达 87.7 的白光[84]。

(3) 激光光强分布不均问题: 由于激光具有高度相干性, 当激光光束照射到粗糙表面后, 通过散射形成不同的子波, 这些不同的子波相互干涉, 形成许多随机分布的颗粒状、明暗不均的强度图样,即散斑现象 [85]。目前常用的抑制散斑的方法有:利用不同波长的光源 [86]、利用脉冲激光的叠加 [87]、移动散射体 [88]、移动孔径光阑 [89]、振动屏幕 [90] 等。另外,由于半导体激光器的光强近似呈高斯分布,强度由中间向边缘逐渐变弱,因此在目标面上能量分布不均匀。

(4) 成本问题:目前半导体激光器的光功率普遍偏小,而且其光电转换效率比较低,实现大功率激光照明则需要集成多颗 LD,这就极大提高了激光照明系统的成本。因此有必要对 LD 的阈值电流、输出功率、工作温度等性能参数进行改进。

8.3.3 白光 LD 光通信技术

近些年,无线通信技术的发展极为迅速,无线通信技术所具有的灵活性以及地区覆盖性,使全球的各个地区和领域可以进行信息的沟通,它已经成为当今世界上最具有吸引力的一种通信方式。但是射频频谱资源是国家的自然资源,为了减少信号干扰和电磁污染,它的使用受到每个国家的管理和约束。随着无线数据传输量需求的增加,移动数据流量呈指数增长,射频频谱更加拥挤,剩下的频谱资源逐渐减小,频谱竞争也更加激烈。而可见光频谱中有约 400THz 带宽的免许可频谱还未开发,是射频频谱 1000 多倍,甚至更高。因此,可见光通信 (VLC) 可以对基于射频的移动通信系统进行补充,设计出高容量的移动数据网络。另外,可见光通信能够同时提供照明和通信,可以作为补充功能并入现有照明基础设施中。可见光通信利用光源作为其发射器,在该发射器中,信息被调制为发射光的强度。调制信号的频率应保持足够高 (通常高于 300Hz),以保证居住者认为光源正常点亮。图 8.15 为可见光通信系统的工作原理图。

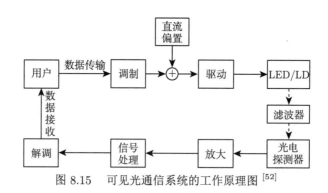

图 8.15 可见光通信系统的工作原理图 [52]

迄今为止,大多数可见光通信系统都将 LED 作为发射器。以 LED 为核心的 VLC 系统通常使用单芯片或多芯片方法,即荧光型白光 LED(pc-LED) 和红绿蓝 (RGB)LED。由于 LED 受自身结电容和载流子自发辐射复合寿命等因素的影响,

8.3 氮化物白光激光技术

LED 本身的物理带宽是有限的，特别是结电容较大的大功率 LED，荧光型白光 LED 带宽只有几兆赫[91-93]，可见光通信系统的带宽和速率主要受白光 LED 带宽的限制。μ-LED 虽然可以以较高的频率进行调制，但是其光功率只有几个毫瓦[94]，无法满足照明需求。另外，在高电流密度下，LED 会因载流子泄露而发生效率下降现象，导致 LED 只能在小电流密度下驱动。这会导致单个 LED 器件只能产生少量光通量，给降低器件成本造成了困难。

而氮化物 LD 作为理想的高能量密度注入光源，由于是受激辐射发光机制，可以有效地避免传统 LED 在大电流密度下的 droop 现象[7-10]。氮化物 LD 作为激发源产生白光存在优势：① 达到阈值之后，载流子浓度被钳制，不会出现效率下降，因此可以通过提高单芯片的输出光功率来降低光源成本；② 光谱半高宽小，易于与荧光材料的最佳激发波长相匹配而实现较高的能量转换效率；③ LD 亮度高、体积小，使得终端照明的设计具有更大的自由度；④ 可以在高电流密度下工作，一般为 10 kA/cm^2[9]。此外，氮化物 LD 本身还具有高调制带宽的特性，高达吉赫兹[52,95,96]。LD 的峰值带宽是 LED 的 100 多倍。因此氮化物 LD 在照明领域发展的同时，也可以发挥在通信领域的特长，使得可见光通信的应用性能得到极大的提升。

综上，LD 相比于 LED 具有两大优势，一是在可以在高电流密度下工作，并且激射状态下不会出现效率下降问题，是一种高功率密度光源；二是具有高达吉赫兹的调制带宽。因此当需要以更高的速度进行通信时，它们更适合用作前端发射器。在与静态可见光通信应用相关的点对点传输的情况下，由于高光功率和光束会聚的特性，LD 也颇具优势。

2015 年 1 月，英国爱丁堡大学的 Tsonev 等将 RGB LD 作为可见光通信的光源，并且证明在标准室内照度水平下，通过波分复用技术 (WDM) 可以实现光学无线接入数据速率 100 Gb/s 以上，如图 8.16 所示[79]。2015 年 7 月，KAUST、台湾大学和交通大学也开始探索 RGB LD 为基础的白色光源同时用于照明与通信的可行性，他们强调蓝光 LD 在复合白光中起着关键性的作用，蓝光 LD 工作电流的大小直接影响色温的高低以及调制带宽的高低[80]。2015 年 11 月，UCSB 和 KAUST 将 pc-LD 作为可见光通信的光源，其 3dB 带宽高达 1.1GHz，并且证明黄色荧光成分基本不会对白光的调制带宽产生影响，这是首次有研究机构考虑以荧光转换激光为基础的白色光源同时用于照明与通信的可行性问题，其可见光通信系统如图 8.17 所示[95]。2015 年 12 月，台湾大学、交通大学、KAUST 和 UCSB 合作用 450nm 蓝光 LD 激发荧光粉薄膜形成激光复合白光，3dB 带宽约为 1GHz，并提出蓝光 LD 频率响应的降低归因于荧光粉–空气界面反射[97]。2015 年 12 月，KAUST、台湾大学和交通大学采用蓝光 LD 远程激发荧光粉薄膜，证明黄色荧光成分基本不会对白光的调制带宽产生影响，反而蓝光滤波片的加入会导致光强的降低，从而使其在通信应用中的信噪比 (SNR) 也降低[98]。2018 年 5 月，台湾大学的林恭如小组采用不同种类的漫射器作为 RGB LD 复合白光的匀化器，虽然白光得到了匀化，但是却严重影响了可见光通信系统的传输速率以及距离[82]。2019 年，台湾大学和台湾省的中山大学通过在 RGB LD 光路中添加 $Lu_3Al_5O_{12}:Ce^{3+}/CaAlSiN_3:Eu^{2+}$ (LuAG:Ce/CASN:Eu) 混合荧光粉，使激光复合白光的显色指数提高到了 80 以上，并研究了荧光层厚度和 RGB LD 功率之间的关系，通过优化获得了兼具高显指 (80.4) 和高传输速率 (10.4 Gb/s) 的 RGB LD 可见光通信系统，如图 8.18 所示[99]。

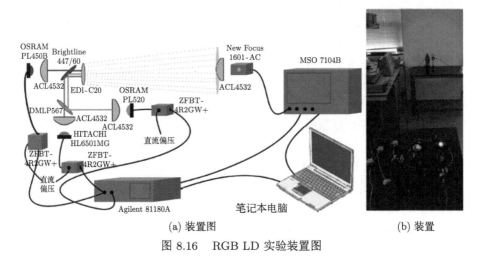

图 8.16　RGB LD 实验装置图

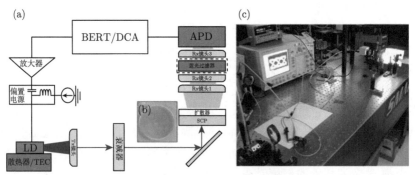

图 8.17　pc-LD 可见光通信实验装置图

表 8.3 总结了白光 LD 可见光通信技术的最新进展，可以看出半导体激光器在高速可见光通信领域的巨大潜力。但目前其发展还受到自身某些特性的限制，需要进一步优化设计。其难点主要体现在以下几个方面。首先，LD 非常脆弱，必须工作在稳定的工作环境中，电压稍高就会被烧坏。其次，半导体光源器件的结

8.3 氮化物白光激光技术

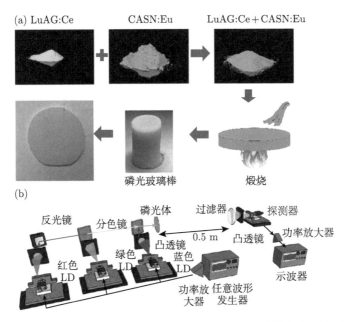

图 8.18　RGB LD+ LuAG:Ce/CaSN:Eu 可见光通信实验装置图

表 8.3　LD 复合白光可见光通信技术的最新进展

发射端	颜色转换器	带宽/Hz	传输距离	传输速率	时间	参考文献
RGB LD	—	230 M/780 M/1 G	30 cm	14 Gb/s	2015	[79]
RGB LD	—	1.18/1.15/1.02 G	20 cm	4.4/4/4 Gb/s	2015	[80]
Blue LD	YAG:Ce	1.1 G	5 cm	2 Gb/s	2015	[95]
Blue LD	YAG:Ce	900 M	60 cm	5.2 Gb/s	2015	[97]
Blue LD	YAG:Ce	1 G	50 cm	4 Gb/s	2015	[98]
RGB LD	—	0.8/1.2/1.5 G	0.5 m	8.8 Gb/s	2017	[81]
Blue LD	LuAG:Ce/CaSN:Eu	1.5 G	0.5 m	4.4 Gb/s	2018	[100]
RGB LD	—	—	0.5 m	11.2 Gb/s	2018	[82]
Blue LD	YAG:Ce	1 G	1 m	1250 Mb/s	2019	[101]
RGB LD	LuAG:Ce/CaSN:Eu	—	0.5 m	10.4 Gb/s	2019	[99]

构决定了其自身频率特性不平坦的特点。对于不同的信号频率分量，半导体光源器件发光强度的响应也不一样。这种器件的非线性会造成电信号转换成光信号过程中的非线性畸变，进而影响可见光通信系统的性能。然后，LD 可见光通信的安全性问题。LD 发出的高强度光会对人体的皮肤或者眼睛造成伤害，所以在使用 LD 作可见光发射机的光源器件时必须要考虑 LD 的安全性。最后，信号频率逐渐增加后，硬件电路不能再看成简单的集总电路，电路的元件特性也会发生巨

大的变化，例如当频率足够高的时候，电容器的感性成分越来越大，甚至大于其容性成分，电感器的容性成分也会逐渐显著。这样不仅电路的设计和分析会变得异常复杂，电路板的制作和电路调试也都会很麻烦，这些都给系统的调试和设计带来很大的困难。除了这些，信道的衰减以及多径效应等问题的存在也在一定程度上影响了系统的性能。

参 考 文 献

[1] 刘建平, 杨辉. 全球氮化镓激光器材料及器件研究现状 [J]. 新材料产业, 2015 (10): 44-48.

[2] Ohta M, Ohizumi Y, Hoshina Y, et al. High-power pure blue laser diodes[J]. Physica Status Solidi (a), 2007, 204(6): 2068-2072.

[3] Kneissl M, Yang Z, Teepe M, et al. Ultraviolet semiconductor laser diodes on bulk AlN[J]. Journal of Applied Physics, 2007, 101(12): 123103.

[4] Yoshida H, Yamashita Y, Kuwabara M, et al. A 342-nm ultraviolet AlGaN multiple-quantum-well laser diode[J]. Nature Photonics, 2008, 2(9): 551-554.

[5] Miyoshi T, Yanamoto T, Kozaki T, et al. Recent status of white LEDs and nitride LDs[C]//Gallium Nitride Materials and Devices III. SPIE, 2008, 6894: 186-192.

[6] Jiang L, Liu J, Tian A, et al. GaN-based green laser diodes[J]. Journal of Semiconductors, 2016, 37(11): 111001.

[7] Wallace J. Laser plus phosphor emits white light without droop[J]. Laser Focus World: The Magazine for the Photonics & Optoelectronics Industry, 2013, 49(11): 24.

[8] Wierer Jr J J, Tsao J Y, Sizov D S. The potential of III-nitride laser diodes for solid-state lighting[J]. Physica Status Solidi (c), 2014, 11(3-4): 674-677.

[9] Wierer Jr J J, Tsao J Y. Advantages of III-itride laser diodes in solid-tate lighting[J]. Physica Status Solidi (a), 2015, 212(5): 980-985.

[10] Basu C, Meinhardt-Wollweber M, Roth B. Lighting with laser diodes[J]. Advanced Optical Technologies, 2013, 2(4): 313-321.

[11] 胡磊, 张立群, 刘建平, 等. 高功率氮化镓基蓝光激光器 [J]. 中国激光, 2020, 47(7):6.

[12] Jovicic A, Li J, Richardson T. Visible light communication: opportunities, challenges and the path to market[J]. IEEE Communications Magazine, 2013, 51(12): 26-32.

[13] Yeh C H, Liu Y L, Chow C W. Real-time white-light phosphor-LED visible light communication (VLC) with compact size[J]. Optics Express, 2013, 21(22): 26192-26197.

[14] Grobe L, Paraskevopoulos A, Hilt J, et al. High-speed visible light communication systems[J]. IEEE Communications Magazine, 2013, 51(12): 60-66.

[15] Oubei H M, Duran J R, Janjua B, et al. 4.8 Gbit/s 16-QAM-OFDM transmission based on compact 450-nm laser for underwater wireless optical communication[J]. Optics express, 2015, 23(18): 23302-23309.

[16] Xue B, Liu Z, Yang J, et al. Characteristics of III-nitride based laser diode employed for short range underwater wireless optical communications[J]. Optics Communications, 2018, 410: 525-530.

[17] Nakamura K, Mizukoshi I, Hanawa M. Optical wireless transmission of 405 nm, 1.45 Gbit/s optical IM/DD-OFDM signals through a 4.8 m underwater channel[J]. Optics Express, 2015, 23(2): 1558-1566.

[18] 张好军, 赵建林. GaN 基紫光激光器应用于激光引信抗干扰探索 [J]. 中国激光, 2011, 38(7): 0702007-1-0702007-6.

[19] Maeda T, Terao M, Shimano T. A review of optical disk systems with blue-violet laser pickups[J]. Japanese Journal of Applied Physics, 2003, 42(2S): 1044.

[20] Dingle R, Shaklee K L, Leheny R F, et al. Stimulated emission and laser action in gallium nitride[J]. Applied Physics Letters, 1971, 19(1): 5-7.

[21] Amano H, Sawaki N, Akasaki I, et al. Metalorganic vapor phase epitaxial growth of a high quality GaN film using an AlN buffer layer[J]. Applied Physics Letters, 1986, 48(5): 353-355.

[22] Amano H, Kito M, Hiramatsu K, et al. P-type conduction in Mg-doped GaN treated with low-energy electron beam irradiation (LEEBI)[J]. Japanese Journal of Applied Physics, 1989, 28(12A): L2112.

[23] Nakamura S, Mukai T, Senoh M S M, et al. Thermal annealing effects on p-type Mg-doped GaN films[J]. Japanese Journal of Applied Physics, 1992, 31(2B): L139.

[24] Nakamura S, Mukai T, Senoh M, et al. $In_xGa_{(1-x)}N/In_yGa_{(1-y)}N$ superlattices grown on GaN films[J]. Journal of Applied Physics, 1993, 74(6): 3911-3915.

[25] Nakamura S, Senoh M, Nagahama S, et al. InGaN-based multi-quantum-well-structure laser diodes[J]. Japanese Journal of Applied Physics, 1996, 35(1B): L74.

[26] Nakamura S, Senoh M, Nagahama S I, et al. Violet InGaN/GaN/AlGaN-Based Laser Diodes with an Output Power of 420 W[J]. Japanese Journal of Applied Physics, 1998, 37(6A): L627.

[27] Nakamura S, Senoh M, Nagahama S, et al. Blue InGaN-based laser diodes with an emission wavelength of 450 nm[J]. Applied Physics Letters, 2000, 76(1): 22-24.

[28] 日亚公司. 日亚开始工程化出货 1W CW 纯蓝光半导体激光器 [EB/OL].[2007-11-20]. http://www.nichia.co.jp/cn/product/laser.html.

[29] 欧司朗.Blue Laser Diode in Multi-Die-Package[EB/OL]. [2016-09-30] https://www.osram.com/ecat/com/en/class_pim_web_catalog_103489/prd_pim_device_4065060.

[30] Avramescu A, Lermer T, Müller J, et al. InGaN laser diodes with 50 mW output power emitting at 515 nm[J]. Applied Physics Letters, 2009, 95(7): 071103.

[31] Miyoshi T, Masui S, Okada T, et al. 510~515 nm InGaN-based green laser diodes on c-plane GaN substrate[J]. Applied Physics Express, 2009, 2(6): 062201.

[32] Enya Y, Yoshizumi Y, Kyono T, et al. 531 nm green lasing of InGaN based laser diodes on semi-polar {2021} free-standing GaN substrates[J]. Applied Physics Express, 2009, 2(8): 082101.

[33] Adachi M, Yoshizumi Y, Enya Y, et al. Low threshold current density InGaN based 520-530 nm green laser diodes on semi-polar {2021} free-standing GaN substrates[J]. Applied Physics Express, 2010, 3(12): 121001.

[34] Strauss U, Somers A, Heine U, et al. GaInN laser diodes from 440 to 530nm: a performance study on single-mode and multi-mode R&D designs[C]//Novel In-Plane Semi-Conductor Lasers XVI. SPIE, 2017, 10123: 19-28.
[35] 欧司朗.reen Laser Diode in TO56 Package [EB/OL]. [2021-07-09]. https://ams-osram.com/products/product-selector.
[36] Nakarmi M L, Nepal N, Ugolini C, et al. Correlation between optical and electrical properties of Mg-doped AlN epilayers[J]. Applied Physics Letters, 2006, 89(15): 152120.
[37] Taniyasu Y, Kasu M, Makimoto T. An aluminium nitride light-emitting diode with a wavelength of 210 nanometres[J]. Nature, 2006, 441(7091): 325-328.
[38] Kaufmann U, Schlotter P, Obloh H, et al. Hole conductivity and compensation in epitaxial GaN: Mg layers[J]. Physical Review B, 2000, 62(16): 10867.
[39] Shatalov M, Gaevski M, Adivarahan V, et al. Room-temperature stimulated emission from AlN at 214 nm[J]. Japanese Journal of Applied Physics, 2006, 45(12L): L1286.
[40] Yoshida H, Takagi Y, Kuwabara M, et al. Entirely crack-free ultraviolet GaN/AlGaN laser diodes grown on 2-in. sapphire substrate[J]. Japanese Journal of Applied Physics, 2007, 46(9R): 5782.
[41] 黄德修. 半导体光电子学 [M]. 2 版. 北京: 电子工业出版社, 2013.
[42] 余金中. 半导体光子学 [M]. 北京: 科学出版社, 2015.
[43] Coldren L A, Corzine S W, Mashanovitch M L. Diode Lasers and Photonic Integrated Circuits[M]. New York: John Wiley & Sons, 2012.
[44] Shan W, Schmidt T J, Yang X H, et al. Temperature dependence of interband transitions in GaN grown by metalorganic chemical vapor deposition[J]. Applied Physics Letters, 1995, 66(8): 985-987.
[45] Feneberg M, Leute R A R, Neuschl B, et al. High-excitation and high-resolution photoluminescence spectra of bulk AlN[J]. Physical Review B, 2010, 82(7): 075208.
[46] Kasap S O. Optoelectronics and Photonics[M]. London: Pearson Education UK, 2013.
[47] Watson S, Viola S, Giuliano G, et al. High speed visible light communication using blue GaN laser diodes[C]//Advanced Free-Space Optical Communication Techniques and Applications II. SPIE, 2016, 9991: 60-66.
[48] Lee C, Zhang C, Becerra D L, et al. Dynamic characteristics of 410 nm semipolar (2021) iii-nitride laser diodes with a modulation bandwidth of over 5 GHz[J]. Applied Physics Letters, 2016, 109(10): 101104.
[49] Shen C, Ng T K, Leonard J T, et al. High-modulation-efficiency, integrated waveguide modulator-laser diode at 448 nm[J]. ACS Photonics, 2016, 3(2): 262-268.
[50] Watson S, Tan M, Najda S P, et al. Visible light communications using a directly modulated 422 nm GaN laser diode[J]. Optics Letters, 2013, 38(19): 3792-3794.
[51] 王海荣. "蓝光之父" 带来全球领先激光照明技术 [N/OL]. 深圳商报, [2016-12-14]. https://stic.sz.gov.cn/xxgk/gzyw/content/post_2911740.html.
[52] Zafar F, Bakaul M, Parthiban R. Laser-diode-based visible light communication: Toward gigabit class communication[J]. IEEE Communications Magazine, 2017, 55(2):

144-151.

[53] Narukawa Y, Nagahama S, Tamaki H, et al. Development of high-luminance white light source using GaN-based light emitting devices[J]. Oyo Butsuri, 2005, 74(11): 1423-1432.

[54] Kozaki T, Nagahama S, Mukai T. Recent progress of high-power GaN-based laser diodes[J]. Novel In-Plane Semiconductor Lasers VI, 2007, 6485: 16-23.

[55] Xu Y, Chen L, Li Y, et al. Phosphor-conversion white light using InGaN ultraviolet laser diode[J]. Applied Physics Letters, 2008, 92(2): 021129.

[56] 徐云, 陈良惠, 胡海峰, 等. 以半导体激光器为基础的固态白光光源 [C]// 第十五届全国化合物半导体材料、微波器件和光电器件学术会议. 广州, 2008.

[57] Xu Y, Hu H, Chen L, et al. Analysis on the high luminous flux white light from GaN-based laser diode[J]. Applied Physics B, 2010, 98(1): 83-86.

[58] Ryu H Y, Kim D H. High-brightness phosphor-conversion white light source using InGaN blue laser diode[J]. Journal of the Optical Society of Korea, 2010, 14(4): 415-419.

[59] Denault K A, Cantore M, Nakamura S, et al. Efficient and stable laser-driven white lighting[J]. AIP Advances, 2013, 3(7): 072107.

[60] Cantore M, Pfaff N, Farrell R M, et al. High luminous flux from single crystal phosphor-converted laser-based white lighting system[J]. Optics Express, 2016, 24(2): A215-A221.

[61] Song Y H, Ji E K, Jeong B W, et al. High power laser-driven ceramic phosphor plate for outstanding efficient white light conversion in application of automotive lighting[J]. Scientific Reports, 2016, 6(1): 1-7.

[62] Zheng P, Li S, Wang L, et al. Unique color converter architecture enabling phosphor-in-glass (PiG) films suitable for high-power and high-luminance laser-driven white lighting[J]. ACS Applied Materials & Interfaces, 2018, 10(17): 14930-14940.

[63] Masui S, Yamamoto T, Nagahama S I. A white light source excited by laser diodes[J]. Electronics and Communications in Japan, 2015, 98(5): 23-27.

[64] Salimian A, Fern G R, Upadhyaya H, et al. Laser diode induced lighting modules[J]. ECS Journal of Solid State Science and Technology, 2016, 5(3): R26.

[65] Lee T X, Chou C C, Chang S C. Novel remote phosphor design for laser-based white lighting application[C]//Fifteenth International Conference on Solid State Lighting and LED-based Illumination Systems. SPIE, 2016, 9954: 96-104.

[66] Zhang X, Yu J, Wang J, et al. All-inorganic light convertor based on phosphor-in-glass engineering for next-generation modular high-brightness white LEDs/LDs[J]. ACS Photonics, 2017, 4(4): 986-995.

[67] Yang Y, Zhuang S, Kai B. High brightness laser-driven white emitter for Etendue-limited applications[J]. Applied Optics, 2017, 56(30): 8321-8325.

[68] Peng Y, Mou Y, Wang H, et al. Stable and efficient all-inorganic color converter based on phosphor in tellurite glass for next-generation laser-excited white lighting[J]. Journal of the European Ceramic Society, 2018, 38(16): 5525-5532.

[69] Peng Y, Mou Y, Sun Q, et al. Facile fabrication of heat-conducting phosphor-in-glass with dual-sapphire plates for laser-driven white lighting[J]. Journal of Alloys and Com-

pounds, 2019, 790: 744-749.

[70] Lee D H, Kim S, Kim H, et al. Highly efficient and highly conductive phosphor-in-glass materials for use in LD-driven white-light lamp[J]. International Journal of Precision Engineering and Manufacturing-Green Technology, 2019, 6(2): 293-303.

[71] Zhang X, Si S, Yu J, et al. Improving the luminous efficacy and resistance to blue laser irradiation of phosphor-in-glass based solid state laser lighting through employing dual-functional sapphire plate[J]. Journal of Materials Chemistry C, 2019, 7(2): 354-361.

[72] Ma Y, Luo X. Packaging for laser-based white lighting: Status and perspectives[J]. Journal of Electronic Packaging, 2020, 142(1): 010801.

[73] Hu F, Li Y. Laser and phosphor hybrid source for projection display[C]//Solid State Lasers XXII: Technology and Devices. SPIE, 2013, 8599: 296-302.

[74] Ma Y, Lan W, Xie B, et al. An optical-thermal model for laser-excited remote phosphor with thermal quenching[J]. International Journal of Heat and Mass Transfer, 2018, 116: 694-702.

[75] Bachmann V, Ronda C, Meijerink A. Temperature quenching of yellow Ce^{3+} luminescence in YAG: Ce[J]. Chemistry of Materials, 2009, 21(10): 2077-2084.

[76] Neumann A, Wierer J J, Davis W, et al. Four-color laser white illuminant demonstrating high color-rendering quality[J]. Optics Express, 2011, 19(104): A982-A990.

[77] Voelkel R, Weible K J. Laser beam homogenizing: Limitations and constraints[C]// Optical Fabrication, Testing, & Metrology III. International Society for Optics and Photonics, 2008,7102: 222-233.

[78] Shin I H, Lee J J, Kang H S. Novel color combiner composed of red, green, and blue laser diodes and a thermally expanded core fiber waveguide[J]. Optical Engineering, 2011, 50(9): 094005.

[79] Tsonev D, Videv S, Haas H. Towards a 100 Gb/s visible light wireless access network[J]. Optics Express, 2015, 23(2): 1627-1637.

[80] Janjua B, Oubei H M, Retamal J R D, et al. Going beyond 4 Gbps data rate by employing RGB laser diodes for visible light communication[J]. Optics Express, 2015, 23(14): 18746-18753.

[81] Wu T C, Chi Y C, Wang H Y, et al. Tricolor R/G/B laser diode based eye-safe white lighting communication beyond 8 Gbit/s[J]. Scientific Reports, 2017, 7(1): 1-10.

[82] Huang Y F, Chi Y C, Chen M K, et al. Red/green/blue LD mixed white-light communication at 6500K with divergent diffuser optimization[J]. Optics Express, 2018, 26(18): 23397-23410.

[83] Yang J, Liu Z, Xue B, et al. Highly uniform white light-based visible light communication using red, green, and blue laser diodes[J]. IEEE Photonics Journal, 2018, 10(2): 1-8.

[84] Janjua B, Ng T K, Zhao C, et al. True yellow light-emitting diodes as phosphor for tunable color-rendering index laser-based white light[J]. ACS Photonics, 2016, 3(11): 2089-2095.

[85] Goodman J W. Some fundamental properties of speckle[J]. JOSA, 1976, 66(11): 1145-1150.

[86] George N, Jain A. Speckle reduction using multiple tones of illumination[J]. Applied Optics, 1973, 12(6): 1202-1212.

[87] Sato K I, Asatani K. Speckle noise reduction in fiber optic analog video transmission using semiconductor laser diodes[J]. IEEE Transactions on Communications, 1981, 29(7): 1017-1024.

[88] Lowenthal S, Joyeux D. Speckle removal by a slowly moving diffuser associated with a motionless diffuser[J]. JOSA, 1971, 61(7): 847-851.

[89] McKechnie T S. Reduction of speckle by a moving aperture-first order statistics[J]. Optics Communications, 1975, 13(1): 35-39.

[90] Rawson E G, Nafarrate A B, Norton R E, et al. Speckle-free rear-projection screen using two close screens in slow relative motion[J]. JOSA, 1976, 66(11): 1290-1294.

[91] Le Minh H, O'Brien D, Faulkner G, et al. 100-Mb/s NRZ visible light communications using a postequalized white LED[J]. IEEE Photonics Technology Letters, 2009, 21(15): 1063-1065.

[92] Li H, Chen X, Guo J, et al. A 550 Mbit/s real-time visible light communication system based on phosphorescent white light LED for practical high-speed low-complexity application[J]. Optics Express, 2014, 22(22): 27203-27213.

[93] Li H, Zhang Y, Chen X, et al. 682 Mbit/s phosphorescent white LED visible light communications utilizing analog equalized 16QAM-OFDM modulation without blue filter[J]. Optics Communications, 2015, 354: 107-111.

[94] Ferreira R X G, Xie E, McKendry J J D, et al. High bandwidth GaN-based micro-LEDs for multi-Gb/s visible light communications[J]. IEEE Photonics Technology Letters, 2016, 28(19): 2023-2026.

[95] Lee C, Shen C, Oubei H M, et al. 2 Gbit/s data transmission from an unfiltered laser-based phosphor-converted white lighting communication system[J]. Optics Express, 2015, 23(23): 29779-29787.

[96] Lee C, Zhang C, Cantore M, et al. 4 Gbps direct modulation of 450 nm GaN laser for high-speed visible light communication[J]. Optics Express, 2015, 23(12): 16232-16237.

[97] Chi Y C, Hsieh D H, Lin C Y, et al. Phosphorous diffuser diverged blue laser diode for indoor lighting and communication[J]. Scientific Reports, 2015, 5(1): 1-9.

[98] Retamal J R D, Oubei H M, Janjua B, et al. 4-Gbit/s visible light communication link based on 16-QAM OFDM transmission over remote phosphor-film converted white light by using blue laser diode[J]. Optics Express, 2015, 23(26): 33656-33666.

[99] Huang Y F, Chi Y C, Cheng C H, et al. LuAG: Ce/CASN: Eu phosphor enhanced high-CRI R/G/B LD lighting fidelity[J]. Journal of Materials Chemistry C, 2019, 7(31): 9556-9563.

[100] Wu T C, Chi Y C, Wang H Y, et al. White-lighting communication with a $Lu_3Al_5O_{12}$:

Ce^{3+}/CaAlSiN$_3$: Eu^{2+} glass covered 450-nm InGaN laser diode[J]. Journal of Lightwave Technology, 2018, 36(9): 1634-1643.

[101] Yeh C H, Chow C W, Wei L Y. 1250 Mbit/s OOK wireless white-light VLC transmission based on phosphor laser diode[J]. IEEE Photonics Journal, 2019, 11(3): 1-5.

第 9 章 太阳能电池研究进展

9.1 引言

太阳能电池是通过光电效应或者光化学效应直接把光能转化成电能的装置。因其存在无污染、无噪声、建设周期短、不受地域限制、规模设计自由度大、易储存等优点，被认为是缓解能源紧张及减少温室气体排放的重要技术之一。所以，目前，开发利用太阳能电池已成为可持续发展的重要战略决策。

9.2 太阳能电池的种类和光伏原理

9.2.1 pn 结型太阳能电池

pn 结型太阳能电池较为普遍，其中，pn 结区吸收入射光子，形成空穴-电子对，并在内建电场的作用下分离，从而形成光电流。其效率较高，开路电压较大，但缺点是制作工艺复杂[1-3]，成本较高。

pn 结光伏效应的基本原理如图 9.1 所示[4]。以单晶硅太阳能电池为例，在 p 型和 n 型硅界面两侧载流子浓度梯度的驱使下，p 区中过剩的空穴向 n 区扩散，n 区中过剩的电子向 p 区扩散。扩散的结果使得靠近 pn 结界面的 p 区出现大量受主离子形成的负电荷，自身电势降低，而靠近 pn 结界面的 n 区出现大量施主离子形成的正电荷，自身电势升高。最终导致能带的弯曲以及 n 区和 p 区费米能级的重叠，形成由 n 区指向 p 区的电场，即内建电场。由于该区域杂质完全离化，可动电荷离去后仅剩下不能移动的离子，该区域又称为耗尽区。当有太阳光照射时，能量大于半导体禁带宽度的光子 ($h\nu > E_g$) 被吸收，激发价带中的电子向导带跃迁，导带中形成自由电子，价带中留下自由空穴，这些电子-空穴对在内建电场的作用下被分离，电子被扫到 n 区，空穴被扫到 p 区，最终在 p 区和 n 区将分别积累正电荷和负电荷，形成以 p 区为正极、n 区为负极的光生电动势，我们将 pn 结吸收光能产生电动势的现象称为 pn 结的光生伏特效应。

如果两个金属电极的外接引线是断路的，被分离的光生载流子会积聚在两侧电极并形成一个与内建电场指向相反的电场，降低内建电场对光生电子-空穴对的分离能力，最终达到平衡状态，并在外电路上表现为开路电压。如果与 pn 结连接的两个金属电极的外接引线是短路的，电池就会向外输出电流 (短路电流)。如果把太阳能电池接上负载,则被结分开的过剩载流子中就有一部分把能量消耗于降低 pn 结势垒,而剩余部分的光生载流子则用来产生光生电流。

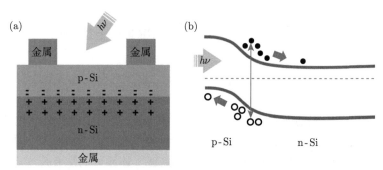

图 9.1　单晶硅 pn 结型太阳能电池的电池结构 (a) 和能带结构 (b) 示意图

9.2.2　肖特基势垒太阳能电池

为避免 pn 结复杂的制备工艺,需要寻找替代的太阳能电池结构。方案之一是利用晶硅与金属接触形成的肖特基结代替 pn 结,实现光生载流子的分离,即肖特基势垒太阳能电池。

肖特基势垒太阳能电池结构非常简单,由半导体 (以硅为例) 与金属的接触而形成,如图 9.2(a) 所示。由于硅 (以 n 型为例) 与金属 (以较大功函数的金属,如 Au 为例) 的功函数存在差异,半导体一侧的电荷将流向金属直至建立热平衡。与 pn 结相似,半导体中被留下的掺杂离子在半导体与金属的交界面处建立内建电场,使半导体和金属的费米能级达到平衡。由于半导体和金属的载流子浓度在数量级上的差别,内建电场完全建立在半导体一侧,如图 9.2(b) 所示。这种具有耗尽区的金属和半导体之间的接触称为肖特基结。与 pn 结相似,在能量大于半导体禁带宽度的光子照射下,肖特基结也会发生光生伏特效应:当光子能量被半导体材料吸收后,在半导体的耗尽区内激发产生电子–空穴对,这些电子–空穴对

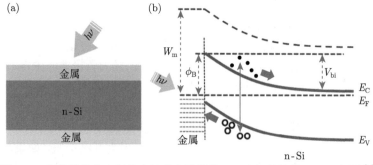

图 9.2　肖特基势垒太阳能电池的电池结构 (a) 和能带结构 (b) 示意图 [5]

在内建电场的作用下被分离,电子和空穴分别被两侧的金属电极收集。在外电路断路和短路的条件下,肖特基结内建电场对光生载流子的分离作用将分别在外电路上输出开路电压和短路电流。

与晶体硅 pn 结型太阳能电池相比,硅肖特基结太阳能电池有很多优点[6]:制备过程简单,具有低能耗、无污染等特点,降低了太阳能电池成本;内建电场建立在硅的表面,一方面能够有效分离光生载流子,提高太阳能电池效率,另一方面可以增强太阳能电池的抗辐射能力。对于这种结构的太阳能电池,金属材料的导电性和光透过性决定了肖特基势垒太阳能电池的性能。然而,目前由于高导电性和光透过性对于一种材料难以同时实现,硅肖特基势垒太阳能并没有受到太多关注。

9.2.3 染料敏化太阳能电池

染料敏化太阳能电池 (dye-sensitized solar cell, DSSC) 的结构示意图如图 9.3 所示,DSSC 由四个部分组成:①由沉积在透明导电玻璃基板上的中孔氧化物层 (通常为 TiO_2) 构成的光电阳极;②单层染料敏化剂共价结合到 TiO_2 层的表面用于收集光并产生光子激发的电子;③有机溶剂中含有氧化还原对 (通常为 I^-/I_3^-) 的电解质,用于在对电极处收集电子并影响染料再生;④由镀铂的导电玻璃基板制成的对电极[7]。

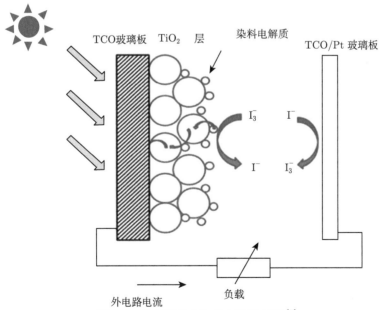

图 9.3 染料敏化太阳能电池原理图[7]

DSSC 由阳极 (或光电极,工作电极) 及阴极 (或对电极) 组成,这两个电极通常由专门设计的涂有导电氧化物 (即铟或掺杂氟的氧化锡) 的透明导电玻璃制成。

一般来说，掺杂铟的氧化锡 (indium doped tin oxide, ITO) 的透射率高于掺杂氟的氧化锡 (fluorine doped tin oxide, FTO) 的透射率；而 FTO 的薄层电阻小于 ITO。在烧结过程中，随着温度的升高，ITO 的薄层电阻急剧增加；而 FTO 的薄层电阻在烧结过程中保持不变 [8]。

工作电极上涂有敏化染料包裹的二氧化钛层，该涂层可使工作电极具有多孔性，因此更多的阳光可以渗透到半导体层中。由于二氧化钛半导体材料对可见光不敏感，因此，需要敏化材料来增强可见光吸收率 [9]。氧化锌具有与二氧化钛类似的带隙、传输性质和电子注入效率，是二氧化钛的替代物。但是，通过使用不同的染料和电解质，很难观察到氧化锌的行为。对电极的功能是将电子注入电解质中从而形成如图 9.3 所示的内部电路。工作电极和对电极从电解质溶液中分离，电解质促进了工作电极和对电极之间的电荷传输。

DSSC 的效率约为 12%。当阳光照射到 DSSC 表面时，染料分子会收集光子并产生激发的电子。敏化剂将激发的电子注入纳米多孔半导体膜的导带中。失去电子的染料分子随后被氧化。注入的电子穿过无孔 TiO_2 薄膜朝着透明导电电极 (工作电极) 移动，并以电能的形式转移而到达负载。当染料从氧化还原介质 (I^-/I_3^-) 接收电子时，被氧化的染料分子就会再生，在该过程中氧化还原介质被氧化。之后，这些被氧化的氧化还原介质 (I_3^-) 扩散到对电极，电子通过外部电路到达对电极而引起的还原作用使 I_3^- 还原，从而形成整个电路 [10]。DSSC 的具体工作原理还可以解释为：染料分子被入射光子激发，激发的染料处于较高的能级，并且将电子释放到 TiO_2(或其他纳米材料，如 ZnO、CuO 等) 纳米带的导带中，从而产生电势差；这些电子可自由移动并通过外部电路到达对电极，并且在对电极和电解质的界面处参与氧化还原反应，然后提供给染料分子。

9.2.4 聚合物太阳能电池

聚合物太阳能电池 (polymer solar cell, PmSC) 是在有机溶剂溶液中加工而成的，而小分子太阳能电池则主要是在高真空环境下通过热蒸发沉积来制备的。聚合物太阳能电池由于具有高的吸收系数、丰富的有机物含量、高效的溶液方案、较低的能源制备要求、低比重、良好的机械柔韧性、可调的材料特性和高透明度等优势而备受关注。典型的光伏发电过程涉及从入射光子中产生自由载流子的过程。以基于聚合物-富勒烯的聚合物太阳能电池为例，该器件的结构和能量模型如图 9.4 所示。外部量子效率 (EQE) 作为波长 (λ) 的函数是所收集的光生电荷与入射光子数量之比，为吸收 (A)、激子扩散 (ED)、电荷分离 (CS) 及电荷收集 (CC) 四个效率 (η) 的乘积，即 $EQE(\lambda) = \eta_A(\lambda) \times \eta_{ED}(\lambda) \times \eta_{CS}(\lambda) \times \eta_{CC}(\lambda)$。光电压 (或开路电压，$V_{OC}$) 与受体的最低未占分子轨道 (LUMO) 能级和施主的最高占据分子轨道 (HOMO) 能级之间的能量差相关，从而提供了电荷分离的主要驱动力。图

9.2 太阳能电池的种类和光伏原理

9.4(b) 比较了具有代表性的聚合物太阳能电池的太阳光谱和 EQE 光谱 (带隙为 1.9 eV)。其中短路电流密度 (J_{SC}) 等于电池响应度与入射太阳光谱辐照度之间乘

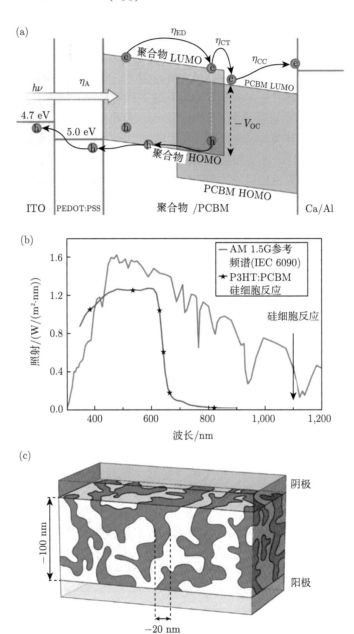

图 9.4 (a) 聚合物太阳能电池的运行机制；(b) 太阳光谱与 P3HT:PCBM 太阳能电池的光响应之间的比较；(c) 聚合物和受体的双连续互穿网络的概念形态模型 [11]

积的积分。因此，有必要利用更宽的太阳光谱，并扩大供体的 LUMO 能级和受体的 HOMO 能级之间的能级差，从而产生更高的 J_{SC} 及 V_{OC}。材料科学的创新为这些目标的实现提供了有效的方法。形态是本体异质结聚合物太阳能电池中的另外一个关键因素。本体异质结的首选形态是图 9.4(c) 所示的双连续互穿网络。施主域和受主域的大小应为激子扩散长度的两倍 (约 10 nm)，从而使激子扩散到施主–受主界面，从而电荷产生获得有效的 $\eta_{ED}(\lambda)$ 和 $\eta_{CS}(\lambda)$。施主–受主界面的电荷分离后，空穴和电子必须分别通过施主和受主网络到达正极和负极。另一个关键因素是有机物–电极界面，其中电荷被提取到外部电路。而电荷收集效率 $\eta_{CC}(\lambda)$ 与网络中的载流子传输和提取步骤有关。因此，聚合物太阳能电池的高效构建要求系统地考虑三个关键领域：材料设计、形态和处理以及界面工程。值得指出的是，纳米结构的氧化锌、氧化钛和镉硒等也可以充当聚合物供体的受体，研究表明其效率超过 3%。

在聚合物太阳能电池的结构中，界面层具有多种功能[12]。首先，它们能减少光敏层和电极之间的能量势垒，从而形成用于有效电荷提取的欧姆接触。其次，它们可以形成用于单一类型载体的选择性接触，例如也可以用作空穴阻挡层的电子传输层。某些界面材料，如 TiO_x 和碳酸铯 ($CsCO_3$)，可以改变电极的功函数，因此可以实现聚合物太阳能电池的反向结构。底层有源层的光场调制和保护也是界面层可以提供的重要功能。

聚合物太阳能电池的底层一般是一层透明导电材料氧化铟锡 (ITO)，其功函数约为 4.7 eV。顶部施加的 PEDOT:PSS 的 p 型界面层 (功函数约 5.0 eV) 与光敏层形成欧姆接触，从而有效地收集电荷，并使 ITO 表面更为光滑以去除潜在的针孔。但 PEDOT:PSS 的酸性性质会影响器件的稳定性。为了解决这个问题，研究人员引入了各种过渡金属氧化物 (如 V_2O_5、MoO_3、WO_3 和 NiO 等) 来替代 PEDOT:PSS。溶液处理的过渡金属氧化物近年来被广泛应用于聚合物太阳能电池中，并取得了较好的结果。此外，在阴极一侧，使用低功函数的金属 (如钙、钡、镁等) 在阴极与聚合物的界面处提供欧姆接触。研究人员引入了无机化合物来代替此类活性金属，从而提高器件的稳定性。

溶液处理的界面层通常比真空处理 (如热蒸发、溅射) 更可取，并有利于实现本体异质结聚合物太阳能电池的全部潜力。n 型无机金属氧化物，例如基于溶胶–凝胶的 TiO_x 和氧化锌 (ZnO_x) 已被证明是制备有机电子产品的理想选择。低温 (150°C) 水解溶胶–凝胶工艺制备的非晶 TiO_x 可有效应用于有机发光二极管 (OLED) 和聚合物太阳能电池的制备中。此外，TiO_x 还可用作空穴阻挡层和光学隔离物，并通过调制光场来增强器件内部的吸收。研究表明，使用 TiO_x 光学垫片可以使 P3HT:PCBM 系统的光电流和 EQE 提高 40%，并使 PCDTBT:PCBM 系统内部的量子效率提高近 100%。

9.2.5 新型太阳能电池

钙钛矿太阳能电池 (perovskite solar cell，PSC)、量子点太阳能电池 (quantum dot solar cell, QDSC) 等都属于新型太阳能电池，下面主要介绍这两种太阳能电池的结构和工作原理。

1. 钙钛矿太阳能电池的结构和工作原理

钙钛矿太阳能电池利用钙钛矿结构的光吸收材料进行光伏活动，类似于染料太阳能电池利用染料收集光。钙钛矿太阳能电池是科学界目前正在研究的主要光伏技术，涉及可再生能源的转化。钙钛矿太阳能电池的结构主要由四个部分组成：基于 TiO_2 纳米粒子的介孔支架；简单的平面和异质结结构；介孔上层结构及倒置的平面异质结的规则构型 (图 9.5)。详细解释就是：导电玻璃 (或塑料箔) 支撑电子提取层 (如 TiO_2、SnO_2 等)，电子提取层上沉积有钙钛矿活性材料；钙钛矿层上方涂有空穴传输材料，电池顶部蒸发有金接触点。太阳光的吸收导致电荷的产生，负电荷和正电荷载流子都通过钙钛矿输运到电荷选择性接触中。钙钛矿太阳能电池的核心为钙钛矿层，钙钛矿化合物的化学通式为 ABX_3，其中 A 为一价阳离子 (如甲基铵 $CH_3NH_3^+$、甲酰胺 $CH_2(NH_2)_2^+$、Cs^+、Rb^+ 等)，B 代表 Pb^{II} 或 Sn^{II}，X 代表 I 或 Br。有机金属卤化物钙钛矿由有机阳离子 (即甲基铵 $CH_3NH_3^+$、乙基铵 $CH_3CH_2NH_3^+$、甲铵 $NH_2CH=NH_2^+$、铯 Cs)，碳原子家族的二价金属阳离子 (即 Ge^{2+}、Sn^{2+}、Pb^{2+}) 和一价卤素阴离子 (即 F^-、Cl^-、Br^-、I^-) 组成。钙钛矿材料具有出色的光电性能、高的吸收系数和迁移率、低的激子结合能以及

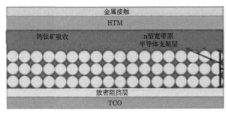
(a) 规则结构，$\eta \approx 20\%$

(b) 平面异质结结构，$\eta \approx 19\%$

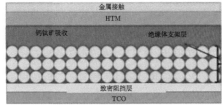
(c) 介孔超结构，$\eta \approx 17\%$

(d) 倒置平面异质结结构，$\eta \approx 18\%$

图 9.5 钙钛矿太阳能电池的结构及最大效率 [13]

大的平衡载流子扩散长度等优势,因此被广泛应用于太阳能电池的制备中。

钙钛矿太阳能电池的工作原理仍在研究中。上述染料敏化太阳能电池 (Dye-sensitized solar cell, DSSC) 的工作机制有助于理解钙钛矿太阳能电池的功能。目前被广泛接受的钙钛矿太阳能电池的简化工作原理为:当光照射到钙钛矿太阳能电池上时,钙钛矿吸光并产生激子。电子和空穴对是由热能产生的,热能分别通过电子和空穴选择性接触扩散和分离 (图 9.6)。一旦阳极和阴极分别含有电子和空穴,就可以通过连接外部电路来为外部负载供电。可比的电荷载流子扩散长度和光吸收长度可实现最佳性能。平面异质结构型的钙钛矿太阳能电池的工作原理清楚地表明,光生电子和空穴共存于钙钛矿吸收膜中,并具有足够的扩散长度以达到选择性接触。了解钙钛矿太阳能电池材料的电子和光学特性的起源对于详细解释器件的工作机理至关重要。研究表明,材料的成分会显著影响钙钛矿太阳能电池的性能。例如,自由电荷载流子的扩散长度在器件性能中起关键作用。与将活性层厚度限制为几百纳米的孔相比,常规的钙钛矿吸收剂 $CH_3NH_3PbI_3$ 具有较低的电子扩散长度。因此,当利用该吸收剂时,器件通常采用中孔结构,而基于复合卤化物的钙钛矿 (如 $CH_3NH_3PbI_{3-x}Cl_x$) 改善了电子扩散长度,因此常采用平面结构。

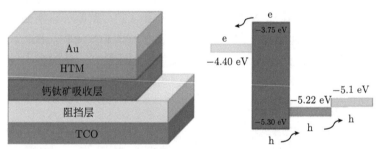

图 9.6 典型 PSC 的结构示意图及相应的能带图 [13]

2. QDSC 的结构和工作原理

典型的 QDSC 主要由三个部分组成:量子点 (quantum dot, QD) 敏化的光电阳极、对电极 (counter electrode, CE) 及含有 S^{2-}/Sn^{2-} 对的氧化还原电解质 (图 9.7(a))。通过在导电玻璃上涂覆一层最佳厚度约为 10 mm、孔隙率为 50%~60% 的介孔宽禁带半导体层 (TiO_2、ZnO) 来制备光电阳极。QD 光敏剂被吸附在介孔半导体纳米结构表面,导致光子吸收和电子注入。如图 9.7(b) 所示,当 QD 受到持续的入射光照射时,它们可以吸收光子并将电子从 QD 的价带 (valence band, VB) 激发到导带 (conduction band, CB)。在 QD 与 TiO_2 之间的导带能量差的驱动力作用下,电子被注入 TiO_2 的导带中,然后沿着 TiO_2 纳米微晶的渗透网络转移到透明导电氧化物 (transparent conductive oxide, TCO) 中，从而实现电荷分离。电子通过 TiO_2 介孔膜传输到 TCO 基板，然后通过外部电路传输到对电极。QD 随后在作为多孔介质的电解质中通过还原再生。最后，空穴被输送到对电极，在该处氧化还原系统的氧化产物被还原。除了这些所需的电荷传输过程外，同时还存在一些非必要的过程，也称为电荷复合，并会严重降低太阳能电池的性能。

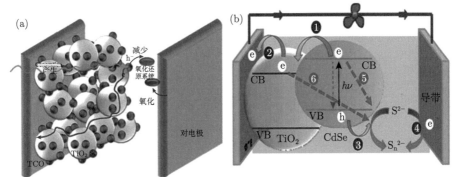

图 9.7 (a) 典型 QDSC 的结构示意图；(b)QDSC 中激发半导体纳米微晶之后的界面电荷转移过程[14]

参 考 文 献

[1] Chapin D M, Fuller C S, Pearson G L. A new silicon p-n junction photocell for converting solar radiation into electrical power[J]. Journal of Applied Physics, 1954, 25(5): 676-677.

[2] Godfrey R B, Green M A. 655 mV open-circuit voltage, 17.6% efficient silicon MIS solar cells[J]. Applied Physics Letters, 1979, 34(11): 790-793.

[3] Blakers A W, Green M A. 20% efficiency silicon solar cells[J]. Applied Physics Letters, 1986, 48(3): 215-217.

[4] 熊绍珍, 朱美芳. 太阳能电池基础与应用 [M]. 北京: 科学出版社, 2009.

[5] 刘鑫. 石墨烯/硅肖特基结太阳能电池的制备与性能研究 [D]. 北京: 中国科学院大学, 2015.

[6] 施敏, 伍国珏. 半导体器件物理 [M]. 耿莉, 张瑞智, 译. 西安: 西安交通大学出版社, 2008.

[7] Gong J, Sumathy K, Qiao Q, et al. Review on dye-sensitized solar cells (DSSCs): Advanced techniques and research trends[J]. Renewable and Sustainable Energy Reviews, 2017, 68: 234-246.

[8] Sima C, Grigoriu C, Antohe S. Comparison of the dye-sensitized solar cells performances based on transparent conductive ITO and FTO[J]. Thin Solid Films, 2010, 519(2): 595-597.

[9] Grätzel M. Photoelectrochemical cells[J]. Nature, 2001, 414(6861): 338-344.

[10] Stathatos E. Dye sensitized solar cells: a new prospective to the solar to electrical energy conversion. Issues to be solved for efficient energy harvesting[J]. Journal of Engineering Science & Technology Review, 2012, 5(4): 9-13.

[11] Li G, Zhu R, Yang Y. Polymer solar cells[J]. Nature Photonics, 2012, 6(3): 153-161.

[12] Chen L M, Xu Z, Hong Z, et al. Interface investigation and engineering-achieving high performance polymer photovoltaic devices[J]. Journal of Materials Chemistry, 2010, 20(13): 2575-2598.

[13] Asghar M I, Zhang J, Wang H, et al. Device stability of perovskite solar cells-A review[J]. Renewable and Sustainable Energy Reviews, 2017, 77: 131-146.

[14] Duan J, Zhang H, Tang Q, et al. Recent advances in critical materials for quantum dot-sensitized solar cells: a review[J]. Journal of Materials Chemistry A, 2015, 3(34): 17497-17510.

第 10 章　集成电路设计与仿真

10.1　引　　言

10.1.1　集成电路的过去、现状和未来

集成电路是指通过一系列特定的加工工艺，将晶体管、二极管等有源器件和电阻、电容等无源器件，按照一定的电路互连，"集成"在一块半导体单晶片（如硅或砷化镓）上，封装在一个外壳内，执行特定电路或系统功能。微电子技术是当代信息技术的一大基石，集成电路技术是微电子技术的核心，芯片是现代社会的基石。

1906 年，第一个电子管诞生，如图 10.1 所示；1947 年，Bardeen、Brattain 和 Shockley 发现了双极型晶体管（BJT），开启了新的信息时代，如图 10.2 所示。1958 年，Jack Kilby 发明了全球第一块集成电路板，于 2000 年获得诺贝尔物理学奖，这标志着世界从此进入集成电路的时代，如图 10.3 所示。2009 年，Intel 酷睿 i 系列全新推出，创纪录地采用了领先的 32 nm 工艺，逐渐向 10 nm、7 nm，并最终走向 3 nm 和 2 nm 工艺。

图 10.1　第一个电子管

集成电路制作工艺的日益成熟以及各个集成电路生产、制造厂商之间的不断竞争，使集成电路发挥了它更大的功能，能够更好地服务于社会。集成电路从产生到成熟大致经历了如下过程：电子管—晶体管—集成电路—超大规模集成电路。集成电路具有体积小、质量轻、寿命长和可靠性高等优点，同时成本也相对低廉，便于大规模生产。

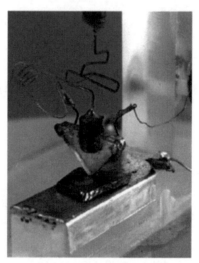

图 10.2　第一个双极型晶体管

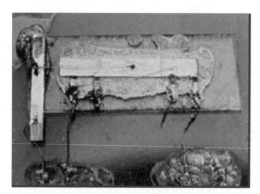

图 10.3　第一块集成电路板

在 20 世纪 80 年代初期，消费类电子产品 (如立体声收音机、彩色电视机和盒式录像机等) 是半导体需求的主要推动力。从 80 年代末开始，个人计算机 (PC) 成为半导体需求强大的推动力。至今，PC 仍然推动着半导体产品的需求。

从 20 世纪 90 年代至今，通信与计算机一起占领了世界半导体需求的 2/3。其中，通信的增长最快。信息技术正在改变我们的生活，影响着我们的工作。信息技术在提高企业竞争力的同时，已成为世界经济增长的新动力。

事实上，早在 20 世纪 50 年代，工程师们就萌生了集成电路的想法，仙童半导体公司、德州仪器以及摩托罗拉、ARM 公司即是其中的典型代表，如图 10.4 所示。如今，半个世纪过去了，从最初的上百微米级，到如今的十余纳米级，集成电路的性能简直犹如指数爆炸般扶摇直上，电子器件的性能更是以知名的"摩尔定律"所描述的奇迹速度飞速提升，甚至 2017 年英特尔 (Intel) 公司声称将实现的 7 nm 工艺——经典工艺于平面集成的技术极限，将传统意义上的集成电路性能发挥到极限。

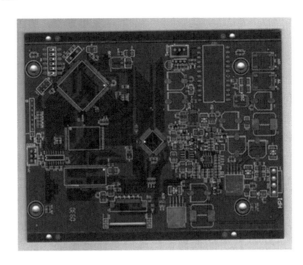

图 10.4　某 ARM 芯片

经典工艺对于平面集成的极限是 7 nm，而摩尔定律并不是真正的自然规律，它是在描述人文世界产业状况的函数模型，随着英特尔公司 7 nm 工艺的实现，摩尔时代将迎来它的黄昏——后摩尔时代，那时的我们，将利用 3D 技术和折叠工艺来实现立体集成，甚至很可能在并不久远且可预见的未来，摩尔时代即将迎来它的终结——同时，第五代计算机，即电子计算机，也将完结它的传奇——硅时代也许将画上句号。

当然，这不意味着信息技术 (IT) 也已经发展到极限，这只是个开始，第五代计算机使命完结之日，也将是第六代计算机重装上阵之时，其中，仿生计算机、光子计算机、量子计算机将是未来 IT 业的主角。

中国的集成电路产业起步于 20 世纪 60 年代中期，1976 年，中国科学院计算机研究所研制成功 1000 万次大型电子计算机，所使用的电路为中国科学院 109 厂研制的发射极耦合逻辑 (emitter coupled logic) 型电路；1986 年，电子工业部 (现工业和信息化部) 提出 "七五" 期间，我国集成电路技术 "531" 发展战略，即推进 5 μm 技术，开发 3 μm 技术，进行 1 μm 技术科技攻关；1995 年，电子工业部提出 "九五" 集成电路发展战略：以市场为导向，以 CAD 为突破口，产学研用相结合，以我为主，开展国际合作，强化投资；在 2003 年，中国半导体占世界半导体销售额的 9%，电子市场达到 860 亿美元，中国成为世界第二大半导体市场，中国中高技术产品的需求将成为国民经济新的增长动力。到现在已经初具规模，形成了产品设计、芯片制造、电路封装共同发展的态势。

但是，值得特别注意的是：由于 20 世纪 50 年代后我国的外交窘境，基础薄弱，人才缺乏，外加雪上加霜的自然及社会问题，中国以集成电路为代表的电子硬件技术行业一度停滞不前，甚至有一种持续至今仍颇具影响的偏激理念认为：对于电子器件而言，软件的进步要远重要于硬件的进步，而硬件上的不足，可以靠软件的不断进步来几乎完全地抵消弥补。

这也导致了当今中国电子科技行业的软肋：软件水平与硬件水平的极端不协调，与硬件水平的巨大短板（直到 2002 年，我国才真正独立开发出本土的 CPU——龙芯）。这也是当今高校的 IT 方面的教材中，软件开发所占比例远大于硬件开发的原因。

直至 21 世纪初，当计算机、手机这类第三次技术革命的产物得以走进千家万户，而我国的电子行业却依旧羸弱无力时，这种幼稚理念终于被推倒，市场规律使得中国电子行业为他们在近 40 年来对于技术进步的错误认知付出了高昂的代价，却也给了那些固步自封者一个极大的教训——实际上，由于意识形态及封建余毒的原因，对此了解甚少的人们在此前一直将电子行业视作"旁门左道"，而较之硬件开发，等程度上软件开发所需的成本低于硬件开发，因此可想而知，当时的国人对于硬件开发并不热忱。

于是，在 21 世纪的头 10 年，我国本土的电子行业只能眼睁睁地看着外国同行在本土的巨大市场赚得满盆金箔，而自己，除了眼红，也只能默默发展实力，穷追猛赶并静待时机了。而值得庆幸的是，我们已经意识到了过往观念中的错误，国家已将推动芯片国产化上升至国家安全的高度，2020 年信息安全政策的重点将落实在硬件领域，特别是对集成电路产业的扶持力度堪称近十年之最。与此同时，随着集成电路发展纲要及地方扶持政策的相继落地，集成电路产业将获得前所未有的发展机遇。

10.1.2　集成电路的分类及主流工艺

集成电路按照不同的标准有不同的分类方法，具体情况如下。按照电路规模分类，可以分为小规模集成电路 (small scale IC，SSI)、中等规模集成电路 (medium scale IC，MSI)、大规模集成电路 (large scale IC，LSI)、超大规模集成电路 (very large scale IC，VLSI)、特大规模集成电路 (ultra large scale IC，ULSI) 和巨大规模集成电路 (gigantic scale IC，GSI)，其分类标准如表 10.1 所示。按照导电载流子类型分类，可以分为双极型集成电路、MOS 型集成电路和 Bi-CMOS 型集成电路。按照电路处理信号的方式分类，可以分为数字集成电路、模拟集成电路和数模混合集成电路。按实现方法分类，可以分为薄膜集成电路、半导体集成电路和混合集成电路。按电路功能分类，可以分为通用集成电路 (general-purpose integrated circuit, GPIC)、专用集成电路 (application specific integrated circuit, ASIC) 和

专用标准集成电路 (application specific standard products，ASSP)。按设计方法分类，可以分为全定制 (full custom) 集成电路、半定制 (semi-custom) 集成电路和可编程 (programmable) 集成电路。

表 10.1 划分集成电路规模的标准

类别	数字集成电路		模拟集成电路
	MOS 集成电路	双极型集成电路	
SSI	<100	<100	<30
MSI	$10^2 \sim 10^3$	100~500	30~100
LSI	$10^3 \sim 10^5$	500~2000	100~300
VLSI	$10^5 \sim 10^7$	>2000	>300
ULSI	$10^7 \sim 10^9$		
GSI	$>10^9$		

集成电路的制造工艺十分复杂，简单地说，就是指在衬底材料上，通过各种方法形成不同的"层"，在选定区域掺入杂质，以改变半导体材料的导电性能，形成半导体器件的过程。这个过程需要通过许多步骤才能完成，从晶圆片到集成电路成品大约需要数百道工序。通过这复杂的一道道工序，就能够在一块微小的芯片上集成成千上万甚至上亿个晶体管。

集成电路的种类繁多，以构成电路基础的晶体管来区分，主要有双极型集成电路和 MOS 集成电路两类，前者以双极型平面晶体管为主要器件，后者以 MOS 场效应晶体管为主要器件。在早期的集成电路生产中双极型工艺曾是唯一可能的工艺，随着工艺的进步，PMOS 和 NMOS 工艺先后出现并超过了双极型工艺，随后 CMOS 工艺出现并开始占主导地位。尽管在集成度方面远落后于 CMOS 工艺，但是双极型工艺凭借其高速度、高跨导、低噪声及较高的电流驱动能力等优势发展依旧很快，目前主要的应用领域是模拟和超高速集成电路。

CMOS 工艺是在 PMOS 工艺和 NMOS 工艺的基础上发展起来的，其特点是将 NMOS 器件和 PMOS 器件同时制作在同一块硅衬底上。CMOS 工艺具有集成度高、设计灵活、抗干扰能力强、低功耗等优点，是当代超大规模集成电路制造广泛采用的主流工艺。CMOS 工艺主要分为 p 阱工艺、n 阱工艺和双阱工艺，其中 n 阱工艺由于工艺简单、电路性能较 p 阱工艺更优而获得广泛应用。双阱工艺是在高阻的硅衬底上，同时形成 p 阱和 n 阱，将 NMOS 管和 PMOS 管分别做在 p 阱和 n 阱中，使 CMOS 电路达到最优的特性，并适合于高密度的集成，但其工艺较复杂[1]。

10.2 数字集成电路

10.2.1 数字集成电路的设计方法

数字集成电路设计是一门不断发展的学科，设计过程较为复杂。就设计方法而言，可以划分为同步设计和异步设计。同步设计中主要存储器件是触发器，由统一的全局时钟触发；异步设计中，用于存储的单元一般为锁存器，用"握手"实现对数据流测控制，异步电路是电平敏感的电路。目前，市场上绝大多数数字电路的设计都采用同步电路的设计方法；随着频率、功耗等方面需求的增大，异步设计以及 GALS(全局异步局部同步) 设计开始不断走进人们的视线 [2]。

"同步"是相对于"异步"而言的，称之为"同步"是因为系统中存储单元 (触发器) 的状态是由统一的时钟触发而改变的。

同步系统中的基本存储单元是触发器 (flip-flop)，存储单元存储状态的改变是在时钟沿的控制下完成的，因此同步电路具有许多优点：

(1) 同步电路比较容易使用触发器的异步清零/置位端口，保证了各个存储单元具有相同的初始态。

(2) 同步电路中各个存储单元的状态只在时钟沿到来时发生改变，之后会保持稳定，这在很大程度上避免了工艺、温度等对电路的影响，并能消除毛刺，使设计稳定可靠。

(3) 同步电路设计容易实现流水线结构。在 CPU 等功能模块的设计中，经常会使用到流水线的设计方法，并以此来提高芯片的效率、大幅提高芯片的运行速度。

(4) 同步电路的设计有着相应的电子设计自动化 (EDA) 软件的支持，比如功能仿真工具 modelsim、Quartus II、NC_Verilog，综合工具 Design Compiler，自动布局布线工具 SOC Encounter、Astro，静态时序分析工具 Prime Time 等，可以说，同步电路的设计方法以及相应的 EDA 软件均较为成熟，为设计提供了很大的便利，在简化设计、加快设计进度的同时，保证了设计的准确性。由于同步电路的种种优点，现在商业化的芯片，大都采用同步设计方案。

同步设计方法是一种相对比较成熟的数字电路设计方法，在对于功耗、频率等特性参数要求越来越高的情况下，同步设计方法也出现了一些问题，比如功耗大、速度慢、电磁噪声辐射大、时钟树插入困难等。在不断对同步设计方法优化的同时，一些设计人员开始寻找别的设计方案，希望能从根本上解决同步系统的问题。

异步电路中的基本存储单元主要是锁存器，它是一种电平敏感的存储器件。在异步系统设计中，相邻存储器之间用"握手"的机制来实现对数据的传输进行控制，常用的协议有捆绑数据协议、4 相双轨协议及 2 相双轨协议等。异步设计方法的缺点是用于实现"握手"功能的控制电路部分占用的芯片面积大、功耗大，且异步系统的 EDA 设计工具相对缺乏。

GALS 设计的设计思想结合了同步设计和异步设计的优点，相互取长补短。该设计方法仍处于理论研发阶段，相应的 EDA 工具还很不成熟，但其发展前景良好。

10.2.2 数字集成电路的基本单元电路

数字集成电路按照其内部有源器件的不同分类可以分为两类：双极型晶体管集成电路和 MOS 晶体管集成电路。因此，这里对基本单元电路的介绍主要包括晶体管–晶体管逻辑 (transistor-transistor logic，TTL) 门、发射极耦合逻辑 (emitter coupled logic，ECL) 门以及 CMOS 逻辑门[3-6]。

1) TTL 门

a. TTL 与非门。

长期以来，数字电路设计者把如图 10.5 所示的电路称为标准的 TTL 与非门电路形式，它是 54/74 系列电路的基本单元。

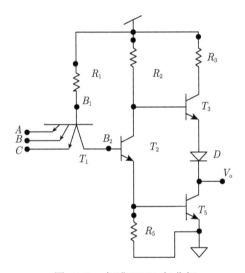

图 10.5 标准 TTL 与非门

该电路的输入极由多发射极晶体管 T_1 和电阻 R_1 组成，其作用是对输入信号进行逻辑与。多发射极晶体管相当于是将多个晶体管的基极和集电极分别短接在一起，用各个发射极作为逻辑门的信号输入端。T_1 的等效电路如图 10.6 所示。

由图 10.6 可知，当输入信号中有低电平时，B_1 节点的电压为逻辑低电平；只有当所有的输入信号都是高电平时，B_1 节点的电压才为逻辑高电平，T_1 晶体管实现了逻辑与的关系。

该电路的中间级由 T_2、R_2 和 R_4 组成。它的作用是从 T_2 的集电极和发射极同时输出两个电压变化方向相反的信号，分别作为输出级 T_3 和 T_5 的驱动信号，同时，T_2 管将输入级的输出电流放大，以给 T_5 管提供足够大的基极驱动电流。

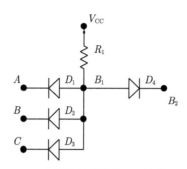

图 10.6　多发射极晶体管 T_1 的等效电路

输出级由 R_3、T_3、二极管和 T_5 组成，其中 R_3、T_3 和二极管组成的连接与 T_5 晶体管形成推挽式输出结构，不仅降低输出阻抗，提高带负载能力，而且还可以提高电路的工作速度，降低功耗。

标准单元 TTL 与非门电路的结构简单，推挽式的输出结构能够有效地降低电路的功耗，缺点是电路中的二极管在导通瞬间会存在大量的超量存储电荷，在由导通态变为截止态时由于本身又没有泄放通路，只能靠自身慢慢复合消失，因此降低了整个电路的工作速度，使电路的电压传输曲线矩形性变差，即抗干扰能力较差，因此产生了改进的五管单元和六管单元结构，这里不再详细解释。

b. TTL 或非门。

图 10.7 为两输入端或非门电路，其中 R_1、T_1、T_2 与 R_2、T_3、T_4 组成了相同的电路结构，只要 A、B 中有一个为高电平，则 T_2、T_4 中至少有一个导通，促使 T_5 导通，T_6 和二极管 D 截止，输出为低电平；只有当 A、B 全部为低电平时，T_2、T_4 都截止，T_5 也截止，T_6 和 D 导通，输出为高电平，因此该电路实现的逻辑关系为或非结构。

值得注意的是，当或非门的输入端并联使用时，总的输入电流等于各个输入端电流之和；而当与非门的输入端并联使用时，总的高电平输入电流等于各个输入端电流之和，总的低电平输入电流等于单个输入端的电流。

c. OC 门。

在 TTL 电路中，与非门是目前大量生产和使用的门电路，但在实际应用中往往需要各种功能的电路结构。为了解决实际需求，一方面在 TTL 与非门中增加扩展器，以实现增加输入端的个数级以及实现输出逻辑功能的扩展；另一方面是生产其他功能的 TTL 门电路，如 OC (open collector) 门等。

上述介绍的推挽式输出结构的 TTL 门电路的输出端不能直接连在一起实现"线与"，因为当两个门直接连接时，如果一个门的输出为高电平，另一个的输出为低电平，就会有一个很大的电流从高电平输出门中的二极管 D 流向低电平输

10.2 数字集成电路

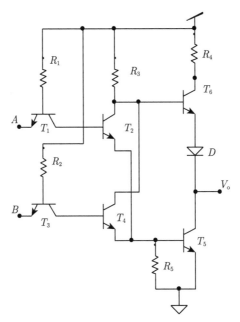

图 10.7　TTL 或非门电路

出门中的晶体管 T_5，这个大电流不仅会使原本的输出高电平降低、输出低电平抬高，而且会使门电路因为电流过大而损坏 T_5 管，甚至造成逻辑混乱。采用 OC 门电路可以有效地解决这个问题。

OC 门是将标准 TTL 与非门单元中的高电平输出部分去掉，低电平输出管 T_5 的集电极是开路的，其电路图和逻辑符号如图 10.8 所示。实际应用时须将 OC 门的输出端用导线连在一起，接到一个公共的上拉电阻 R_1 上，以实现"线与"。

d. 三态门。

普通的 TTL 门的输出只有两个状态——高电平和低电平。三态门 (third state logic gate) 除了有高电平输出和低电平输出外，还有第三种状态的输出——禁止态，这时输出端相当于悬空，所以三态门的特点是允许把多个三态门的输出端连接在一条公共母线上，使总线结构分时多路通信得以实现。

三态门的应用有两个，其中一个是总线结构，三态门在计算机系统中得到了广泛的应用，其中一个重要用途就是构成数据总线。利用分时传送原理，可以实现多组三态门挂在同一个数据总线上进行数据传送，在某一时刻只允许一组三态门的输出可以传输到数据总线上。三态门的另外一个应用是双向传输，在不同的使能信号的控制下，可以现实数据从 A 端传送到 B 端，也可以从 B 端传送到 A 端。

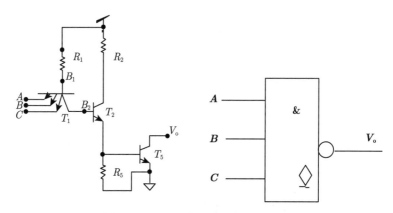

图 10.8　OC 门的电路图和逻辑符号

图 10.9 是一种三态门的电路图和逻辑符号图。从电路图中可以看出,当 EN = 1 时,二极管 D_1 截止,此时电路与原来的 TTL 电路的工作原理相同,输出取决于输入信号 A 和 B,实现与非的逻辑功能。当 EN = 0 时,二极管 D_1 导通,T_2 的集电极为低电平,T_3 和二极管 D 截止,同时,多发射输入管中由于 EN = 0,T_1 的基极电压为逻辑低电平,T_2、T_5 管也截止,输出 V_o 为高阻态。

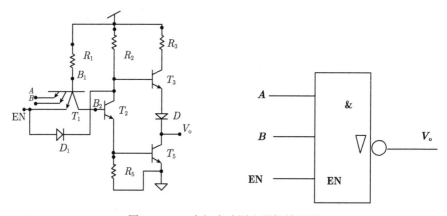

图 10.9　三态门电路图和逻辑符号图

2) ECL 门

TTL 电路中的晶体管都会进入饱和态,因此使得电路的工作速度受到了一定的限制,为了适应数字系统在超高速方面的需求,人们以非饱和形式对开关电路进行了研究。1962 年,美国摩托罗拉公司制成第一个电流型逻辑电路——发射极耦合逻辑 (ECL) 集成电路。其中的晶体管只工作在截止区和放大区,不会进入饱和区,因而晶体管的基区没有多余的存储电荷,晶体管基本没有存储时间,且电路的输入/输出逻辑幅度小，因而 ECL 电路的速度很高，这种电路的平均延迟时间可以达到几纳秒甚至亚纳秒数量级。

从某种意义上来说，ECL 电路开关速度的提高是以牺牲功耗为代价的，它空载时每门的平均功耗约为 25 mW，比 TTL(10 mW) 电路要大很多。近年来，通过改进电路结构和采用新的工艺，ECL 电路的平均延迟时间保持在亚纳秒数量级的同时，电路的功耗可以下降至几毫瓦的数量级，使 ECL 电路成为数字系统中无以匹敌的重要角色。

典型的 ECL 门电路图和逻辑符号图如图 10.10 所示，ECL 的基本逻辑门是或/或非门，由电流开关、射极跟随器和参考电压源三部分组成。电流开关部分由 T_{1A}、T_{1B}、T_2、R_{C1}、R_{C2} 和 R_E 组成射极耦合电流开关，它是 ECL 电路的核心部分，T_2 为定偏管。T_{1A}、T_{1B} 并联电路完成"线或"的逻辑功能，即 A、B 中有一个及以上为高电平时，T_{1A}、T_{1B} 的发射极为相对高电平 (或逻辑)，T_{1A}、T_{1B} 的集电极为相对低电平 (或非逻辑)，定偏管 T_2 的集电极为相对高电平 (或逻辑)。T_5、D_1、D_2，$R_1 \sim R_3$ 构成参考电压源，为定偏管 T_2 的基极提供偏置电压，它决定着电路逻辑电平的位置、阈值电压和抗干扰能力。T_3 和 T_4 为射极开路的射极输出器，作为整个电路的输出级，保持输出的相位和逻辑关系不变，并进行电平位移，以使电路的输入和输出电平相匹配。

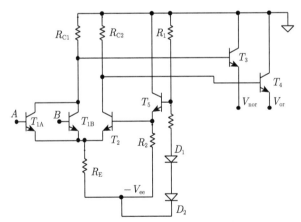

图 10.10　ECL 门电路结构和逻辑符号图

与 TTL 电路相比，ECL 电路具有以下优点：

(1) 由于 ECL 门电路中的晶体管为非饱和态，且电路中的电阻取值较小，逻辑高低电平变化幅度小，因而其工作速度是各种集成门电路中最高的；

(2) 同时具有或和或非两个互补性的输出，使用方便、灵活；

(3) 输出端采用射极跟随器，输出电阻低，带负载能力强，扇出系数可以达到

25~100;

(4) 电路内部的开关噪声低。

ECL 电路的主要缺点是噪声容限低、电路功耗大以及输出电平的稳定性较差。

3) CMOS 逻辑门

MOS 管是金属–氧化物–半导体场效应管 (metal-oxide-semiconductor field-effect-transistor) 的简称，属于单极型晶体管。以 MOS 管作为开关元件的门电路称为 MOS 门电路。MOS 门电路与 TTL 门电路能够完成相同的逻辑功能，但性能却迥异，MOS 逻辑门的功耗低，便于大规模集成，但速度低于 TTL 逻辑门电路。现在，单片机中几乎都是 MOS 器件。

a. CMOS 反相器。

CMOS 反相器由一对增强型的 NMOS 和 PMOS 管构成，其电路如图 10.11 所示。其中 M_n 为驱动管，M_p 为负载管。两个 MOS 管的栅极连在一起作为反相器电路的输入端，漏极连在一起作为反相器的输出端，NMOS 管的源极接电路的低电平，PMOS 管的源级接电路的高电平 V_{dd}，NMOS 管的衬底接电路的最低电平，PMOS 管的衬底接电路的最高电平，V_{dd} 需要大于 M_n 管和 M_p 管的开启电压绝对值之和。

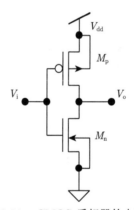

图 10.11　CMOS 反相器的电路图

如果 M_n 和 M_p 的阈值电压分别为 V_{tn} 和 V_{tp}。当输入电压为低电平 0 V 时，M_n 管中 $V_{gsn}=0$ V，小于其阈值电压 V_{tn}，因此 M_n 截止；此时 $|V_{gsp}| > |V_{tp}|$，M_p 管导通，输出电压为高电平 V_{dd}。当输入电压为高电平 V_{dd} 时，M_n 管中 $V_{gsn} > V_{tn}$，M_n 导通，此时 $|V_{gsp}| < |V_{tp}|$，M_p 管截止，输出电压为低电平 0 V。该电路实现了逻辑非的功能。用 Spice 软件对该电路进行模拟仿真，其网单文件如下所示：

M1 Vo Vi Gnd Gnd NMOS L=2u W=22u AD=66p PD=24u AS=66p PS=24u

10.2 数字集成电路

M2 Vo Vi Vdd Vdd PMOS L=2u W=22u AD=66p PD=24u AS=66p PS=24u

Vdd Vdd gnd 5

Vi Vi gnd pulse(0 5 20n 1n 1n 20n 40n)

.include ml2_125.md

.tran 1n 80n

.print tran Vi Vo

该 Spice 网单文件中表明：电源电压 V_{dd} 为 5 V，低电平为 0 V，输入信号 V_i 为脉冲信号，其高电平为 5 V，低电平为 0 V，延迟时间为 20 ns，上升时间和下降时间都是 1 ns，脉冲宽度为 20 ns，脉冲周期为 40 ns。MOS 管的模型参数文件为 ml2_125.md 文件，采用的是 1.25 μm 的 CMOS 工艺，查看输入、输出信号随时间的变化波形其仿真波形如图 10.12 所示。

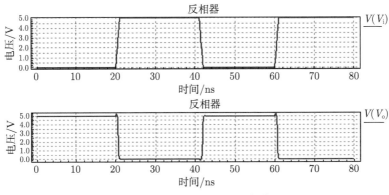

图 10.12 CMOS 反相器的仿真波形图

b. CMOS 与非门电路。

CMOS 与非门电路图如图 10.13 所示，其中 M_{n1} 和 M_{n2} 是串联的 (相当于两个开关串联)，M_{p1} 和 M_{p2} 是并联的 (相当于两个开关并联)。其中一对管 M_{n1} 和 M_{p1} 的栅极短接作为输入端 A，另一对管 M_{n2} 和 M_{p2} 的栅极短接作为输入端 B，V_o 为输出端。

其工作原理如下：

(1) 当 $A=B=0$ 时，M_{n1} 和 M_{n2} 截止，M_{p1} 和 M_{p2} 导通，输出为高电平；

(2) 当 $A=0$，$B=1$ 时，M_{n2} 导通，M_{n1} 截止，同时 M_{p1} 导通，M_{p2} 截止，输出为高电平；

(3) 当 $A=1$，$B=0$ 时，M_{n1} 导通，M_{n2} 截止，同时 M_{p2} 导通，M_{p1} 截止，输出为高电平；

(4) 当 $A=B=1$ 时，M_{n1} 和 M_{n2} 都导通，M_{p1} 和 M_{p2} 都截止，输出为低电平。

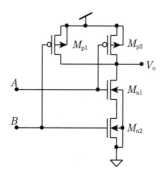

图 10.13　CMOS 与非门电路图

综上所述，该电路实现了与非门的功能。其 Spice 网单文件如下所示：
M1 Vo A N2 Gnd NMOS L=2u W=22u AD=66p PD=24u AS=66p PS=24u
M2 N2 B Gnd Gnd NMOS L=2u W=22u AD=66p PD=24u AS=66p PS=24u
M3 Vo A Vdd Vdd PMOS L=2u W=22u AD=66p PD=24u AS=66p PS=24u
M4 Vo B Vdd Vdd PMOS L=2u W=22u AD=66p PD=24u AS=66p PS=24u
VA A gnd pulse(0 5 0n 1n 1n 20n 40n)
VB B gnd pulse(0 5 0n 1n 1n 50n 100n)
Vdd Vdd gnd 5
.include ml2_125.md
.tran 1n 90n
.print tran A B Vo

该 Spice 网单文件中表明：电源电压 V_{dd} 为 5 V，低电平为 0 V，输入信号 V_A 为脉冲信号，其高电平为 5 V，低电平为 0 V，延迟时间为 0 ns，上升时间和下降时间都是 1 ns，脉冲宽度为 20 ns，脉冲周期为 40 ns，输入信号 V_B 也是脉冲信号，其高电平为 5 V，低电平为 0 V，延迟时间为 0 ns，上升时间和下降时间都是 1 ns，脉冲宽度为 50 ns，脉冲周期为 100 ns。MOS 管的模型参数文件为 ml2_125.md 文件，采用的是 1.25 μm 的 CMOS 工艺，查看输入、输出信号随时间的变化波形，其仿真波形如图 10.14 所示。

c. CMOS 或非门电路。

CMOS 或非门电路图如图 10.15 所示，其中 M_{n1} 和 M_{n2} 是并联的（相当于两个开关并联），M_{p1} 和 M_{p2} 是串联的（相当于两个开关串联）。其中一对管 M_{n1} 和 M_{p1} 的栅极短接作为输入端 A，另一对管 M_{n2} 和 M_{p2} 的栅极短接作为输入端 B，V_o 为输出端。

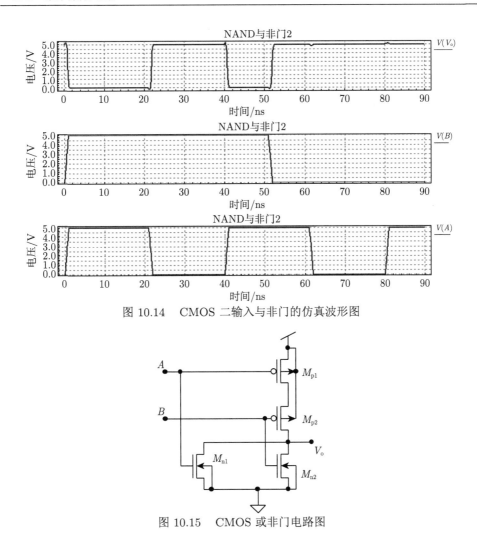

图 10.14　CMOS 二输入与非门的仿真波形图

图 10.15　CMOS 或非门电路图

其工作原理如下：

(1) 当 $A=B=0$ 时，M_{n1} 和 M_{n2} 截止，M_{p1} 和 M_{p2} 导通，输出为高电平；

(2) 当 $A=0$，$B=1$ 时，M_{n2} 导通，M_{n1} 截止，同时 M_{p1} 导通，M_{p2} 截止，输出为低电平；

(3) 当 $A=1$，$B=0$ 时，M_{n1} 导通，M_{n2} 截止，同时 M_{p2} 导通，M_{p1} 截止，输出为低电平；

(4) 当 $A=B=1$ 时，M_{n1} 和 M_{n2} 都导通，M_{p1} 和 M_{p2} 都截止，输出为低电平。

综上所述，该电路实现了或非门的功能。其 Spice 网单文件如下所示：

M1 Vo A Gnd Gnd NMOS L=2u W=22u AD=66p PD=24u AS=66p PS=24u

M2 Vo B Gnd Gnd NMOS L=2u W=22u AD=66p PD=24u AS=66p PS=24u
M3 Vo B N4 Vdd PMOS L=2u W=22u AD=66p PD=24u AS=66p PS=24u
M4 N4 A Vdd Vdd PMOS L=2u W=22u AD=66p PD=24u AS=66p PS=24u
VA A gnd pulse(0 5 0n 1n 1n 20n 40n)
VB B gnd pulse(0 5 0n 1n 1n 50n 100n)
Vdd Vdd gnd 5
.include ml2_125.md
.tran 1n 90n
.print tran A B Vo

该 Spice 网单文件中电源电压、输入信号、使用的 MOS 管模型参数等都与二输入与非门的相同，这里不再赘述，查看输入、输出信号随时间的变化波形，其仿真波形如图 10.16 所示。

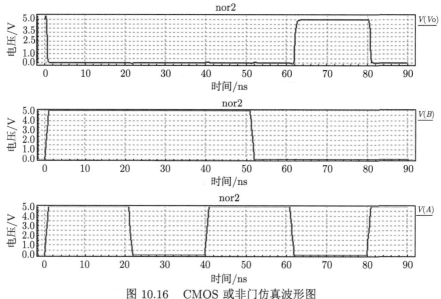

图 10.16　CMOS 或非门仿真波形图

d. CMOS 与门。

CMOS 与门可以由与非门和反相器级联构成，如图 10.17 所示。

e. CMOS 或门。

CMOS 或门可以由或非门和反相器级联构成，如图 10.18 所示。

f. CMOS 与或非门。

如果想要实现 $V_o = \overline{AB + CD}$ 所示的与或非逻辑关系，可以采用图 10.19 所示的电路结构。该电路中标识相同节点名称的节点连接在一起，只有输入 A、B

都为高电平或 C、D 都为高电平时,输出才为低电平;否则,输出都为高电平,实现了与或非的逻辑关系。

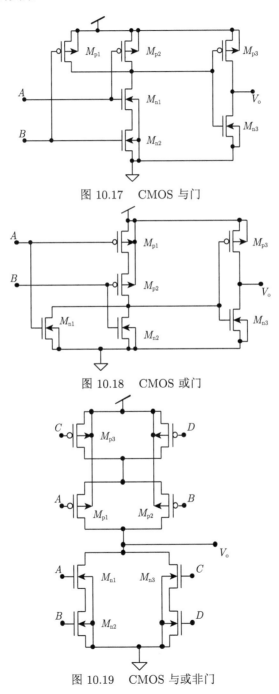

图 10.17　CMOS 与门

图 10.18　CMOS 或门

图 10.19　CMOS 与或非门

g. CMOS 传输门。

CMOS 传输门与反相器一样，是构成各种逻辑电路的一种基本单元电路，其电路如图 10.20 所示。在控制信号 C 和 $CB(CB$ 为 C 的非信号) 的作用下实现信号的传送或中断。由于 MOS 管的源、漏极是可以互换的，因此信号可以由 A 向 B 传送，也可以由 B 向 A 传送。传送的信号可以是数字信号，也可以是模拟信号，又被称为模拟开关。

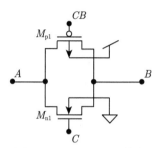

图 10.20 CMOS 传输门

当 $C=1$ 时，$CB=0$，M_{n1} 和 M_{p1} 都导通，可以得到 $B=A$，A 可以是 V_{dd} 与 V_{gnd} 之间的任意电压值；或者得到 $A=B$，B 是 V_{dd} 与 V_{gnd} 之间的任意电压值。

当 $C=0$ 时，$CB=1$，M_{n1} 和 M_{p1} 都截止，输入与输出之间断开，不能实现信号的传送。

h. CMOS 漏极开路门 (open drain，OD)。

与 TTL 电路中的 OC 门一样，CMOS 门的输出电路结构也可以是漏极开路的连接形式。它可以实现"线与"的功能，常用作输出缓冲/驱动器，或者用作输出电平转换器。图 10.21 所示的是一个二输入与非缓冲/驱动器电路，其输出是一个漏极开路的 NMOS 管，须外接上拉电阻 R 和电源 V_{dd2} 才能正常工作，实现的逻辑关系为：$V_o = \overline{A \cdot B}$。

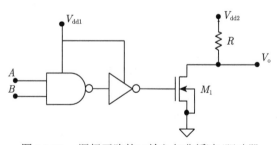

图 10.21 漏极开路的二输入与非缓冲/驱动器

i. CMOS 三态门。

普通逻辑门只能输出高电平和低电平两种逻辑状态。而三态门，除了可以输出正常的高低电平外，还有第三种状态输出，这种状态输出被称为高阻态。典型的 CMOS 三态门电路如图 10.22 所示。

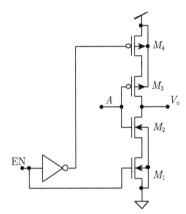

图 10.22　典型的 CMOS 三态门电路

该电路中，当 EN = 1 时，M_1 和 M_4 都导通，M_2 和 M_3 构成 CMOS 反相器，实现 $V_o = \bar{A}$ 的逻辑功能；当 EN = 0 时，M_1 和 M_4 都截止，输出与电源电压和低电平之间都断开，输出为高阻态。

10.2.3　数字集成电路中的时序电路

1. 时序电路的分类

数字电路的核心组成部分除了基本的门电路单元 (如与门、或门、非门等) 外，还包含更为复杂的时序电路。时序电路与组合逻辑门电路不同，他的输出不仅取决于输入信号，还依赖于电路的历史状态，这种状态的记忆功能是通过存储电路来实现的。在时序电路中，触发器是最基本的存储单元，用于保存电路的当前状态。每个触发器能够存储一个二进制位 (0 或 1)，并且可以在时钟信号的控制下更新其状态。因此，一个触发器本身就是最简单的时序逻辑电路，它能够实现最基本的状态记忆功能。通过将多个触发器与门电路组合，可以构建更复杂的时序电路。

按照功能划分，时序电路可以分为计数器、寄存器、移位寄存器、随机存储器和顺序脉冲发生器等。

按照触发器状态变化的特点，时序电路可以分为同步时序电路和异步时序电路。其中，同步时序电路中所有触发器的始终端由同一时钟脉冲驱动，所有触发器的状态变化都与所施加的时钟脉冲信号同步；异步时序电路中没有统一的时钟脉冲，时钟脉冲信号只驱动一部分触发器，其他触发器靠输入信号或时序逻辑内部产生的信号去驱动，因此各触发器的翻转不是同时进行的。

按照输出信号的特性，时序电路可以分为米利 (Mealy) 型和摩尔 (Moore) 型。米利型时序电路的输出状态与输入变量和触发器状态有关；摩尔型时序电路的输出状态与触发器状态有关，与输入变量无关。

时序逻辑电路还有许多种分类方法，比如，按照构成的元器件类型可以分为 TTL 和 CMOS 时序电路；按照集成度可以分为 SSI、MSI、LSI 及 VLSI 等；按照是否编程分为可编程和不可编程时序电路等。

2. 时序电路的分析方法

时序电路的分析就是根据给定的时序逻辑电路图，通过电路分析，确定电路的状态表、状态图或时序波形，从而确定电路的逻辑功能及工作特点。

分析时序电路的一般步骤为：

(1) 根据电路写出时钟方程、驱动方程和输出方程；
(2) 求解各触发器的状态方程；
(3) 对电路状态和输出进行计算；
(4) 根据计算结果，列写状态表，画出状态图或时序图；
(5) 进行电路功能说明。

3. 计数器电路的设计仿真示例

这里以十三进制异步计数器为例来说明时序电路的设计及仿真过程。计数器是数字系统中具有记忆功能的一种电路，它用来累计输入脉冲的个数，实现计数的功能。由于触发器具有记忆功能，因此利用触发器可以构成多种形式的计数器[7]。

异步计数器的设计不同于同步计数器的设计，在选择计数器中每个触发器的时钟信号，以及化简次态卡诺图，求解状态方程时有所区别。该计数器的触发器的时钟条件为异步，计数器的增减趋势默认为加法计数器。

(1) 明确触发器类型、数目及选择状态编码。

这里选用的触发器为 JK 触发器。设 Q^n 为次态，Q 为现态，其中的 J 和 K 为两个控制端口，则 JK 触发器的特性方程为：$Q^n = J\overline{Q} + \overline{K}Q$。

当 $J=1$，$K=1$ 时，Q^n 与 Q 是互补的，所以输出状态会由 0 变为 1，或者由 1 变为 0，即触发器的状态发生翻转。

当 $J=1$，$K=0$ 时，$Q^n = Q + \overline{Q} = 1$，所以触发器的状态始终为 1。

当 $J=0$，$K=1$ 时，$Q_n = 0 + 0 = 0$，所以触发器的状态始终为 0。

当 $J=0$，$K=0$ 时，$Q^n = Q$，触发器还保持原来的状态。

10.2 数字集成电路

JK 触发器的功能表如表 10.2 所示。

表 10.2 JK 触发器功能表

J	K	Q^n	Q	备注
0	0	0	0	保持
		1	1	
0	1	0	0	置0
		1	0	
1	0	0	1	置1
		1	1	
1	1	0	1	翻转
		1	0	

由于计数器是由触发器和控制门构成的，而每个触发器可以存储一位二进制信息，所以十三进制计数器需要 4 个触发器 ($13<16=2^4$)，采用 8421 编码。

13 进制有 13 个状态，在这里分别用 $S_0 \sim S_{12}$ 表示，并建立原始状态图，如图 10.23 所示。

$$
\begin{array}{cccccccc}
& /0 & /0 & /0 & /0 & /0 & /0 & \\
S_0 & \to S_1 & \to S_2 & \to S_3 & \to S_4 & \to S_5 & \to S_6 & \\
/1\uparrow & & & & & & \downarrow/0 & \\
S_{12} & \leftarrow S_{11} & \leftarrow S_{10} & \leftarrow S_9 & \leftarrow S_8 & \leftarrow S_7 & & \\
& /0 & /0 & /0 & /0 & /0 & &
\end{array}
$$

图 10.23 十三进制异步计数器原始状态图

(2) 确定十三进制异步计数器的状态图及状态转换表。

状态编码选用 8421BCD 编码，1101~1111 四个为未用码组。编码后的状态图如图 10.24 所示。

$$
\begin{array}{cccccccc}
& /0 & /0 & /0 & /0 & /0 & /0 & \\
0000 & \to 0001 & \to 0010 & \to 0011 & \to 0100 & \to 0101 & \to 0110 & \\
/1\uparrow & & & & & & \downarrow/0 & \\
1100 & \leftarrow 1011 & \leftarrow 1010 & \leftarrow 1001 & \leftarrow 1000 & \leftarrow 0111 & & \\
& /0 & /0 & /0 & /0 & /0 & &
\end{array}
$$

图 10.24 8421BCD 编码装换图

将图 10.24 的状态转换图用真值表表示，其中十三进制异步计数器的 4 位现态的原始状态分别用 Q_3、Q_2、Q_1、Q_0 表示，进位输出用 C 表示，真值表如表 10.3 所示。

表 10.3　十三进制计数器真值表

状态变化顺序	状态编码				进位输出 C	等效十进制数
	Q_3	Q_2	Q_1	Q_0		
S_0	0	0	0	0	0	0
S_1	0	0	0	1	0	1
S_2	0	0	1	0	0	2
S_3	0	0	1	1	0	3
S_4	0	1	0	0	0	4
S_5	0	1	0	1	0	5
S_6	0	1	1	0	0	6
S_7	0	1	1	1	0	7
S_8	1	0	0	0	0	8
S_9	1	0	0	1	0	9
S_{10}	1	0	1	0	0	10
S_{11}	1	0	1	1	0	11
S_{12}	1	1	0	0	1	12
S_0	0	0	0	0	0	0

(3) 选择各触发器的时钟脉冲信号。

该计数器选用下降沿触发的边沿 JK 触发器。根据电路的工作情况，画出时序图，其中，设时钟脉冲信号为 CP，4 个触发器的时钟脉冲信号分别为 CP_0、CP_1、CP_2 和 CP_3 表示，如图 10.25 所示。

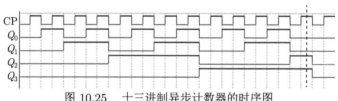

图 10.25　十三进制异步计数器的时序图

从时序图上可以看出，当时钟信号 CP 下降沿到来时，Q_0 发生翻转，所以 Q_0 的时钟信号 CP_0=CP。Q_1 的时钟信号 CP_0 选择 Q_0，这是因为，每当 Q_0 的下降沿到来时，Q_1 发生翻转，而且时钟脉冲的选择是越少越好。Q_2 的时钟信号 CP_2 选择 CP，这是因为，Q_2 一个计数周期内翻转的次数为 4 次，而第 4 次翻转时只有时钟信号 CP 处于下降沿。Q_3 在一个计数周期内翻转了两次，每次翻转都有 Q_2 处于下降沿，因此，Q_3 的时钟信号 CP_3 选择 Q_2。

(4) 求出计数器的状态方程和输出方程。

计数器的 4 位原始状态为 Q_3、Q_2、Q_1、Q_0，在这里我们设计数器的次态分别为 Q_3^n、Q_2^n、Q_1^n、Q_0^n。进位输出 C 的值由 Q_3、Q_2、Q_1、Q_0 决定。为了方便得到状态方程和输出方程，因而画出计数器电路的次态卡诺图。其中，计数器在计数过程中计不到的项作为约束项，在卡诺图中用 X 来表示。化简异步计数器的卡诺图不同于同步计数器的卡诺图，在没有时钟信号的情况下，电路状态也要当作约束项，再加上计数器中的无效状态，这两种情况均要当作约束项处理。其次态卡诺图如图 10.26 所示。

$Q_3Q_2 \backslash Q_1Q_0$	00	01	11	10
00	0001	0010	0100	0011
01	0101	0110	1000	0111
11	0000	XXXX	XXXX	XXXX
10	1001	1010	1100	1011

图 10.26　十三进制计数器的次态卡诺图

分别画出 Q_3^n、Q_2^n、Q_1^n、Q_0^n 四个逻辑函数的次态卡诺图，如图 10.27 所示。

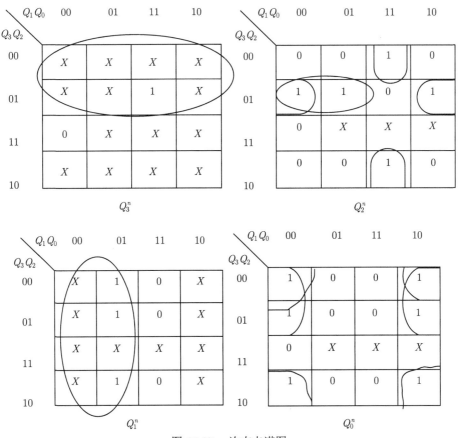

图 10.27　次态卡诺图

上述四个卡诺图分别表示 Q_3^n、Q_2^n、Q_1^n、Q_0^n 四个逻辑函数。根据卡诺图的化简，得到电路的状态方程为

$$Q_3^n = \overline{Q_3}$$

$$Q_2^n = \overline{Q_3}Q_2\overline{Q_0} + \overline{Q_3}Q_2\overline{Q_1} + \overline{Q_2}Q_1Q_0$$

$$Q_1^n = \overline{Q_1}$$

$$Q_0^n = \overline{Q_3Q_0} + \overline{Q_2Q_0}$$

进位 C 同样可以利用卡诺图化简的方法得到，如图 10.28 所示。

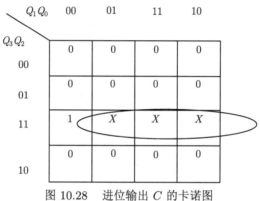

图 10.28　进位输出 C 的卡诺图

得出 C 输出方程为：$C = Q_3Q_2$。

(5) 求解驱动方程。

由于已经决定选用 JK 触发器，它是在下降沿有效。而 JK 触发器的特性方程为 $Q^n = J\overline{Q} + \overline{K}Q$，其中 Q^n 为次态，Q 为现态，J，K 为两个控制端。把电路的状态方程转化为电路的 JK 触发器特性方程的形式，即得到驱动方程：

$$Q_3^n = \overline{Q_3}$$

$$Q_2^n = Q_1Q_0\overline{Q_2} + (Q_3 + Q_0Q_1)Q_2$$

$$Q_1^n = \overline{Q_1}$$

$$Q_0^n = (\overline{Q_3} + \overline{Q_2})\overline{Q_0}$$

从而可以分别可求出各个触发器的 J、K 值：

$$J_3 = 1, \quad K_3 = 1$$

$$J_2 = Q_1Q_0, \quad K_2 = Q_3 + Q_1Q_0$$

10.3 模拟集成电路

$$J_1 = 1, \quad K_1 = 1$$
$$J_0 = \overline{(\overline{Q_3} + \overline{Q_2})} = \overline{Q_3 Q_2}, \quad K_0 = 1$$

(6) 画出计数器的逻辑电路图。

根据得到的各触发器的 J、K 值，得到该计数器的电路图，如图 10.29 所示。由图可知该电路包含有二输入与门、二输入与非门、二输入或门和 JK 触发器。

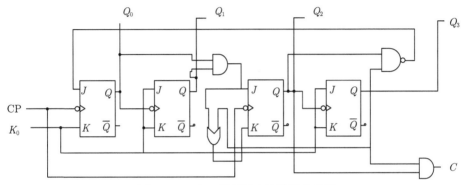

图 10.29 十三进制异步计数器电路图

(7) 自启动问题。

由于有三个无效状态，即 1101、1110、1111，把这些值代入上述驱动方程中，会分别得到 0010、0110、1000 三个结果，而这三个结果均为有效状态，故电路能够自启动。

10.3 模拟集成电路

10.3.1 模拟集成电路的设计方法

超大规模集成电路技术已经发展到可以在一块芯片上集成数百万个晶体管的水平。芯片中组成子系统的电路，尤其是数模接口部分的电路，现在已经能够以数模混合的方式集成在一起形成一个片上系统。数字系统设计自动化方面已经非常成功，但这并不适用于模拟电路设计。一般来说，模拟电路设计仍然需要"手工"进行，也就是使用全定制设计方法来实现[8,9]。

全定制方法是利用人机交互图形系统，由版图设计人员从每个半导体器件的图形、尺寸开始设计，直至整个版图的布局和布线。

全定制设计方法中，当确定了芯片的性能、允许的芯片面积成本以后，对逻辑电路等各层次进行精心的设计，对不同的方案进行反复比较，特别是对影响性能的关键因素做出分析，一旦确定以后便会进入全定制版图设计阶段。全定制版图设计的特点是针对每个晶体管进行电路和版图优化，获得最佳的性能以及最小的芯片面积[10]。

10.3.2 模拟集成电路的主要电路单元

模拟集成电路中的主要单元电路有：恒流源、带隙基准电路、运算放大器、开关电容电路、比较器，以及数模和模数转换器等[6,11,12]。下面将对其进行分开展述。

1. 恒流源电路

恒流源电路广泛地应用在模拟集成电路中，其电路设计的基本思路是从一个参考电流源"复制"电流。恒流源电路可以分为双极型电路和 MOS 型电路。由于 npn 型恒流源电路使用更加广泛，因此这里主要介绍 npn 型的电路及其改进形式。

1) 基本的双极型恒流源电路

双极型恒流源电路可以分为 npn 型和 pnp 型。当 BJT 管工作在放大区时，其集电极电流具有恒流特性，基本的双极型恒流源结构如图 10.30 所示，其中 (a) 图为基本的 npn 恒流源电路，(b) 图为基本的 pnp 恒流源电路。

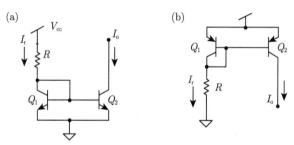

图 10.30 双极型基本恒流源结构

设定上述两个电路中 Q_1 和 Q_2 具有相同的发射结压降 V_{BE}、相同的电流放大倍数 β，相同的尺寸和参数，则有

$$I_{B1} = I_{B2}$$

$$I_{C1} = I_{C2} = I_o$$

可以得到输出电流 I_o 与输入电流 I_r 之间的逻辑关系：

$$I_o = I_r - 2I_{B2} = I_r - 2\frac{I_{C2}}{\beta} = I_r - 2\frac{I_o}{\beta}$$

有

$$I_\text{o} = \frac{I_\text{r}}{1 + \dfrac{2}{\beta}}$$

由上式可知：当 β 足够大时，$I_\text{o} \approx I_\text{r}$，因此基本的恒流源又称镜像恒流源。当 β 较小时，I_o 与 I_r 之间匹配性差，需要对电路结构进行改进，以提高其匹配性。

2) 改进的双极型恒流源电路

为了减小 β 的影响，提高 I_o 与 I_r 之间的匹配度，产生了改进的恒流源电路，如图 10.31 所示。

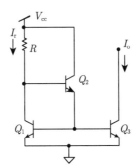

图 10.31　改进的恒流源电路

该电路中 Q_1 和 Q_3 具有相同的发射结压降 V_BE、相同的电流放大倍数 β，相同的尺寸和参数；Q_1 和 Q_2 和 Q_3 的发射结压降 V_BE 相同，电流放大倍数 β 相同，有

$$I_\text{B1} = I_\text{B3}$$

$$I_\text{C1} = I_\text{C3} = I_\text{o}$$

可以得到输出电流 I_o 与输入电流 I_r 之间的逻辑关系：

$$I_\text{o} = I_\text{r} - I_\text{B2} = I_\text{r} - \frac{I_\text{E2}}{1+\beta} = I_\text{r} - \frac{2}{1+\beta} \cdot \frac{I_\text{C3}}{\beta} = I_\text{r} - \frac{2I_\text{o}}{\beta(\beta+1)}$$

有

$$I_\text{o} = \frac{I_\text{r}}{1 + \dfrac{2}{\beta(\beta+1)}}$$

可见，与基本的恒流源电路相比，在相同的 β 下，改进的电路中输出电流 I_o 更接近于输入电流 I_r。

此外，双极型的恒流源电路还有比例恒流源、温度恒定的恒流源、小电流恒流源和威尔逊恒流源等结构形式。比例恒流源可以实现多个输出电流与输入电流之间呈一定的比例关系；温度恒定的恒流源可以实现输出电流只与供电电压有关，与温度无关；小电流恒流源可以输出一个比输入电流小很多的输出电流；威尔逊恒流源具有对称性更好、输出电阻更大、稳定性更高、输入输出电流匹配性更高等优点。这里不再详细介绍。

3) 基本的 MOS 恒流源电路

MOS 恒流源电路也分为 NMOS 恒流源和 PMOS 恒流源。根据 MOS 管的输出特性曲线可知，当 MOS 管工作在饱和区时，I_{DS} 电流具有恒流特性。基本的 NMOS 恒流源和 PMOS 恒流源电路如图 10.32 所示。

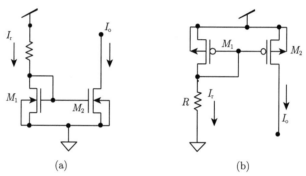

图 10.32　(a)NMOS 管和 (b)PMOS 管恒流源电路

图 10.32 中 M_1 和 M_2 构成的电路结构称为"电流镜"。M_1 管中，$V_{GS1} = V_{DS1}$，所以 M_1 工作在饱和区。假设 M_2 的 $V_{GS} - V_{th} \leqslant V_{DS}$，则 M_2 也工作在饱和区。MOS 管的栅极上没有电流，是一个电压控制器件，且只有工作在饱和区时才具有恒流特性。

根据 MOS 管的萨氏方程，有

$$I_o = I_{DS2} = \frac{\mu_N C_{OX}}{2} \left(\frac{W}{L}\right)_2 (V_{GS2} - V_{th2})^2$$

$$I_r = I_{DS1} = \frac{\mu_N C_{OX}}{2} \left(\frac{W}{L}\right)_1 (V_{GS1} - V_{th1})^2$$

由图 10.32 可知：$V_{GS1} = V_{GS2}$，取 $V_{th1} = V_{th2}$，则有 $\dfrac{I_o}{I_r} = \dfrac{(W/L)_2}{(W/L)_1}$。

当 M_1 和 M_2 的沟道宽长比相同时，输出电流 I_o 与输入电流 I_r 相等，此时的恒流源称为电流镜；当 M_1 和 M_2 的沟道宽长比不相同时，输出电流 I_o 与输入电流 I_r 呈一定的比例关系，此时的恒流源称为比例恒流源，在模拟电路中有着广泛的应用。

如果考虑 MOS 管沟道长度调制效应 (λ 为沟道长度调制系数)，则有

$$\frac{I_\mathrm{o}}{I_\mathrm{r}} = \frac{(W/L)_2}{(W/L)_1} \cdot \frac{(1+\lambda V_{\mathrm{DS}2})}{(1+\lambda V_{\mathrm{DS}1})}$$

此时，输出电流 I_o 实际上并不是一个只与 MOS 管的宽长比有关的恒流源。改善 I_o 的恒流特性的方法有以下两种：一种是减小以至于消除 M_2 的沟道长度调制效应；另一种是设定 $V_{\mathrm{DS}1} = V_{\mathrm{DS}2}$，则 I_o 与 I_r 只与 MOS 管的沟道宽长比有关，从而得到具有很好的恒流特性的电流源。

2. 基准电压源

基准电压和基准电流模块是模拟电路中的常用模块，常用来提供运放的直流偏置、模数和数模电路中的参考电平等。

常用的基准电压源电路可以分为双极型结构和 Bi-CMOS 型结构，这里主要介绍双极型基准电压源的结构和工作原理。

模拟集成电路中对偏置电压源和基准电压源的要求有：

(1) 输出电压稳定；

(2) 与温度和电源电压无关。

在集成电路中，常用的与电源电压无关的标准电压有以下三类：

(1) 三极管 BE 结的正向压降：$V_{\mathrm{BE}} = 0.6 \sim 0.8\mathrm{V}$，它的温度系数是负的，约为 $-2\mathrm{mV \cdot {}^\circ C^{-1}}$；

(2) 齐纳二极管的击穿电压 V_Z，$V_\mathrm{Z} = 6 \sim 9\mathrm{V}$，它的温度系数是正的，约为 $+2\mathrm{mV \cdot {}^\circ C^{-1}}$；

(3) 等效热电压，$V_\mathrm{t} = 26\mathrm{mV}$，它的温度系数是正的，约为 $+0.086\mathrm{mV \cdot {}^\circ C^{-1}}$。

可见，这三种标准电压的温度系数有正有负，可将这三种电压源进行不同的组合来得到不同的对电源电压和温度都不敏感的电压源和基准电压。下面我们来举例介绍。

1) 双极型三管能隙基准源

双极型三管能隙基准源的电路如图 10.33 所示，其中，Q_1、Q_2 及 R_3 构成小电流恒流源，各管的 β 都足够大，可以忽略 I_B 电流的影响。

由图 10.33 可知

$$V_\mathrm{B} = V_{\mathrm{BE}1} = V_{\mathrm{BE}2} + I_2 R_3$$

$$I_2 R_3 = \Delta V_{\mathrm{BE}}$$

忽略 I_B 的影响，有

$$V_2 = I_2 R_2 = \frac{R_2}{R_3} \cdot \Delta V_{\mathrm{BE}}$$

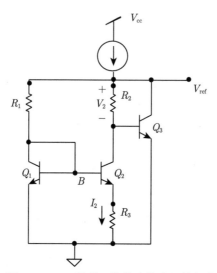

图 10.33 双极型三管能隙基准源的电路

根据 pn 结的 I-V 方程，Q_1 和 Q_2 的发射结都正偏，有

$$I_{E1} = I_{ES1}\left(e^{V_{BE1}/V_t} - 1\right) \approx A_{E1} \cdot J_R \cdot e^{V_{BE1}/V_t}$$

$$I_{E2} = I_{ES2}\left(e^{V_{BE2}/V_t} - 1\right) \approx A_{E2} \cdot J_R \cdot e^{V_{BE2}/V_t}$$

其中，A_E 为双极型三管的有效发射结面积；J_R 为双极型三管的发射极反向饱和电流密度。上述两式相除得到

$$\Delta V_{BE} = V_t \cdot \ln\frac{I_{E1}/A_{E1}}{I_{E2}/A_{E2}} = V_t \cdot \ln\frac{J_1}{J_2}$$

J 为双极型三管的发射极电流密度。由图 10.33 可知

$$V_{ref} = V_{BE3} + V_2 = V_{BE3} + \frac{R_2}{R_3} \cdot V_t \cdot \ln\frac{J_1}{J_2}$$

可见 V_{ref} 与 V_{BE} 和 V_t 有关，二者具有相反的温度系数，通过适当组合，可使输出电压 V_{ref} 基本不随温度的变化而变化。室温下，V_{ref} 的值约为 1.19 V，近似于硅的能隙电压 (1.12 V)，因此将之称为能隙基准源。

2) 双极型两管能隙基准源

双极型三管能隙基准源的输出电压是硅的能隙电压的整数倍，如要得到不是硅的能隙电压整数倍的输出电压，则可采用双极型两管能隙基准源，其电路如图 10.34 所示。

10.3 模拟集成电路

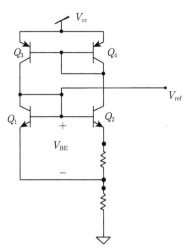

图 10.34 双极型两管能隙基准源

分析过程如下：假设各管的 β 都足够大，可以忽略 I_B 电流的影响，并设定：

$$\frac{I_{C3}}{I_{C4}} \approx \frac{I_{C1}}{I_{C2}} \approx \frac{I_{E1}}{I_{E2}} = P$$

有

$$V_{R2} = (I_{E1} + I_{E2})R_2 = (1+P)I_{E2} \cdot R_2$$

又

$$V_{BE} = V_{BE1} = V_{BE2} + I_{E2}R_1$$

得到

$$I_{E2} = \frac{\Delta V_{BE}}{R_1}$$

由上文分析可知

$$\Delta V_{BE} = V_t \cdot \ln\frac{J_1}{J_2}$$

得到该电路的输出基准电压为

$$V_{ref} = V_{BE} + (1+P) \cdot \frac{R_2}{R_1} \cdot V_t \cdot \ln\frac{J_1}{J_2}$$

3. 运算放大器

运算放大器是电子系统的通用功能模块，它能用来处理各种模拟信号，完成放大、振荡、调制和解调，以及模拟信号的相除、相乘、相减和比较等功能，广泛应用于脉冲电路，是目前应用最广、产量最大的模拟集成电路。

运算放大器 (简称运放) 实质上是一种高增益的放大器，主要由差分输入级、中间增益级、推挽式输出级和各级的偏置电路组成。

运算放大器的主要参数如下所述。

(1) 输入失调电压 V_{OS}。

当运放的输出直流电压为 0 时，在运算放大器输入端之间所加的直流补偿电压为输入失调电压。V_{OS} 是标志运算放大器对称性的一个重要参数，对于 MOS 管的输入级，它的大小一般在 ±1mV 至几十毫伏。

(2) 最大共模输入电压 V_{ic}。

使运算放大器的共模抑制比相较于规定电压下的共模抑制比下降 6 dB 时加在输入端的共模输入电压。

(3) 最大差模输入电压 V_{id}。

运算放大器两输入端所能承受的最大差模输入电压。超过此电压，差分管将进入非线性区，而使运放的性能显著恶化。

(4) 开环直流电压增益 A_{ud}。

运算放大器工作在线性区时，输出电压变化与差分输入电压变化的比值，一般用 dB(分贝) 表示，dB 被定义为 $20\lg A_{ud}$。

(5) 共模抑制比。

运算放大器差模电压增益 A_{ud} 与共模电压增益 A_{uc} 之比：

$$\text{CMRR} = 20\lg \frac{A_{ud}}{A_{uc}}$$

共模抑制比 (CMRR) 是标准运放的一个不对称参数，且有

$$\frac{dV_{OS}}{dV_{ic}} = -\frac{A_{uc}}{A_{ud}} = -\frac{1}{\text{CMRR}}$$

(6) 电源电压抑制比。

电源电压抑制比 (PSRR) 反映了运放对噪声的抑制能力，一般全差分结构具有很好的电源抑制。它定义为

$$\text{PSRR} = \frac{\text{从运放的输入到输出的开环增益}}{\text{从电源到运放输出的增益}}$$

(7) 输出压摆。

运算放大器的输出电压的最大范围，大多数带有运放的系统都需要很大的压摆以适应宽范围的信号幅度。

(8) 输出电压转换速率 S_r。

也称为压摆率，是指运放在闭环增益为 1 时，在额定条件下，当输入为大信号阶跃脉冲时，输出电压的最大变化率，反映了运放对于任意输入波形的大信号瞬态特性。

(9) −3dB 带宽。

运放的开环电压增益 A_{ud} 的半功率点的频率,即运放的开环增益下降到低频电压增益的 $\frac{1}{\sqrt{2}}$ 时所对应的频率。这是运放小信号工作时的频率。

(10) 单位增益带宽。

运放的开环增益下降到 0 dB(1 倍) 时所对应的频率。这也是运放小信号工作时的频率。

下面以一个简单的五管差分单元放大器为例来说明运放的设计及仿真过程。其电路如图 10.35 所示。该电路仿真时采用的是 0.5μm 的 CMOS 工艺,其工艺参数有:$k'_n = \mu_n C_{ox} = 110 \mu A \cdot V^{-2}$, $k'_p = \mu_p C_{ox} = 59 \mu A \cdot V^{-2}$, $\lambda_n = 0.1V$, $\lambda_p = 0.2V$, $V_{tn} = 0.755V$, $V_{tp} = -1.02V$; $V_{dd} = 2.5 V$, $V_{ss} = -2.5V$, $V_b = -1.62V$。

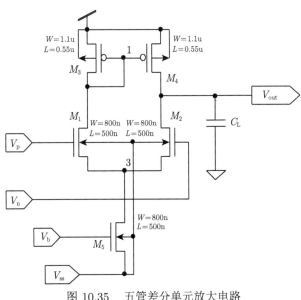

图 10.35 五管差分单元放大电路

设计技术指标要求:

放大倍数 A_v=20 V/V;

负载电容 C_L=5 pF;

转换速率 $S_R \geqslant 5V/\mu s$;

增益带宽 GB$\geqslant$200 kHz;

功耗 $P_{diss} \leqslant 1$ mW;

共模电压输入范围为:−1 V$\leqslant$ICMR$\leqslant$2 V。

计算:

(1) 计算电路中所有 MOS 管的沟道 $\left(\dfrac{W}{L}\right)$ 的值；

(2) 利用计算结果对电路用进行 Spice 模拟仿真，查看电路的放大倍数。

该电路的分析计算过程如下所述。

(1) 根据转换速率的要求，确定 M_5 的漏电流 I_5 的范围。

$$\mathrm{SR} = \dfrac{I_5}{C_L} > 5\mathrm{V}/\mu\mathrm{s}$$

$$I_5 > 25\mu\mathrm{A}$$

(2) 根据功耗的要求，确定 M_5 的漏电流 I_5 的范围。

$$P_{\mathrm{DISS}} = U_{总} * I_5 \leqslant 1\mathrm{mW}$$

$$I_5 \leqslant 20\mu\mathrm{A}$$

(3) 根据单位增益带宽，确定 M_5 的漏电流 I_5 的范围。

$$\mathrm{GB} = \dfrac{1}{R_{\mathrm{OUT}}C_L} \geqslant 2\pi \times 200 \times 10^3$$

$$R_{\mathrm{OUT}} = r_{\mathrm{ds}2} + r_{\mathrm{ds}4} = \dfrac{1}{\lambda_\mathrm{n} I_{\mathrm{DS}2}} + \dfrac{1}{\lambda_\mathrm{p} I_{\mathrm{DS}4}}$$

$$I_5 \geqslant 41.87\mu\mathrm{A}$$

根据以上三步的计算结果，这里取 $I_5 = 50\mu\mathrm{A}$。

(4) 根据最大输入共模范围，计算 M_3 和 M_4 的沟道宽长比。

可知

$$V_{\mathrm{IC,MAX}} = V_{\mathrm{DD}} - V_{\mathrm{SG}3} + V_{\mathrm{THN}1}$$

$$V_{\mathrm{SG}3} = 2.5 - 2 + 0.755 = 1.255(\mathrm{V})$$

M_3 工作在饱和区有

$$I_3 = \dfrac{I_5}{2} = \dfrac{\mu_\mathrm{P} C_{\mathrm{OX}}}{2} \left(\dfrac{W}{L}\right)_3 (V_{\mathrm{SG}3} - |V_{\mathrm{THP}3}|)^2$$

代入已知的参数：$\left(\dfrac{W}{L}\right)_3 = \left(\dfrac{W}{L}\right)_4 = 15.34$，这里取 $\left(\dfrac{W}{L}\right)_3 = \left(\dfrac{W}{L}\right)_4 = 16$。

(5) 根据电路的放大倍数，确定 M_1 和 M_2 的沟道宽长比。

该电路的电压放大倍数为：$A_V = g_{\mathrm{m}1} R_{\mathrm{OUT}} = 20$，则有：$g_{\mathrm{m}1} = 150\mu\mathrm{s}$。

10.3 模拟集成电路

又 $g_{m1} = \sqrt{2\mu_N C_{OX}\left(\dfrac{W}{L}\right)_1 \dfrac{I_5}{2}}$，代入参数计算得到 $\left(\dfrac{W}{L}\right)_1 = \left(\dfrac{W}{L}\right)_2 = 4.09$，这里取 $\left(\dfrac{W}{L}\right)_1 = \left(\dfrac{W}{L}\right)_2 = 5$。

(6) 根据最小共模输入电压范围，确定 M_5 的沟道宽长比。

对于 M_1 管有

$$V_{GS1} = \sqrt{\dfrac{I_5/2}{\mu_N C_{OX}/2 \left(\dfrac{W}{L}\right)_1}} + V_{THN1} = 1.06(\text{V})$$

由最小共模输入电压范围可知

$$V_{IC,MIN} = V_{SS} + V_{DS5} + V_{GS1}$$

$$V_{DS5} = 0.44\text{V}$$

M_5 工作在饱和区，有

$$I_5 = \dfrac{\mu_N C_{OX}}{2}\left(\dfrac{W}{L}\right)_5 (V_{GS5} - V_{THN5})^2 = \dfrac{\mu_N C_{OX}}{2}\left(\dfrac{W}{L}\right)_5 V_{DS5}^2$$

计算得到 $\left(\dfrac{W}{L}\right)_5 = 4.695$，这里取 $\left(\dfrac{W}{L}\right)_5 = 5$。

(7) 将上述计算结果代入电路，对电路进行仿真，其 spice 网单文件如下：

C1 Vout Gnd 5pF
M1 1 Vp 3 Vss NMOS L=1u W=5u
M2 Vout Vn 3 Vss NMOS L=1u W=5u
M5 3 Vb Vss Vss NMOS L=1u W=5u
M4 Vout 1 Vdd Vdd PMOS L=1u W=16u
M3 1 1 Vdd Vdd PMOS L=1u W=16u
Vdd Vdd gnd 2.5
Vss Vss gnd -2.5
Vb Vb gnd -1.62
Vin Vp Vn 0.0 AC 1.0 0.0
Vcm Vn Gnd 0
.include 0.5um.md
.ac dec 20 0.1 1g
.print ac Vm(Vout)

其仿真波形如图 10.36 所示,由图可知,在以上计算结果条件下,其放大倍数约为 41,超过了设计指标,符合设计要求。

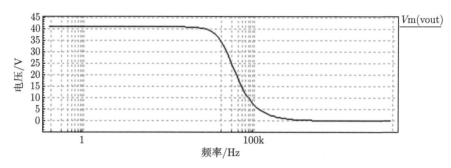

图 10.36　五管放大电路仿真波形

4. 开关电容电路

开关电容电路 (switched capacitor circuits,简称 SC 电路) 是由受时钟信号控制的开关与电容组成的电路,它利用电容器电荷的存储与转移原理来实现电路功能。自 20 世纪 70 年代末开关电容滤波器进入实用阶段以来,随着大规模和超大规模集成工艺技术的发展和 SC 电路理论研究的拓展,SC 电路不仅在模拟数据采集系统中的信号滤波方面占有重要的地位,而且已进一步应用到其他模拟信号处理领域。

MOS 开关电容电路是由 MOS 模拟开关和 MOS 电容组成,电路在时钟信号的控制下,完成电荷的存储和转换。

开关电容电路的主要优点是:①与 CMOS 工艺兼容;②高精度的时间常数;③良好的电压线性度;④良好的温度特性。主要的缺点是:①时钟馈通;②需要不相重叠的时钟信号;③信号的宽度必须小于时钟频率。

基本开关电容单元可以等效为电阻,这恰恰是实现 SC 滤波器的机理。下面来分类介绍。

1) 并联开关电容等效电阻电路

基本的并联型 MOS 等效电阻电路如图 10.37 所示。图中两个 MOS 模拟开关管 M_1、M_2 分别受两相时钟 clk 和 clkb 控制。clk 和 clkb 具有同频、相位相反、振幅相等而不重叠的特性。V_1、V_2 为两个直流电压源。

其工作原理分析如下所述。

当 clk 为高时,clkb 为低,M_1 导通,M_2 截止,电压 V_1 向电容 C_1 充电至 V_1,此时 C_1 上的电荷为:$Q_1 = C_1 \cdot V_1$。

10.3 模拟集成电路

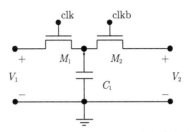

图 10.37 并联型 MOS 开关电容电路

当 clkb 为高时，clk 为低，M_2 导通，M_1 截止，电容 C_1 通过 M_2 放电到 V_2，此时 C_1 上的电荷为：$Q_1' = C_1 \cdot V_2$。

在这个过程中，通过电容 C_1 从左侧传送到右侧的电荷量 $\Delta Q = Q_1 - Q_1' = C_1(V_1 - V_2)$，假设 clk 和 clkb 的频率为 f_c(周期为 T_C)，当 f_c 远高于信号频率时，在一个周期 T_C 内，左端向右端传递的平均电流 $\bar{I}$ 为

$$\bar{I} = \frac{\Delta Q}{T_C} = \frac{C_1(V_1 - V_2)}{T_C} = \frac{V_1 - V_2}{\dfrac{1}{f_c \cdot C_1}} = \frac{V_1 - V_2}{R_{\text{eff}}}$$

可见，该电路可等效为一个电阻，其等效电阻 R_{eff} 为：$R_{\text{eff}} = \dfrac{1}{f_c \cdot C_1}$。

2) 串联开关电容等效电阻电路

基本的串联型 MOS 等效电阻电路如图 10.38 所示。同样地，两个 MOS 模拟开关管 M_1、M_2 分别受两相时钟 clk 和 clkb 控制。clk 和 clkb 具有同频、相位相反、振幅相等而不重叠的特性。V_1、V_2 为两个直流电压源。

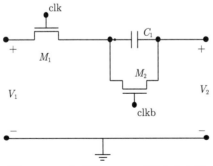

图 10.38 串联型 MOS 开关电容电路

其工作原理分析如下所述。

当 clk 为高时，clkb 为低，M_1 导通，M_2 截止，电容 C_1 上存储的电荷量为：$Q_1 = C_1 \cdot (V_1 - V_2)$。

当 clkb 为高时，clk 为低，M_2 导通，M_1 截止，电容 C_1 通过 M_2 放电，此时 C_1 上的电荷变为零。

在这个过程中，通过电容 C_1 从左侧传送到右侧的电荷量 $\Delta Q = C_1(V_1 - V_2)$，假设 clk 和 clkb 的频率为 f_c(周期为 T_C)，当 f_c 远高于信号频率时，在一个周期 T_C 内，左端向右端传递的平均电流 $\bar{I}$ 为

$$\bar{I} = \frac{\Delta Q}{T_C} = \frac{C_1(V_1 - V_2)}{T_C} = \frac{V_1 - V_2}{\dfrac{1}{f_c \cdot C_1}} = \frac{V_1 - V_2}{R_{\text{eff}}}$$

该电路也可等效为一个电阻，其等效电阻 R_{eff} 为：$R_{\text{eff}} = \dfrac{1}{f_c \cdot C_1}$。

5. 比较器

电压比较器是模拟集成电路中应用较多的电路之一，用于在模拟信号处理中比较和判断两个信号的大小，输出的是一个逻辑值 (0 或者 1)，电压比较器的电路结构、性能等与运放基本相同，所用的符号也与运放一致，如图 10.39 所示。比较器的输入端口分为同相 V_p、反相 V_n 两端，当 $V_p \geqslant V_n$ 时，比较器输出为高电平，反之则输出为低电平。

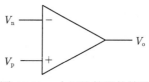

图 10.39 电压比较器的符号

比较器被广泛应用于模拟信号到数字信号的转换过程中。在模–数转换过程中，首先必须对输入进行采样。接着，经过采样的信号，通过比较器以决定模拟信号的数字值。在最简单的情况下，比较器可以作为一个 1 位模–数转换器。

电压比较器的主要参数如下所述。

(1) 开环电压增益 A_u：开环电压增益直接决定了电压判别的灵敏度，增益越高，灵敏度也越高。

(2) 输入失调电压 V_{os}：输入失调电压的大小也直接影响电压判别的灵敏度，失调越小，灵敏度越高。

(3) 输出高、低电平：一般情况下，输出高、低电平分别为电源电压的 70% 与 30%。

(4) 瞬态响应：反映了比较器的动态特性，决定了电压比较器的输出电压的高低压之间的转换时间。

10.3 模拟集成电路

(5) 传输时延 t_p：是指比较器的输入与输出响应之间存在的时延。传输时延随比较器输入幅度的变化而变化。

常用的 CMOS 电压比较器结构主要有级联反相器结构和差分放大结构。级联反相器结构多用于静态工作模式，差分放大结构既可以用作静态比较器，又可以用作动态比较器。

差分比较器的内部电路本质上是一个具有很高电压增益的运算放大器，这里以图 10.40 为例来说明比较器的工作原理。其中，参考电压 V_ref 施加在反相输入端，输入信号 V_i 施加到同相输入端。只有当 V_i 低于 0 V 的参考电压时，输出为低电平 (接近于 -15 V)；当 V_i 增大到高于 0V 时，输出为高电平 (接近于 $+15$ V)。通常情况下，参考电压 V_ref 可以是任意的电压值，可以接在反相输入端，可以接在同相输入端，输入信号接在另外一个输入端上。

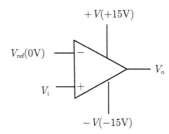

图 10.40 比较器的典型应用电路

6. 数模和模数转换器

我们生活的世界是一个模拟的世界，人类也是用模拟的方式对外界进行感知。因此，需要一座连接模拟世界和数字世界的桥梁，也就是数据转换器。能够把模拟信号转换为数字信号的电子系统统称为模数转换器 (ADC)；能够把数字信号转换为模拟的信号的电子系统统称为数模转换器 (DAC)。

1) 数模转换器

数模转换电路的作用是把数字信号转变为模拟信号，所采用的原理是线性叠加。适合于单片集成电路的基本数模转换电路，根据其工作原理可以分为三类：电流定标电路、电压定标电路和电荷定标电路。

一个电压输出的 DAC 电路的结构框图如图 10.41 所示。它由一个 N 位的码字 $(b_0, b_1, b_2, b_3, \cdots, b_{N-2}, b_{N-1})$ 和一个基准电压 V_ref 组成。b_0 被称为 "最高位 MSB"，b_{N-1} 被称为 "最低位 LSB"。输出电压的表达式为

$$V_\text{o} = K \cdot V_\text{ref} \cdot D$$

其中，K 为比例因子。码字 D 可表示为

$$D = \frac{b_0}{2^0} + \frac{b_1}{2^1} + \frac{b_2}{2^2} + \cdots + \frac{b_{N-1}}{2^{N-1}}$$

其中，N 是码字的总位数；b_{i-1}（i 从 0 到 $N-1$）为第 i 位的系数，取 0 或者 1。结合上述两式，得到 DAC 的输出电压表达式为

$$V_\text{o} = K \cdot V_\text{ref} \cdot \left(\frac{b_0}{2^0} + \frac{b_1}{2^1} + \frac{b_2}{2^2} + \cdots + \frac{b_{N-1}}{2^{N-1}} \right)$$

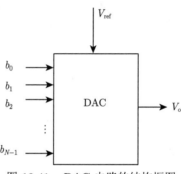

图 10.41　DAC 电路的结构框图

DAC 的特性可以分为静态特性和动态特性。先来了解 DAC 的静态特性。DAC 的精度等于实际输入码字的位数，用 N 位表示，N 为位数。对于一个理想的 3 位 DAC 而言，有 8 个可能的码字，8 个码字中每一个都有其特定的模拟输出电压，这些电平用可分辨的最小值（LSB）来区分。LSB 的值定义为：$\text{LSB} = \dfrac{V_\text{ref}}{2^N}$。

DAC 的精度是有限的，所以其最大的模拟输出电压不等于 V_ref，这个特性用 DAC 的满刻度值（FS）来描述。满刻度值定义为最大码字（$111\cdots$）和最小码字（$000\cdots$）对应的模拟输出量之差，其表达式为

$$\text{FS} = V_\text{ref} - \text{LSB} = V_\text{REF}\left(1 - \frac{1}{2^N}\right)$$

如果考虑 DAC 的平移特性，则其满刻度范围（FSR）定义为

$$\text{FSR} = \lim_{N \to \infty} \text{FS} = V_\text{ref}$$

基于以上分析，可以定义几个对于 DAC 非常重要的参数。第一个参数是量化噪声。量化噪声是指在用一个有限精度的转换器将模拟值进行数字化的过程中所存

在的固有不确定性。这个噪声是 DAC 的一个基本特性，它代表了转换器的精度限制。

DAC 的动态范围 (DR) 等于 FSR 和 LSB 的比，其表达式为

$$\mathrm{DR} = \frac{\mathrm{FSR}}{\mathrm{LSB}} = \frac{\mathrm{FSR}}{\mathrm{FSR}/2^N} = 2^N$$

用分贝的形式表示，其表达式为：$\mathrm{DB}(\mathrm{dB}) = 6.02N\mathrm{dB}$。

DAC 的信噪比 (SNR) 定义为满刻度值和量化噪声均方根值的比。量化噪声的均方根值可以通过取量化噪声平方的均值的平方根得到，其表达式为

$$\mathrm{rms}\,(量化噪声) = \sqrt{\frac{1}{T}\int_0^T \mathrm{LSB}^2 \left(\frac{t}{T} - 0.5\right)^2 \mathrm{d}t} = \frac{\mathrm{LSB}}{\sqrt{12}} = \frac{\mathrm{FSR}}{2^N\sqrt{12}}$$

因此，信噪比可以表示为

$$\mathrm{SNR} = \frac{V_\mathrm{o} \cdot \mathrm{rms}}{\mathrm{FSR}/\left(2^N\sqrt{12}\right)}$$

对于正弦波，V_o 的最大可能的均方根值等于 $(\mathrm{FSR}/2)/\sqrt{2}$ 或者 $V_\mathrm{ref}/\left(2\sqrt{2}\right)$，因此，DAC 的信噪比的最大值为

$$\mathrm{SNR} = \frac{\mathrm{FSR}/\left(2\sqrt{2}\right)}{\mathrm{FSR}/\left(2^N\sqrt{12}\right)} = \frac{2^N\sqrt{6}}{2}$$

下面来了解 DAC 的动态特性。时域或动态特性也是 DAC 特性的重要组成部分。DAC 最基本的动态特性是转换速度。转换速度是指当输入码字改变时 DAC 提供相应的模拟输出所需的时间。决定 DAC 速度的主要因素有寄生电容、增益带宽积和运算放大器的摆率。

2) 模数转换器

ADC 是 DAC 的逆过程，主要区别在于 ADC 必须进行采样，因此 ADC 是一个数据采样电路。ADC 的输入是模拟信号，通常是模拟电压，输出是数字编码。模拟输入可以是 $0 \sim V_\mathrm{ref}$ 的任意值，而数字编码被限制在固定或离散的幅度上。ADC 的结构框图如图 10.42 所示。

ADC 的主要性能参数如下所述。

(1) 精度：精度表示一个 ADC 把模拟信号转换成数字信号的有效位数。

(2) 采样速率：采样速率是指能够连续转换的采样速度，一般而言，其采样速度与转换时间的倒数相等。

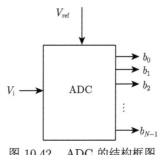

图 10.42　ADC 的结构框图

(3) 转换时间：转换时间是指包括输入信号的捕获时间在内的完成一个信号的转换所需的时间。

(4) 信噪比：ADC 的信噪比与它的精度密切相关，其关系如下：

$$SNR = 6.02N + 1.76$$

其中，N 为实测的可以分辨的有效位数；SNR 则为测量出的信噪比。

(5) 积分非线性 (INL)：积分非线性表示实际有限精度特性曲线与理想有限精度特性曲线在垂直方向上的最大差值。

(6) 微分非线性 (DNL)：微分非线性表示实际有限精度特性曲线在每个垂直台阶上测量的相邻编码之间的距离。

单片集成 ADC 主要有以下几类：积分型、伺服型、并行比较器型和逐次逼近型，其主要变换误差有量化误差和孔阑误差。

量化误差：由于模拟信号是连续的，数字信号是离散的，两者之间不可能完全对应，而会存在一定的误差，这种误差习惯上称为量化误差。

孔阑误差：由于完成一次变换需要一定时间，等到变换完成时，模拟信号的值已经不是刚开始时的值了，因此会造成一定的误差，这种误差称为孔阑误差。

10.4　Bi-CMOS 型集成电路

10.4.1　Bi-CMOS 型集成电路的设计方法

Bi-CMOS 型集成电路就是指使用 Bi-CMOS 型工艺制造出来的电路，它结合了双极型工艺和 CMOS 工艺的优点。Bi-CMOS 型集成电路是把双极型器件和 CMOS 器件同时制作在同一个芯片上，它综合了双极型器件高跨导、强负载驱动能力和 CMOS 器件高集成度、低功耗的优点，使两者互相取长补短，发挥各自的优点。Bi-CMOS 型集成电路多为模拟集成电路，因此其设计方法也多采用全定制设计方法。

10.4.2 Bi-CMOS 型集成电路的典型电路的设计及应用

Bi-CMOS 型集成电路的典型电路就是带隙 (bandgap) 基准电路。产生基准的目的是建立一个与电源和工艺无关,具有确定的温度特性的直流电压或电流[11]。由于大多数工艺参数是随温度变化的,所以如果一个基准是与温度无关的,那么通常它也是与工艺无关的。

如何产生一个对温度变化保持恒定的量呢?我们假设,如果将两个具有相反温度系数的量以适当的权重相加,那么结果就会显示出零温度系数。在半导体工艺中的各种不同器件参数中,双极型晶体管的特性参数被证实具有最好的重复性,并且具有提供正温度系数 (ΔV_{BE}) 和负温度系数 (V_{BE}) 严格定义的量。尽管 MOS 器件的许多参数已被考虑用于基准产生,但是双极型电路还是形成了这类电路的核心。

1. 负温度系数电压

V_{BE} 电压或者说 pn 结二极管的正向电压,具有负温度系数。由相关分析可知

$$\frac{\partial V_{BE}}{\partial T} = \frac{V_{BE} - (4+m)V_T - E_g/q}{T}$$

其中,$m \approx -\frac{3}{2}$, $E_g \approx 1.12\text{eV}$, V_T 具有正的温度系数。可见,V_{BE} 的温度系数与 V_{BE} 本身的大小有关,且温度系数本身与温度有关。

2. 正温度系数电压

早在 1964 年,人们就知道当两个双极型晶体管在不相等的电流密度下时,它们的 V_{BE} 之差就与热力学温度成正比,如图 10.43 所示。

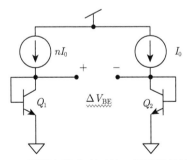

图 10.43 不同电路密度下的双极型管的输出特性

设定图 10.43 中 Q_1 和 Q_2 具有相同的参数,完全对称,且 $\beta \gg 1$(可以忽略基极电流的影响),有

$$I_{E1} \approx I_{C1} = nI_0 \approx I_{S1} \cdot e^{V_{BE1}/V_T}$$

$$I_{E2} \approx I_{C2} = I_0 \approx I_{S2} \cdot e^{V_{BE2}/V_T}$$

$$I_{S1} = I_{S2}$$

可以得到

$$\Delta V_{BE} = V_{BE1} - V_{BE2} = V_T \cdot \ln(n)$$

可见 ΔV_{BE} 表现出正温度系数，且这个温度系数与温度或集电极电流的特性无关。

$$\frac{\partial \Delta V_{BE}}{\partial T} = \frac{K}{q} \cdot \ln(n)$$

利用上面的正、负温度系数的电压，可以设计出一个零温度系数的基准电压 V_{ref}。

$$V_{\text{ref}} = a_1 V_{BE} + a_2 \Delta V_{BE}$$

如 $\dfrac{\partial V_{BE}}{\partial T} \approx -1.5\text{mV} \cdot {}^\circ\text{C}^{-1}$，$\dfrac{\partial V_T}{\partial T} \approx 0.086\text{mV} \cdot {}^\circ\text{C}^{-1}$，$a_1 = 1$，则有 $a_2 \ln(n) = \dfrac{1.5}{0.086} \approx 17.2$。这表明，零温度系数基准为：$V_{\text{ref}} = V_{BE} + 17.2 V_T$，如 $V_{BE} = 0.75\text{V}$，$V_T = 0.026\text{V}$，则 $V_{\text{ref}} \approx 1.19\text{V}$，约为 Si 的带隙电压（1.12 V），因此被称为带隙基准。

如何来实现 V_{BE} 和 ΔV_{BE} 的组合呢？一个基本的电路如图 10.44 所示。在该电路中，Q_2 管的发射结有效面是 Q_1 管的 n 倍，两管的 $\beta \gg 1$，使用某种方法强制使 $V_{o1} = V_{o2}$，由电路可知

$$V_{o1} = V_{BE1}$$

$$V_{o2} = V_{BE2} + IR$$

有 $IR = \Delta V_{BE} = V_{BE1} - V_{BE2}$。

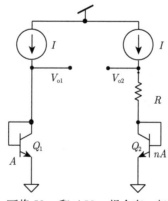

图 10.44　可将 V_{BE} 和 ΔV_{BE} 组合在一起的电路结构

10.4 Bi-CMOS 型集成电路

根据 pn 结的 I-V 方程可知

$$I_{E1} \approx I = A \cdot J_R e^{V_{BE1}/V_T}$$

$$I_{E2} \approx I = nA \cdot J_R e^{V_{BE2}/V_T}$$

式中，J_R 为两管的发射结反相饱和电流密度，设定两者相等，两式相除得到

$$\Delta V_{BE} = V_T \ln(n)$$

此时，从 V_{o2} 处得到的输出电压即为带隙基准电压：

$$V_{ref} = V_{o2} = V_{BE2} + \Delta V_{BE} = V_{BE2} + V_T \ln(n)$$

这意味着，当 $\ln(n) = 17.2$ 时，输出基准与温度无关。上述电路存在两个问题，一个是如何实现 $V_{o1} = V_{o2}$，另外一个是当 $\ln(n) = 17.2$ 时，n 将很大，Q_2 管的有效发射结面积太大，成本高。因此，需要对上述电路进行改进，如图 10.45 所示。

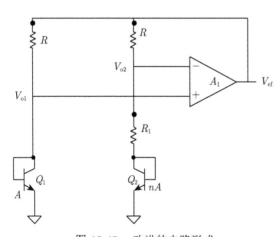

图 10.45 改进的电路形式

该电路中，应用运算放大器 A_1 来强制使 $V_{o1} = V_{o2}$，增加电阻 R 以增大 V_T 的系数，达到减小 n 的目的。由图可知

$$V_{o1} = V_{o2} = V_{BE1} = V_{BE2} + V_{R_1}$$

由之前的分析可知

$$\Delta V_{BE} = V_{BE1} - V_{BE2} = V_T \ln(n) = I_{R_1} R_1$$

有
$$I_{R_1} = \frac{V_\text{T} \ln(n)}{R_1}$$

该电路的输出基准电压为
$$V_\text{ref} = V_\text{BE2} + \frac{R + R_1}{R_1} \cdot V_\text{T} \cdot \ln(n)$$

考虑到双极型工艺与 CMOS 工艺相兼容的问题，这里用衬底 PNP 管来代替 NPN 管。由于电路中仅仅使用了双极型管的发射结，采用衬底 PNP 管后，其基极与集电极短接后接电路的低电平。这不仅更好地应用了衬底 PNP 管本身的工作特点，又没有增加多余的工艺步骤，完全应用 CMOS 工艺便可制作出所需的带隙基准，其电路如图 10.46 所示，其分析过程与图 10.45 类似，这里不再赘述。

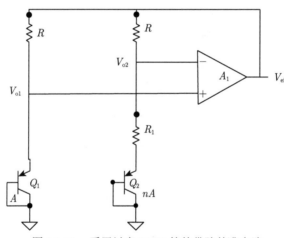

图 10.46　采用衬底 PNP 管的带隙基准电路

10.5　能量回收型集成电路的设计

10.5.1　能量回收型集成电路的原理

能量回收型集成电路又称绝热电路[13-22]。随着 CMOS 电路集成密度和速度的不断提高，功耗已经成为集成电路设计中除速度和面积之外需要考虑的另一个重要问题。低功耗设计问题已引起广泛的关注，它贯穿于从系统级到器件级的整个数字系统设计的全过程，低功耗技术已成为集成电路设计的一个紧急技术需要。近年来，研究者在电路中引入物理学中的绝热原理，从而产生了采用交变电压供电的低功耗绝热电路。此类电路在对输出节点电容充放电时，总是使 MOS 开关器件两端保持较低的电压，因此在 MOS 开关器件沟道电阻上产生的能耗是极小的一部分，而节点电容上的大部分能量将恢复至电源，以便在下一个周期重新使用，实现能量恢复，从而使电路表现出明显的低功耗特点。

绝热逻辑的原理，实际上是利用开关时间的增加来换取电路功耗的降低。下面以静态反相器来具体说明，如图 10.47 所示。

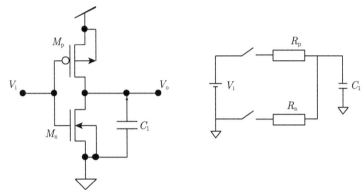

图 10.47　静态 CMOS 反相器及其等效电路

当输入为低电平时，M_p 管导通，此时电源通过 M_p 管对负载电容 C_1 进行充电，若初始状态时负载电容上不存在电荷，即电容两端的电压为零，那么在 M_p 导通瞬间，相当于电源电压全部加到了 M_p 管的沟道电阻上。随着充电过程的进行，负载电容 C_1 中积累电荷，M_p 管的沟道电阻 R_p 两端的电压差减小，最后电容充电到电源电压 V_{dd}。此过程中能量的损耗，电源提供的能量有两个去处，一部分是负载电容 C_1 中存储的电荷能量，另一部分是在对电容充放电过程中损失在 M_p 管的沟道电阻 R_p 上的热能。由电路基础知识可知，电源对负载电容 C_1 的充电过程中充电电流为

$$i(t) = \frac{dq}{dt} = \frac{V_{dd}}{R_p} e^{-\frac{t}{R_p C_1}}$$

即流过 R_p 的电流为 $i(t)$，假设充电时间为 T，那么充电过程中消耗在沟道电阻 R_p 上的能量为

$$E_R = \int_0^T i(t)^2 R_p dt = \frac{1}{2} C_1 V_{dd}^2 \left(1 - e^{-\frac{2T}{R_p C_1}}\right)$$

当 $R_p C_1 \ll T$ 时，可近似为

$$E_R = \frac{1}{2} C_1 V_{dd}^2$$

同理，可以计算出负载电容上存储的能量为

$$E_{C_1} = \frac{1}{2} C_1 V_{\mathrm{dd}}^2$$

对于传统的 CMOS 逻辑电路来说，电容 C_1 上存储的这部分能量会在输入为高电平时，M_n 管导通，负载电容 C_1 对地放电，而消耗在 M_n 管的沟道电阻 R_n 上，以热量形式散失，所以反相器翻转一次的总功耗为

$$E = E_R + E_C = C_1 V_{\mathrm{dd}}^2$$

如若用另一线性电源取代静态 CMOS 反相器中的地节点，当输入为高电平、负载电容 C_1 通过 M_n 管放电时，使这一线性电源由 V_dd 缓慢变化到 0，从而使负载电容 C_1 中的能量重新回收至电源。在能量回收过程中，M_n 管沟道电阻 R_n 两端的电压始终保持在一个很小的范围，即能量回收的放电过程也是一个绝热过程，若 M_n 的沟道电阻为 R_n，且 $R_\mathrm{p} = R_\mathrm{n}$，则放电过程的能量损耗与绝热充电的能量损耗相等。

绝热电路突破了传统 CMOS 电路中能量传输模式的局限性，使得对能量的使用不再是由 V_dd 到信号节点到地的一次性使用方式，而是由电源到信号节点再到电源的重复利用方式，这使电路表现为具有能量恢复的特点。

10.5.2 能量回收型电路的结构分类

如果按照绝热电路实现能量恢复的具体方案划分，则目前提出的绝热电路大体可以分为三大类：逐级级联收缩 (retractile cascode) 结构、可逆逻辑 (reversible logic) 结构及存储 (memory) 结构。前两种方案存在严重缺陷，因此重点讨论具有交叉耦合结构 (即存储结构) 的绝热电路。这里以四相功率时钟的高效电荷恢复逻辑 (efficient charge recovery logic, ECRL) 和功率时钟绝热电路 (power clock adiabatic circuit, CPAL) 电路为例来加以说明。

1. ECRL 电路

ECRL 结构形式简单，优点多，越来越多地受到人们的重视。图 10.48 是其反相器的电路图及其功率时钟图。

ECRL 反相器包含两个下拉求值管 M_n1、M_n2，以及两个交叉耦合的负载管 M_p1、M_p2，它是一个双端输入双端输出的结构。电路工作状态分为四个阶段：预充阶段、保持阶段、回收阶段和等待阶段。

下面来分析 ECRL 反相器的工作原理。首先假设输入端 in 为逻辑高电平，相应地，inb 为逻辑低电平。

10.5 能量回收型集成电路的设计

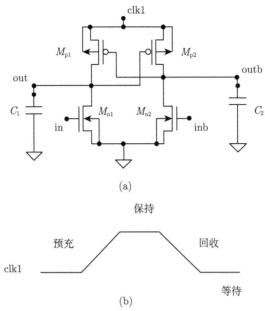

图 10.48 (a)ECRL 反相器的电路图及 (b) 其功率时钟图

当时钟进入预充阶段，clk1 由 0 缓慢上升到 V_{dd}，由于 in 为高电平，M_{n1} 导通，M_{n2} 截止，使得 out 输出端与地相连，其节点电压为 0。当 $V_{clk1} - V_{out} > |V_{TP}|$ 时，M_{p2} 开始导通，功率时钟 clk1 通过 M_{p2} 管对 outb 节点电容进行充电，随着 clk1 上升到 V_{dd} 时，outb 节点电压上升到 V_{dd}。由于在此期间 out 节点一直与地相连，其电压保持为逻辑低电平，随后电路进入保持阶段。

在保持阶段输出端的逻辑状态不变，从而在级联时为下一级提供了有效的输入。当电路进入能量回收阶段时，clk1 由 V_{dd} 线性缓慢下降，clk1 通过 M_{p2} 对 out 负载电容进行回收，当 $V_{clk1} - V_{out} < |V_{TP}|$ 时，M_{p2} 管截止，回收过程结束，电路进入等待阶段，此时输入端等待新的逻辑状态的输入，电路接着进入下一个工作周期。

在 0.5 μm 伯克利短沟道 IGFET 模型 (Berkeley short-channel IGFET model, BSIM)3v3 工艺参数下，对该电路进行模拟仿真，其仿真波形如图 10.49 所示。由图可知，该电路可以实现反相器的功能，优点是电路结构简单，面积小；缺点是在输出端存在悬空状态，能量回收阶段回收不完全，因此需要对电路进行改进。

2. CPAL 电路

针对 ECRL 电路存在悬空的问题，研究者提出了多种不多的改进结构。CPAL 是其中的一种，它不仅解决了输出悬空的问题，还具有非常简单而规则的电路形式，易于简化版图设计，其电路如图 10.50 所示。

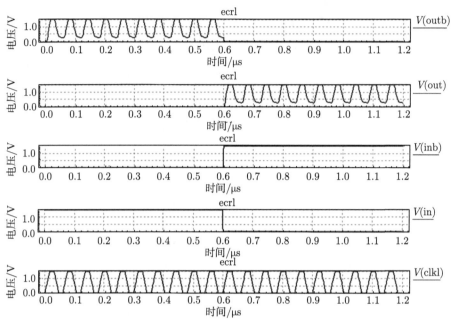

图 10.49　ECRL 反相器的仿真波形

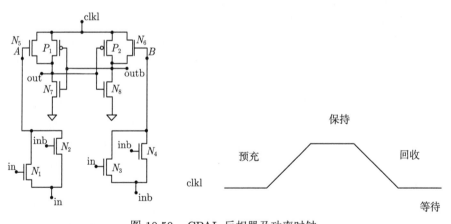

图 10.50　CPAL 反相器及功率时钟

该电路由两部分组成：逻辑复制部分和负载驱动部分。逻辑赋值部分由 N_1、N_2、N_3 和 N_4 组成，负载驱动部分由两对传输门 N_5、P_1 和 N_6、P_2 组成，它的功能主要是实现能量回收，而 N_7 与 N_8 两个 NMOS 管由输出交叉控制，它们的作用是防止输出节点悬空而造成能量损耗。输出节点的能量通过两对传输门回收到功率时钟 clkl 中去，使得输出节点的非绝热电路能量损耗完全消失，因此 CPAL 电路具有更高效的能量回收。CPAL 电路结构简单，与传统 CMOS 相比，所需晶体管较少，因此总的输入栅电容及充放电电荷减少，降低了功耗，提高了速度。

对 CPAL 反相器进行工作原理分析：CPAL 可分为预充求值、保持、回收、等待四个工作阶段。假设输入 in 为逻辑高电平 "1"，输入 inb 为逻辑低电平 "0"。

预充求值阶段：功率时钟 clk1 从低电平逐步上升到高电平 V_{dd}，由于 N_5 处于导通状态，则输出端 out 会随 clk1 的变化而变化，同时结点 A 也会跟随功率时钟变化，当结点 A 随功率时钟 clk1 上升到高电平 V_{dd} 时，N_8 此时已经导通，互补输出端 outb 通过 N_8 箝位至低电平状态。

保持阶段：当输出结点 outb 为 0 时，此时 out 保持高电平 V_{dd}。

回收阶段：由于 N_5 处于导通状态，输出端 out 会随功率时钟 clk1 的变化逐步从 V_{dd} 下降至低电平状态，通过这个过程存储在输出结点 out 的电荷逐步回收至电源端，最终实现对能量的有效回收。

等待状态：输入端 in 和 inb 分别为高电平和低电平，则输入端 N_1 和 N_3 导通，N_2 与 N_4 截止，输入信号 in 通过 N_1 对结点 A 进行充电至 $V_{dd}-V_{th}$ 电位，其中 V_{th} 为 NMOS 的阈值电压，而结点 B 则通过 N_3 放电至低电平 0V。电路等待下一轮信号的到来。

在 0.5 μm BSIM 3v3 工艺参数下，对该电路进行模拟仿真，得到其仿真波形如图 10.51 所示，由图可知，该电路有效地解决了 ECRL 电路中输出端的悬空问题。

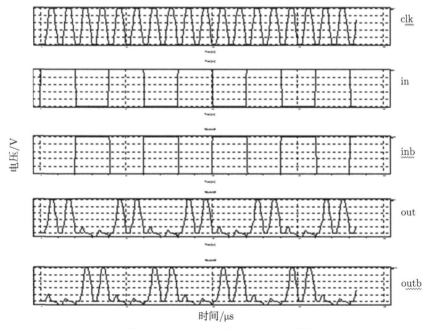

图 10.51　CPAL 反相器仿真波形图

参 考 文 献

[1] 王志功, 朱恩, 陈莹梅. 集成电路设计 [M]. 北京: 电子工业出版社, 2006.
[2] 孔德立. 数字集成电路设计方法的研究 [D]. 西安: 西安电子科技大学, 2012.
[3] 刘培植, 胡春静, 郭琳, 等. 数字电路与逻辑设计 [M]. 北京: 北京邮电大学出版社, 2009.
[4] 宁帆, 张玉艳. 数字电路与逻辑设计 [M]. 北京: 人民邮电出版社, 2003.
[5] 张锁良, 陈雷, 黄亚丽, 等. 数字电子技术基础 [M]. 北京: 北京邮电大学出版社, 2011.
[6] 朱正涌, 张海洋, 朱元红. 半导体集成电路 [M]. 北京: 清华大学出版社, 2008.
[7] 李秀群. 异步计数器设计的新探讨 [J]. 河北广播电视大学学报, 1999(1):4.
[8] Allen P E, Holberg D R. CMOS 模拟集成电路设计 [M]. 冯军, 李智群, 译. 北京: 电子工业出版社, 2009.
[9] 吴建辉. CMOS 模拟集成电路 [M]. 北京: 电子工业出版社, 2011.
[10] 王卫东. 现代模拟集成电路原理及应用 [M]. 北京: 电子工业出版社, 2008.
[11] 拉扎维. 模拟 CMOS 集成电路设计 [M]. 陈贵灿, 程军, 张瑞智, 等译. 西安: 西安交通大学出版社, 2003.
[12] 洪志良, 等. 模拟集成电路分析与设计 [M]. 2 版. 北京: 科学出版社, 2011.
[13] 张亚朋. CMOS 绝热逻辑及其功率时钟电路的分析与研究 [D]. 天津: 河北大学, 2010.
[14] Chandrakasan A P, Brodersen R W. Low Power Digital CMOS Design[M]. Boston: Kluwer Academic Press, 1996.
[15] Rabaey J M, Pedram M. Low Power Design Methodo-Logics[M]. Boston: Kluwer Academic Press, 1996.
[16] 邬杨波, 李宏, 胡建平. 低功耗互补传输门绝热逻辑和时序电路的设计 [J]. 宁波大学学报(理工版), 2008,21(2):195-200.
[17] Lim J, Kim D G, Chae S I. A 16-bit carrylookahead adder using reversible energy recovery logic for Ultralow-energy systems[J]. IEEE J. of Sol. Sta. Circ., 1999,34(6):898-903.
[18] 杭国强, 吴训威. 采用二相功率时钟的无悬空输出绝热 CMOS 电路 [J]. 半导体学报, 2001,22(3):366-372.
[19] Moon Y, Jeong D K. An efficient charge recovery logic circuit[J]. IEEE J. Sol. Sta. Circ., 1996,31(4):514-522.
[20] Kamer A, Dener J S, Flower B, et al. Second-order adiabatic computation with 2N-2P and 2N-2N2P logic circuits[C]//Proc. of the international symposium on low power design. Canada: Dana Point, 1995:191-196.
[21] Oklobdzija V C, Maksimovic D, Lin F. Pass-transistor adiabatic logic using single power-clock supply[J]. IEEE Trans Circuits and Systems II:Analog and Digital Signal Processing, 1997,44(10):842-846.
[22] Liu F, Lau K T. Pass-transistor adiabatic logic with NMOS pull-down configuration[J]. Electronics Letters, 1998,34(8):739-741.